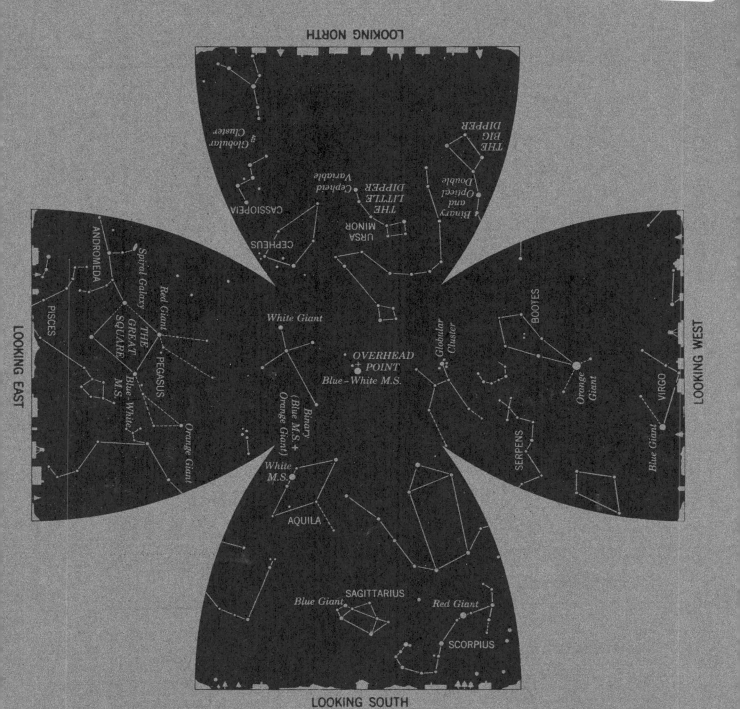

LOOKING NORTH

LOOKING EAST

LOOKING WEST

LOOKING SOUTH

THE BIG DIPPER

Binary and Optical Double

THE LITTLE DIPPER

Cepheid Variable

URSA MINOR

CEPHEUS

CASSIOPEIA

Globular Cluster

ANDROMEDA

Spiral Galaxy

Red Giant

THE GREAT SQUARE

PEGASUS

PISCES

Blue-White M.S.

Orange Giant

White Giant

OVERHEAD POINT
Blue-White M.S.

Globular Cluster

BOOTES

Orange Giant

VIRGO

Blue Giant

SERPENS

Binary (Blue M.S. + Orange Giant)

White M.S.

AQUILA

SAGITTARIUS

Blue Giant

Red Giant

SCORPIUS

ASTRONOMY:
FUNDAMENTALS
AND FRONTIERS

ROBERT JASTROW · MALCOLM H. THOMPSON

# ASTRONOMY: FUNDAMENTALS AND FRONTIERS
Second Edition

John Wiley & Sons, Inc.
New York   London
Sydney   Toronto

Cover: The eclipse of June 30, 1973, photographed
from Kenya by Serge A. Korff and Fred Ayer.

Copyright © 1972, 1974, by Robert Jastrow

**Fourth printing, July, 1975**

**Library of Congress Cataloging in Publication Data:**

Jastrow, Robert, 1925-
  Astronomy: fundamentals and frontiers.
  1. Astronomy. I. Thompson, Malcolm H., joint
author. II. Title.
QB43.2.J37 1974        520        74-687
ISBN 0-471-44078-7

Printed in the United States of America

# FOREWORD

In the introduction to the present textbook the authors briefly discuss the Copernican revolution. We are reminded that the ideas of Copernicus, Kepler, Galileo and Newton profoundly influence human culture, and we recall how long it took for these ideas to be generally understood and accepted. The addition of knowledge that is taking place in our day in astronomy occurs at a spectacularly rapid pace. There can be no doubt that the task of making the results of astronomical research generally available is a most important one. The challenge is strongly felt by the astronomical community.

Professors Jastrow and Thompson have written an astronomy textbook with the liberal arts student in mind at every point of the development. There is a very good balance between the discussion of basic methods and description of the results of astrophysical research, including the most recent advances.

I believe that the authors have chosen wisely in emphasizing the central problems of modern astrophysics, dealing only briefly with some of the chapters of classical astronomy. In placing the discussions of stellar structure and evolution of galaxies, and large scale cosmology before the discussion of the solar system and problems of structure and history of planets, the authors are guided by considerations of evolution in time rather than location in space. This way of arranging the subject material has obvious advantages.

If this textbook is as widely used and read as it deserves the authors will have made a very significant contribution in the area of communication between scientist and the community. For the book tells compellingly about basic research, important to us all, and of its results, its spirit and its excitement.

**Bengt Stromgren**

# PREFACE TO THE FIRST EDITION

Astronomy, more than any other physical or behavioral science, offers the nonscience student a mind-expanding educational experience. The steadily increasing enrollments in introductory astronomy courses reflect a growing awareness of this fact on the part of liberal arts students. The historical and philosophical elements in astronomy, always a large factor in the appeal of this subject for the nonscience major, have been strengthened by new discoveries in stellar evolution and cosmology. Progress in these fields during the last twenty years has filled in many details of the sequence of events that led from the explosive beginnings of the Universe through the birth of innumerable stars and planets to the formation of the sun and the earth. When the latest advances in astronomy are combined with developments in the life sciences, the result is a chain of cause and effect that stretches back over 10 billion years and links the earth and its life forms to events that occurred early in the history of the Universe. At that point—on the threshold of the appearance of life on our planet—the direct contribution of astronomy ends, and the story is taken up by other branches of science.

The advances in astronomical knowledge provide many points of contact between this subject and other scientific disciplines. Modern astronomy, fascinating in itself, seems increasingly to be a fragment of a mosaic that, when viewed from a distance, forms an image of the human observer. The appeal of astronomy to the nonscientist is further strengthened by the fact that its subject matter forces the imagination to contemplate larger expanses of space and time than fall within the province of any other scientific discipline. These qualities make the study of astronomy a uniquely attractive means of introducing the liberal arts student to the physical sciences.

We have focused on the needs and interests of the liberal arts student in our choice of topics, as well as in the style of writing and in the level of required mathematical skills. No mathematics is used beyond the level of elementary algebra, and technical terms are avoided. Each chapter opens with the statement of a central theme to which the previous chapters of the book are clearly related. The remainder of the chapter is an explicit development of this central theme.

The central problems of twentieth century astronomy are emphasized. Full chapters are included on the Hertzsprung-Russell diagram, nuclear reactions in stars, stellar evolution, galactic structure and evolution, radio galaxies, Seyfert galaxies and quasars, cosmology, the history of the moon, and the evolution of planetary atmospheres. Much discussion is devoted to recently opened areas of research such as infrared astronomy, x-ray and gamma-ray astronomy, gravitational waves, pulsars, and black holes in space. The choice of topics and the allotment of space to each topic reflect much of contemporary research publication. An unusual feature is the inclusion of a final chapter on the evolution of life in the Cosmos.

A complete discussion of these topics in a one-semester textbook necessitated the omission of some areas, such as astrometrics, which are centers of active research in contemporary astronomy but are not as directly related to the book's central line of development. Celestial mechanics is treated very briefly in the chapter on the solar system. The motions of the earth in space, tides, eclipses, and celestial coordinates are described in an introductory, separately paginated section.

A basic innovation in the book is its presentation of material on stars and galaxies before the discussion of the solar system. This is the reverse of the traditional presentation, in which astronomical knowledge is given in the order in which it was acquired in human history, starting with the earth and then radiating outward to the moon, planets, stars, and galaxies. Our book embeds the study of the solar system in the context of a general study of stars and planets, and more accurately reflects the impact of the Copernican Revolution on the history of astronomy.

The new organization of material has the advantage that it permits the instructor to use astrophysical knowledge when discussing the structure, chemical composition, and origin of the earth, moon, and planets. Astrophysics reveals how elements are made in the stars; why some are more abundant than others; and how these elements condensed to form the clouds out of which the sun, moon, and planets were born. An astrophysical background is required to discuss conditions at the beginning of the solar system, when the earth and other planets were newly formed. Instructors who prefer the traditional organization, but like other features of the book, can start their course as usual with "The Solar System," whose opening chapters summarize the astrophysical background needed for studying the solar system.

The rapid pace of change in modern astronomy makes the task of the textbook author very difficult if he wishes to present a balanced view of recent developments. We are deeply indebted to a number of friends and colleagues, closely associated with these developments, who have been willing to spare time from their research for careful reviews and detailed criticisms of portions of the manuscript dealing with subjects of which they have a profound knowledge. We are particularly grateful to Dr. Richard Stothers for many informative discussions on stellar evolution and numerous detailed criticisms of Chapters 1 to 8; to Professor Lodewijk Woltjer for a careful commentary on the Chapters dealing with galaxies and cosmology; to Professors Paul Gast and Robert Phinney and to Dr. Vivien Gornitz for their comments on the Chapters relating to the solar system, the earth, the moon and the planets; and to Dr. S. I. Rasool for illuminating discussions of the planets in general and critical review of Chapter 17 in particular. We also profited greatly from conversations with Dr. Patrick Thaddeus on topics in radio astronomy and interstellar chemistry. Professor Neville Woolf gave us the benefit of his reading and criticism of the entire manuscript, and offered valuable comments on the balance of the contents between traditional and contemporary areas of astronomy. Drs. J. W. Hogan, R. Stewart, and Dennis Hegyi reviewed the manuscript for pedagogical effectiveness and offered many helpful suggestions based on classroom testing of the materials.

The completion of the manuscript would not have been possible without the devoted editorial and secretarial assistance of Misses Judith Silverman and Ruth McCarthy. Finally, with particular pleasure we express our thanks to Donald Deneck and Dennis Hudson in particular, and to the extremely capable Wiley editorial, production, picture research, illustration and design departments for their enthusiastic support and cooperation in bringing the raw material of our text into finished form.

**Robert Jastrow**
**Malcolm Thompson**

# PREFACE TO THE SECOND EDITION

The second edition contains several major additions and revisions. A new Chapter 12, inserted at the beginning of the section on the solar system, gives a detailed account of the properties of the sun. The first edition treated the sun only as a case study in stellar evolution, following the Copernican viewpoint, but in doing this we lost the opportunity to describe the fascinating variety of phenomena that are included under the general description of solar surface activity. These events, comprising mainly sunspots, flares, surges, and prominences, have a striking visual impact when photographed in light of suitably chosen wavelengths. They also have profound scientific importance, because they provide insights into the behavior of a gas of electrically charged particles in strong interaction with magnetic fields, under unusual circumstances in nature's laboratory that have not been matched on the earth thus far. The addition of this chapter constitutes a major effort to improve on the first edition. The chapter concludes with a set of plates showing some of the more striking examples of solar surface phenomena, accompanied by detailed captions that expand on the discussions of the corresponding topics in the text.

The chapter on the moon has been completely rewritten to incorporate the scientific results from the final missions in the Apollo program. The discussion of Mars in the following chapter also has been revised to include remarkable evidence acquired in the Mariner 1971 mission, suggesting that at one time the surface of Mars was the scene of extensive volcanic activity with substantial amounts of liquid water present and, possibly, favorable conditions for the chemical evolution of life.

An appendix has been added to describe methods for the measurement of stellar masses, distances and luminosities, as well as distances to extragalactic objects. This material, somewhat more difficult than the remainder of the book, will be of considerable interest to the reader who wishes to know how astronomers acquire the critical observational results that underlie the discussion of stellar properties. A second, brief appendix extends the discussion in Chapter 7 on nuclear reactions in Main-Sequence stars.

Numerous short sections throughout the text have been rewritten and expanded for improved clarity, and discussions of frontier topics have been updated to reflect the current status in these rapidly advancing fields of research.

Several friends and colleagues have been kind enough to devote their time to critical reviews of the new material. We are particularly indebted to Dr. Richard Stothers for his careful reading of all topics relating to stellar structure and evolution; to Professor Ludwig Woltjer for his review of the account of extragalactic distance measurements in Appendix A, and the updated discussions of quasars and cosmology in Chapters 10 and 11; to Professor Edward A. Spiegel and Dr. Richard Defouw for their criticisms of the chapter on the sun; and to Dr. Vivian Gornitz for her comments on the presentation of the Apollo and Mariner results. We are also grateful to several astronomers who are active in solar physics, including Drs. Richard Dunn, Sara Martin, Robert Howard, Richard Newkirk, Harold Zirin and Edward Gibson, for providing unusually fine photographs illustrating solar surface phenomena and for their assistance in guiding us through the interpretation of these materials. We are indebted to Professors Sally Deike, Hal I. Heaton, Stephen Hill, and Michael M. Shurman for compiling valuable lists of errata from earlier printings.

The publication of this edition would not have been possible without the capable assistance of Diana Birmingham, Patricia Dingcong, Vivian Landa, and Yola Schlosser. Finally, in publishing both editions the experience of working with the Wiley staff was exceptionally rewarding and pleasant. With particular pleasure we acknowledge our thanks to our editor, Donald Deneck, and the Manager of College Production, Dennis Hudson. We also thank John Balbalis for illustrations, Stella Kupferberg for picture research, and Jerry Wilke for design. The favorable reception accorded our book thus far has been due in substantial measure to their creative talent, efficiency, and warm support.

Robert Jastrow
Malcolm Thompson

January, 1974

PREFACE TO THE SECOND EDITION

# CONTENTS

**Photo Credits 505**

**Index 507**

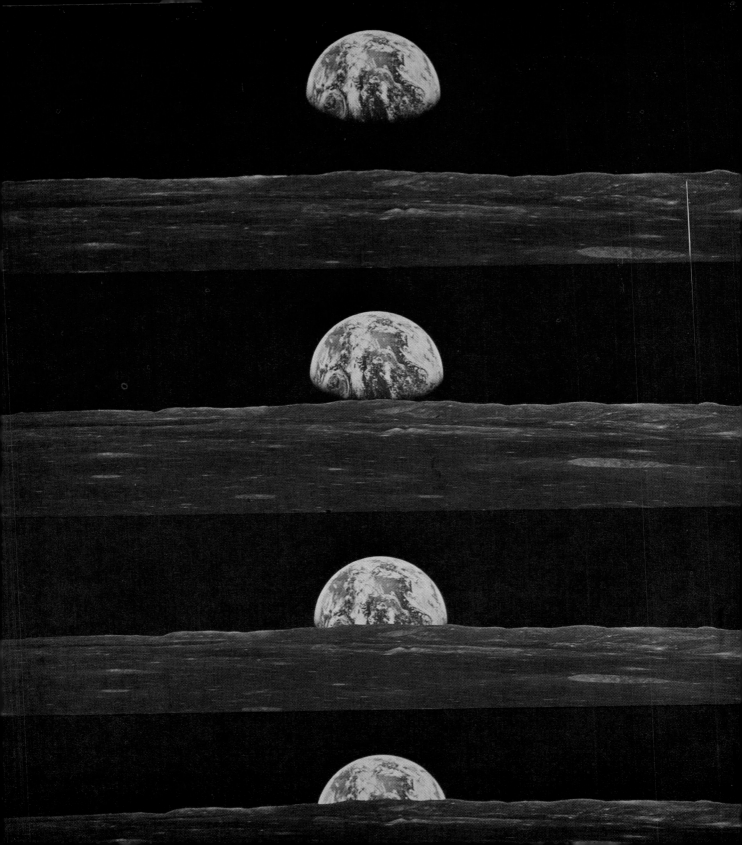

# Introduction: The Earth in Space

The earth seems vast and immobile; throughout two million years in the prehistory of man, it has provided the stage for all human experience, with the heavens seemingly no more than a backdrop of moving lights. Modern astronomy was born in the contrary realization that space is vast, and the world of men is small. Astronomical discoveries have their greatest impact on human thought through man's continually evolving awareness of the modest position of the earth in space.

## THE COPERNICAN REVOLUTION

### The Ancient Ideas

To anyone who follows the motion of the sun day after day, and the motions of the moon and the stars night after night, it is obvious that the earth is the hub of the universe and that all heavenly bodies revolve about it daily, paying homage to the abode of man. It violates the evidence of

*The Setting Earth (Apollo 14)*

the eyes to assert that the earth moves instead.

But there is one difficulty in accepting the picture of the earth as the fixed center of the universe: in this picture a complicated system has to be invented to describe the motions of the planets. Since ancient times, observers of the heavens had noted that from night to night the stars moved across the sky in a regular procession, from east to west, following the motions of the sun and moon. In this nighttime procession the stars maintained the same position relative to one another; the stars of the Big Dipper moved across the sky at that time, and still do today, as a unit.

But five objects in the sky did not hold the same relative positions; they looped forward and backward from west to east and then from east to west. The Greeks, observing this remarkable fact, called each of these five celestial bodies the Greek word for wanderer, *planetes,* from which we derive the word "planet."

Of course, the wandering motion was not observable in one night's time. Over the course of a night, the planets, like the stars, always move from east to west across the sky, reflecting the rotation of the earth about its own axis in the opposite direction to their apparent motion. It was only when the observations of planets were extended for a year or more, that their positions relative to their nearest neighbors among the fixed stars were seen to change in a complicated fashion.

*Figure I.1  A planetarium photograph of the apparent paths of five planets.*

Figure I.1 is a photograph taken in a planetarium that shows the changes in the apparent position of five planets—Mars, Venus, Mercury, Jupiter and Saturn—over a period of many months, as viewed from the earth. The planets seem to go through a series of loops as they move across the background of fixed stars. The looping motion of each planet could be duplicated if the planet were anchored to the outer rim of a wheel which was located on a track moving across the sky. When the observed positions of the planets were plotted, they agreed fairly well with this description of wheels rolling across the sky. But if the ancient astronomers tried to make the calculated motions of the planets fit the observations accurately, the system became very complicated and cumbersome. It had to be assumed that some of the wheels themselves rolled on the rims of other wheels. The best job of fitting was done by Ptolemy, who lived in the second century A.D. He and his followers found 80 wheels within wheels were needed just for the five planets known at that time. Some planets required wheels within wheels within wheels.

## The Theory of Copernicus

In the sixteenth century, a Polish churchman named Nicholas Copernicus proved that he could explain the complicated motions of the planets in a much simpler way than Ptolemy, if he made the assumption that all planets, including the earth, revolved around the sun. Copernicus proposed further that each planet revolved around the sun at a different speed. As a result, when a planet was viewed against the background of fixed stars, sometimes it appeared to be moving forward and sometimes backward, all depending on the relative speeds and positions of the two planets in their orbits.

In this way, Copernicus showed that his theory could account for the looping motions of the planets. However, he did not *prove* that the earth and other planets revolved around the sun; he showed only that his theory provided a simpler explanation of the facts than the alternative theory of Ptolemy that the sun revolved around the earth.

But after Newton discovered the universal law of gravity in the seventeenth century, many people became convinced that all the planets, including the earth, were in fact moving in orbits around the very massive sun, held captive by its gravitational attraction. As yet, there was still no direct proof that the earth moved around the sun, but the growing belief in a universal force of gravity helped to make this idea more plausible.

## Proof that the Earth Moves

During the course of a year, as the earth moves around its orbit, stars nearer to the sun should shift their position relative to the more distant

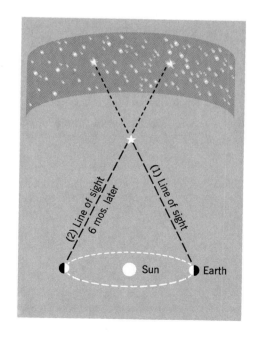

Figure I.2 *The change in the apparent position of a nearby star against the background of distant stars.*

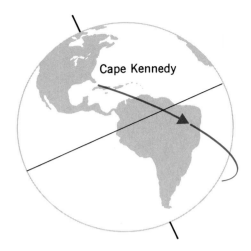

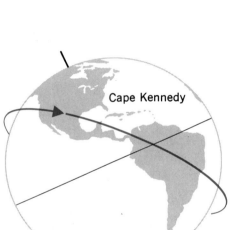

*Figure I.3 Successive orbits of a satellite provide proof of the earth's rotation.*

stars (Figure I.2). This shift in position is known as *parallax*.[1] One of the strongest arguments against the theory of Copernicus had been that no parallax was observed. Actually the parallax was there but was too small to be detected with the naked eye or even with early telescopes, because all stars, even the closest, are exceedingly far away in comparison with the diameter of the earth's orbit. It was not until 1838 that the telescope reached a state of perfection such that in the hands of a very skillful observer—the German astronomer, Bessel—the shift in the position of a nearby star could finally be measured. Bessel's proof of the earth's motion was obtained in 1838, two-hundred and ninety-five years after the death of Copernicus.

## Rotation of the Earth

The clearest proof of the earth's rotation is provided by artificial satellites. Suppose a satellite is launched from Cape Kennedy into an orbit 100 miles above the earth carrying it toward the southeast, at an angle of 30 degrees to the equator (Figure I.3a). This would be a typical launch trajectory from the Cape. Once the satellite is launched, the plane of its orbit stays fixed in space, since no forces are exerted on the satellite to change it. (This is only approximately true, but the change is small enough to be ignored in the present discussion.) Therefore, if the earth is not rotating, the satellite should pass over Cape Kennedy once in every orbit. But it does not; it passes over Alabama on the completion of its first orbit, over Louisiana at the end of the third, and so on (Figure I.3b).

Because the earth is rotating beneath the satellite, it passes over places in the United States that are in this case 1000 miles farther west on each orbit. This fact, which has been observed in all of the hundreds of satellites that have been launched, directly proves the rotation of the earth.

## Tilt of the Earth's Axis

The earth's axis of rotation is pointed in a direction which makes an angle of 23.5 degrees with the perpendicular to the plane of the earth's orbit (Figure I.4).

This fact does not make the earth an unusual planet in the solar system. The direction of the axis of spin of a planet is determined by the particular swirling motions of the gases out of which this planet condensed at the beginning of the solar system's history, when the sun was

---

[1] The parallax can be seen clearly if you hold up your finger vertically with arm extended and see it against the background closing first one eye, then another. You will notice that the position of the finger seems to change markedly.

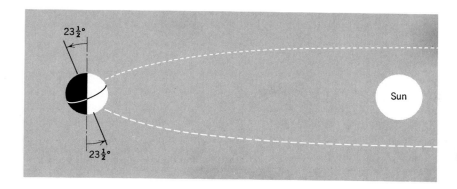

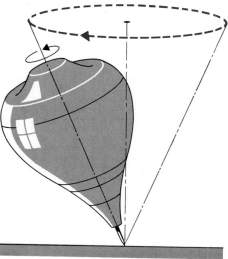

Figure I.4 *The tilt of the earth's axis (left).*

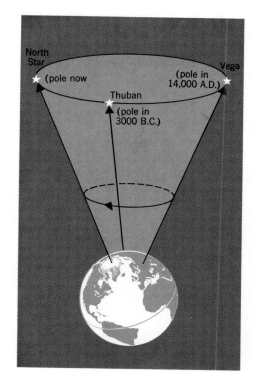

*Figure I.5 The precession of the earth's axis.*

*Figure I.6 Changes in the direction of the axis of rotation.*

still surrounded by a dense, turbulent cloud of gas and dust. As one might expect, this situation produced a wide range in the angles of tilt of the spin axes of the planets. The axes of rotation of Mars and Saturn are tilted at nearly the same angles as the earth's axis (24° and 27°, respectively). On the other hand, Jupiter's axis of rotation is almost exactly perpendicular to the plane of its orbit, while Uranus spins lying on its side, with its axis nearly in the plane of its orbit.

*Precession of the Axis.* The earth's axis shifts its direction during the course of time, always retaining, however, the same angle of tilt to the ecliptic. Its movements are very similar to the movements of a spinning top or gyroscope, rotating very rapidly about its axis while the axis itself revolves slowly about the vertical. This slow change in the direction of the axis of a spinning object, whether the earth or a gyroscope, is called the *precession* of the axis (Figure I.5). The precession of the earth's axis is produced by the gravity of the sun and moon, which tug at the equatorial bulge of the earth. Under the influence of these forces, the axis of rotation revolves in a cone, completing a circle once every 26,000 years.

At present, the axis is pointed toward the North Star; 5000 years ago it pointed in the direction of the Thuban in the Constellation Draco, and 12,000 years from now it will point in the direction of Vega, which will be the replacement for the North Star at that point in the distant future (Figure I.6).

*Nutation.* In addition to the slow movement of the axis of rotation, completed once in every 26,000 years, there is also a smaller movement of the axis, something like a wobble superimposed on the steady 26,000-year drift. The wobble of the axis of rotation of a spinning object is called *nutation* (Figure I.7). As a result of the nutation of the earth's axis of rotation, the axis varies in its inclination to the plane of the earth's orbit by somewhat more than one four-hundredth of a degree every 19 years.

## Other Motions of the Earth

The stars in the Galaxy move randomly, each star buzzing around in space like an atom in a gas of particles at a high temperature. The sun's

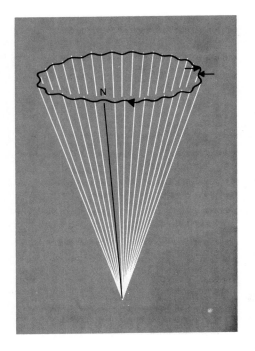

random velocity at this moment is carrying it toward Vega—a bright star directly overhead in the autumn sky around 9 P.M.—at a speed of 12 miles a second. The earth accompanies the sun in its flight toward Vega, tracing a spiral through space (Figure I.8).

The sun and the other stars in the neighborhood of our solar system revolve slowly around the center of the Galaxy. In the course of this motion, our solar system completes one full turn around the Galaxy in 200 million years, covering a distance of nearly one million trillion miles in that time at an average speed of 200 miles per second.

The entire Galaxy, carrying the earth with it, is moving through space relative to other nearby galaxies. As a consequence of this motion, our Galaxy and the Andromeda Galaxy—our nearest large galactic neighbor—are approaching one another at a speed of 180 miles per second. The earth is carried along with our Galaxy in this motion.

Thus the motion of the earth is very complicated. It rotates about its axis, revolves around the sun, moves with the sun around the center of the Galaxy, and moves with the Galaxy on its journey toward Andromeda. Perhaps, the earth participates in still other movements, at even higher speeds. If it does, we probably will never be able to detect the motion. We cannot say that we know all the motions of the earth; we can never define an absolute space through which the earth moves, heading in some ultimate direction at a definite speed.

*Figure I.7    Nutation.*

*Figure I.8    The earth's spiraling motion through space.*

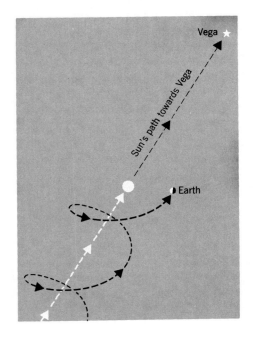

## THE MOON AND THE EARTH

The moon accompanies the earth on its journey through space, revolving around our planet every 27.3 days in a nearly circular orbit. The relative positions of the sun, earth, and moon change continually as a result of the motion of the moon around the earth and the earth around the sun, creating phenomena of great beauty as well as practical importance.

The serene passage of the moon's silver orb across the heavens has excited the admiration of poets since the dawn of language:

"The starry host, rode brightest, till The moon,
Rising in clouded majesty, at length
Apparent queen, unveil'd her peerless light,
And o'er the dark her silver mantle threw."

*Milton*

But the moon also exerts a strong influence over the practical affairs of men because of its exceptional size and mass in comparison to its parent planet. No other satellite in the solar system approaches the moon in this respect. The large size of the moon enables it to cast a giant shadow over parts of the earth on occasion, when the movements of the moon and the earth bring the three bodies into a straight line, in the phenomenon known as a *solar eclipse*. The massiveness of the moon creates a powerful gravitational pull that causes large areas of the land and the oceans to bulge outward along the earth-moon line by amounts ranging from a few inches to many feet, in the phenomenon known as the *tides*.

## The Phases of the Moon

The light we see coming from the moon is reflected sunlight. Half the lunar surface is always in sunlight, but to us the moon's face appears to change shape because we see varying portions of the sunlit side as the moon revolves around the earth. The moon is invisible when it is between the earth and the sun, partly because the side facing the earth is in shadow, but also because in this position the moon, being near the

Figure I.9   *The phases of the moon.*

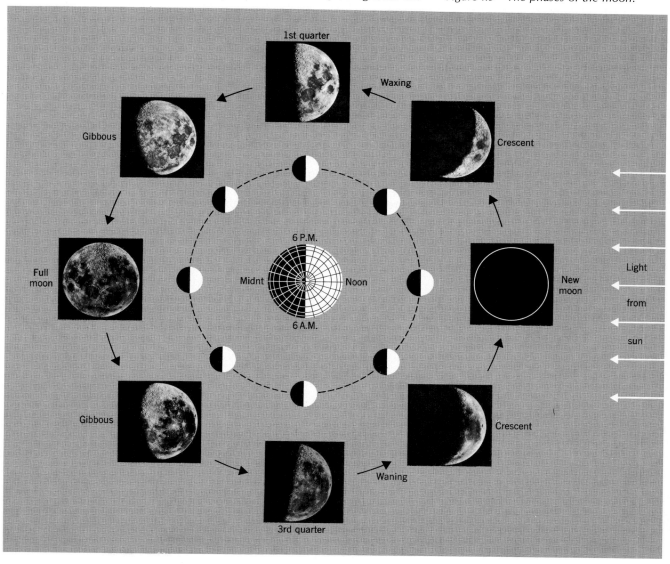

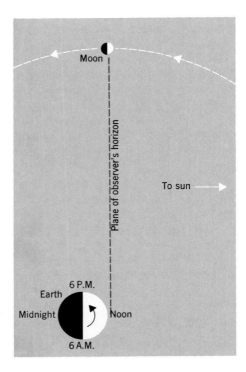

Figure I.10   Rise of the first-quarter moon.

sun in the sky, is lost in the solar brilliance. When the moon is a little off the earth-sun line, we can see the thin crescent of the illuminated edge. When the moon is on the opposite side of the earth from the sun, its fully lighted face dominates the nighttime sky.

The shapes successively assumed by the moon are referred to as its *phases*. The phases of the moon proceed from new moon to crescent, to first quarter, to gibbous, to full moon, and then, in reverse order for the remaining half of the cycle, through gibbous, third quarter, and crescent to new moon again (Figure I.9). In the quarter phases, exactly one-half of the moon's visible face is illuminated. For this reason the moon in its quarter phases is often described as being half full, or as a half moon.

As the moon progresses through the first half of its cycle from new to full, it is called a *waxing* moon. In the second half of the cycle it is called a *waning* moon. The cycle takes 29.5 days.

To astronauts looking back at earth from the moon, the earth goes through the same sequence of phases that we observe the moon going through. Any particular phase of the earth, however, will be opposite the phase of the moon. If we see the moon in the third quarter, the astronauts see a first quarter earth.

*Moonrise and Moonset.* The times at which the moon rises and sets can be worked out from studying Figure I.9. These times depend on the phase of the moon. As an example, suppose the moon is in its first quarter. Suppose an observer is standing on the earth at a point in which the sun is directly overhead, that is, the local time is noon, as shown in Figure I.10. In this figure the earth is viewed from above the North Pole, and the earth is therefore rotating in a counterclockwise direction. As the earth rotates, the observer sees the moon rising above his eastern horizon. The moon appears to be half full with the illuminated half up. This same observer sees the moon directly overhead at 6 P.M. Around midnight the moon sets on his western horizon with the illuminated half facing down. The actual time of setting is delayed somewhat by the fact that the moon's own motion, which is counterclockwise around the earth, carries it backward toward the east with respect to the stars during the night.

The times of moonrise and moonset for all phases of the moon are given in the table below.

| Phase | Rise | Set |
|---|---|---|
| New | 6:00 A.M. | 6:00 P.M. |
| New crescent | 9:00 A.M. | 9:00 P.M. |
| First quarter | Noon | Midnight |
| First gibbous | 3:00 P.M. | 3:00 A.M. |
| Full | 6:00 P.M. | 6:00 A.M. |
| Last gibbous | 9:00 P.M. | 9:00 A.M. |
| Last quarter | Midnight | Noon |
| Old crescent | 3:00 A.M. | 3:00 P.M. |

These times are not precise because they neglect the moon's motion as noted above and also because they assume that the moon's orbit is a perfect circle in the same plane as the earth's orbit and that the observer is standing on the equator. The actual times of the moon's rising and setting found in newspapers may differ from the times in the table by more than one hour at the latitudes included within the continental United States.

## Solar Eclipses

When the moon passes between the sun and the earth, it blocks all or a part of the sun's rays from some regions of the earth. This event, called a solar eclipse, is the most awesome of all astronomical phenomena if the rays of the sun are entirely blocked from the observer. As the moon slides over the disk of the sun, an unnatural darkness descends on the earth, and the birds cease their chatter. For several minutes the moon blots out the sun entirely, the temperature of the air drops, and the solar corona—the sun's outer atmosphere—appears as a pearly, shimmering halo around the moon's black shadowing disc. The brighter stars and planets become visible in the sky at this time. During the next hour the face of the sun reveals itself again, growing from a thin crescent to its full brilliance as the moon continues on its way.

The ancients regarded eclipses as harbingers of important events. Many ancient societies kept careful eclipse records, but the Babylonian astronomers were exceptional among ancient peoples in the attention that they devoted to the compilation of these events. References to eclipses can be found in cuneiform inscriptions on Babylonian tablets that date back to the second millenium B.C. An example of a reference to what appears to be an eclipse is found in a tablet describing unusual events, such as the appearance of a bearded woman, a four-horned sheep, and a talking corpse. Figure I.11a shows a tablet in this series.

The fourteenth line of the tablet (Figure I.11b) says that,

On the twenty-sixth day of the month Sivan in
the seventh year the day was turned to night,
and fire in the midst of heaven (. . . . .).

The event it describes could equally well be interpreted as a thunderstorm or an eclipse; but because the rest of the tablet is concerned with truly exceptional events, it is very likely that this passage described a total eclipse, in which the "fire in the midst of heaven" was the flickering solar corona during the few minutes of totality.

Astronomers can compute the dates of occurrence of eclipses with great precision—for thousands of years in the past as well as in the future. They know the orbits of the earth and the moon with sufficient accuracy to be able to calculate the times at which the sun and the moon will be in line with a particular point on the earth. Starting from

(a)

Figure I.11(a)   A Babylonian tablet recording an unusual event interpreted as a total solar eclipse. (b) Lines from the tablet including the statement, "the day was turned to night."

(b)

the observed positions of the earth and the moon at the present moment, and using their knowledge of the orbits, they trace the motion of the earth and the moon forward and backward in time, predicting the exact moments and the locations on the earth at which an eclipse is to be visible. The astronomical calculations confirm the occurrence of an eclipse in eleventh century B.C. Babylon in the early afternoon of July 31, 1062 B.C.

*Total Eclipses.* The moon's path around the earth is an ellipse which carries it out to a maximum distance of 253,000 miles and in to a minimum distance of 221,000 miles once each month. If an eclipse occurs when the moon is close to the earth so that its apparent diameter is great enough to block all the rays of the sun, the resultant eclipse is said to be *total* (Figure I.12). This combination of circumstances occurs once every two or three years, on the average, at some point on the earth. A total eclipse can be seen at a given spot on the earth only once in every 360 years.

The period of totality of a solar eclipse is usually two to three minutes, and cannot last for more than seven minutes. The last total solar eclipse in the United States occurred on March 7, 1970, and was visible along an arc running across the Gulf of Mexico and up the east coast (Figure

Figure I.12 *Positions of the earth and moon during a total eclipse.*

I.13). The duration of totality was three minutes along the central line of the eclipse swath. The next solar eclipse visible from the United States will occur in 1979. Total solar eclipses visible from other parts of the world occurred in 1973 and others occur in 1974, 1976, 1977, 1979, and 1980.

*The Annular Eclipse.* If a solar eclipse occurs when the moon is at its greatest distance from the earth, the moon's apparent diameter will be smaller than the diameter of the sun, and the outer rim of the sun's disc will remain visible throughout the eclipse, even for a location on the earth that lies exactly on the line between the centers of the moon and the sun (Figure I.14). From such a location, at the midpoint of the moon's passage between the sun and the earth, the moon will appear to be a black disc surrounded by the bright ring of the sun's outer layers. This type of eclipse is called *annular.* Annular eclipses occur with a slightly higher average frequency than total eclipses. Figure I.15 shows an annular eclipse, with an uneven pattern of illumination around the edges of the moon's disc, caused by rays of sunlight streaming through the lunar valleys.

*The Partial Eclipse.* A partial eclipse occurs when the moon is close to the sun-earth line, but not close enough to completely block the sun's rays from any region on the earth. On the regions of the earth that lie within the path of the partial eclipse, the moon's disk will bite into the sun but will never cover it entirely. Partial eclipses occur more frequently than total or annular eclipses. Figure I.16 shows the sequence of events in a partial eclipse. The moon begins to block out the sun at (A), reaches maximum coverage at (B), and recedes to a small bite again at (C). The total time span from (A) to (C) is about two hours.

1:10 pm

1:50 pm

## Lunar Eclipses

Lunar eclipses occur at full moon when the earth, moon, and sun are in a straight line and the moon lies behind the earth, that is, in the full-moon phase (Figure I.17). They are visible at any place on the earth at which the moon is also visible, in contrast to the solar eclipse, which usually is visible only in a narrow band on the earth's surface, less than a hundred miles across. During a total lunar eclipse the moon receives no light except that which is bent around the edges of the earth by refraction in the earth's atmosphere, which acts like a lens, focusing sunlight to the moon. The blue component in this refracted light is reduced in intensity by scattering in the atmosphere of the earth, and the moon is illuminated by sunlight depleted in the blue wavelength, making it appear a dull coppery-red color. As the lunar eclipse develops, the earth's shadow sweeps across the face of the moon at a speed of about 2000 miles an hour.

The earth's shadow as it appears on the moon is clearly circular. The Greeks noticed this fact 2400 years ago and concluded then that the earth is spherical and not flat.

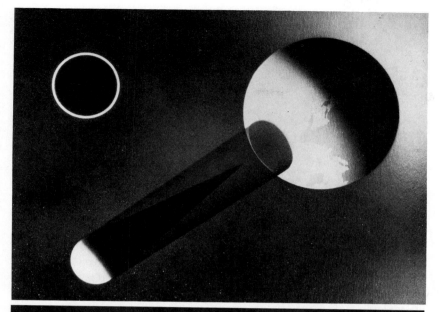

Figure I.14 *Positions of the earth and moon during an annular eclipse.*

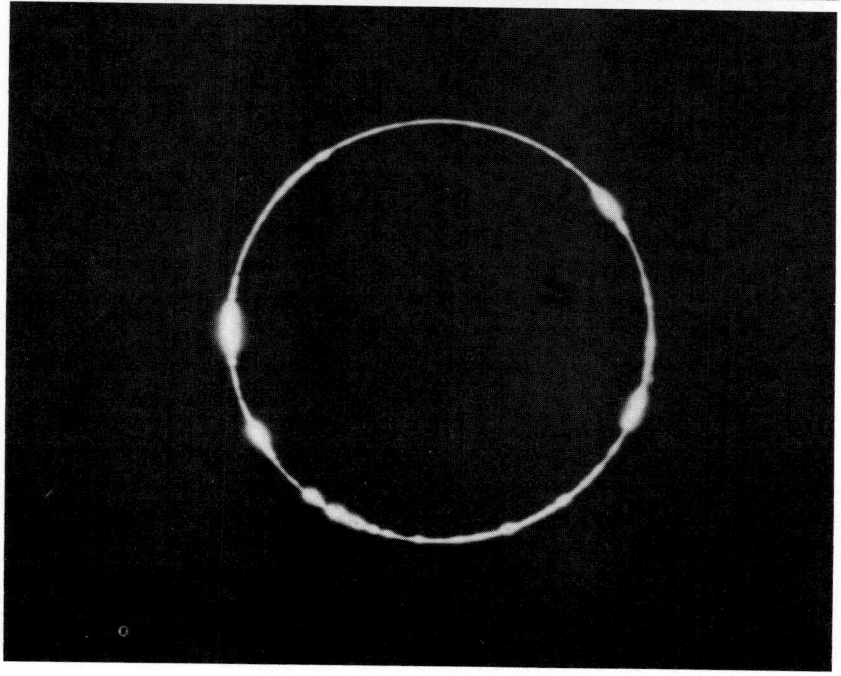

Figure I.15 *The annular eclipse of April 28, 1930.*

## The Eclipse Seasons

From the study of the phases of the moon, you would expect a solar eclipse to occur every month at the new moon, and a lunar eclipse to occur every month at the full moon. In fact, they only occur on the

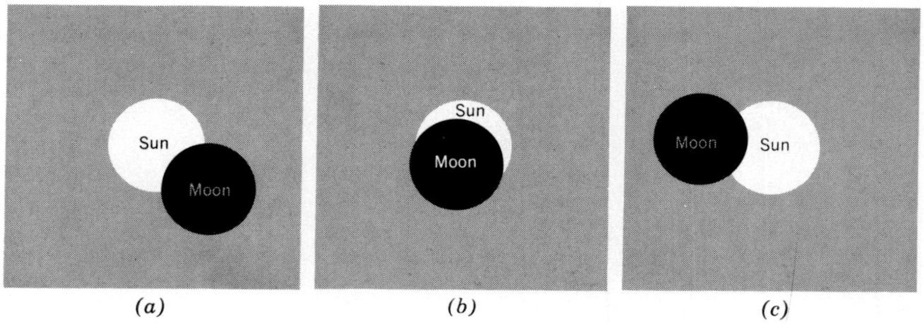

Figure I.16 *Appearance of the sun as the moon crosses its face during a partial eclipse.*

(a)  (b)  (c)

average every six months, during the semiannual *eclipse seasons*. The explanation for their failure to occur at other times is connected with the fact that the moon's orbit is tilted at an angle of 5.2 degrees to the plane of the earth's orbit (Figure I.18). As a result, during most of the year the moon is out of the earth's orbital plane at the new moon as well as the full moon, and an eclipse cannot take place (Figure I.19).

Twice each month, the moon passes through the earth's orbital plane. The points at which it passes through this plane are marked by A and B in Figure I.18. These points are called the nodes of the orbit, and the line AB is known as the line of nodes. As the earth and moon together revolve around the sun, the plane of the moon's orbit stays fixed in space,[2] and the line of nodes also keeps a fixed direction in space (Figure I.20). Twice each year, when the line of nodes points to the sun, an eclipse can occur. At other times of the year, when the line of nodes does not point toward the sun, solar and lunar eclipses cannot occur.

[2] This, is not precisely true; actually, the plane of the moon's orbit precesses slowly about the normal to the ecliptic, completing one revolution in about 19 years. In one year the shift in orientation is small enough so that the description in the text is accurate.

Figure I.17 *A lunar eclipse.*

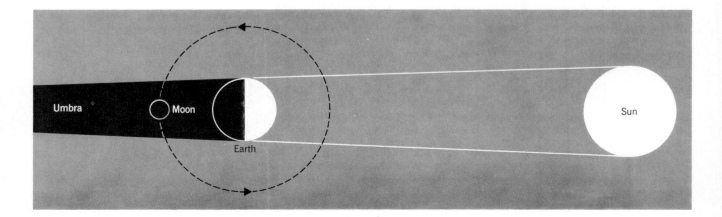

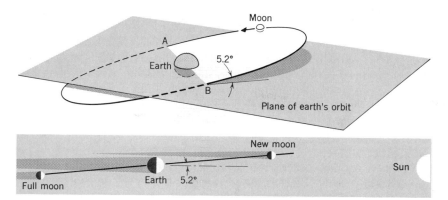

*Figure I.18   The tilt of the plane of the moon's orbit to the plane of the earth's orbit.*

*Figure I.19   Unfavorable conditions for an eclipse.*

## Tides

*The Lunar Tide.* The moon's gravity pulls on all parts of the earth as our satellite circles in its orbit. Different regions of the earth feel the pull of lunar gravity to a slightly different degree, however, because they are located at slightly different distances from the moon. The side of the earth facing the moon, for example, is 8000 miles closer than the side facing the other way. Since the force of gravity becomes weaker with increasing distance, the moon's pull on the near side of the earth is greater than its pull on the rest of the earth. The side of the earth facing

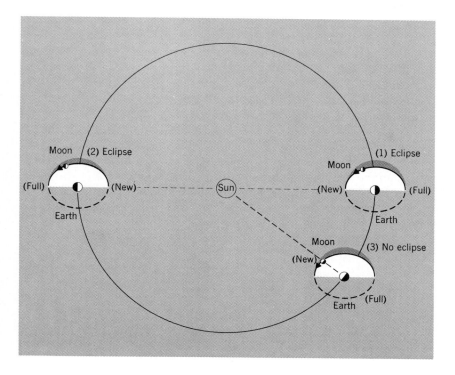

*Figure I.20   Eclipse seasons occur semi-annually at (1) and (2), when the moon's line of nodes points to the sun. No eclipses can occur at other times of the year (3).*

**I-15**

the moon bulges out as a result of this effect. The bulge toward the moon is called the lunar tide.

The height of the lunar tide is greatest when a large body of water, such as one of the major oceans, faces in the direction of the moon, since water can flow freely in response to the extra force. Tides in the open ocean or sea are approximately two feet high. If the ocean or sea tide is channeled into a narrow opening, for example, the mouth of a river, the height of the tide can be much greater than two feet. In the Bay of Fundy, on the border between Maine and Canada, the tide sometimes reaches a height of 50 feet.

When a continent faces the moon, the solid rock resists the moon's gravity. The tide is not zero, however, because even the solid rocks of the earth's interior yield to some degree when a force is applied to them. The response of the continents to the moon's pull, called a *land tide*, is usually no more than a few inches in height. Nonetheless, it can be easily measured with modern instruments, which are able to detect changes in the force of gravity as small as one part in a billion.

There is also a tidal bulge on the side of the earth facing away from the moon, equal in height to the tide facing toward the moon. This fact seems puzzling at first, but has a simple explanation. As mentioned above, the moon pulls most strongly on the side of the earth facing it; it pulls less strongly on the interior of the earth; and it is weakest of all on the side facing away from it. In Figure I.21, the moon's pull is represented by the arrows *A*, *B*, and *C*, whose lengths are proportional to the strength of the force of lunar gravity at these three places. Since force *A* is stronger than force *B*, it tends to pull the ocean away from the solid body of the earth, causing the tidal bulge on the near side. Since force *B* is stronger than force *C*, it tends to pull the solid body of the earth away from the ocean on the far side, causing the tidal bulge on that side. The difference in the moon's gravitational force between the far side and the center of the earth is the same as the difference in force between the center and the near side and, therefore, the height of the tide is the same on the far side and the near side (Figure I.21).

*The Solar Tide.* The sun's gravity also pulls on the part of the earth facing the sun more strongly than it pulls on the other parts of the planet. Thus, like the moon, the sun tends to raise a tide on the earth. Although the sun is much more massive than the moon, it is also much further away. As a result, the tide-raising force of the sun is less than half as strong as that of the moon. However, twice each month the sun and

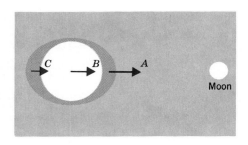

Figure I.21 *Explanation of the tides: the lengths of the arrows indicate the relative strength of the moon's gravitational pull on various parts of the earth.*

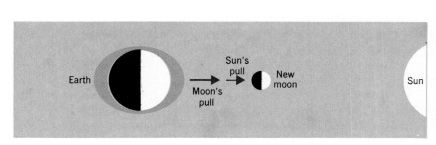

moon work together to produce exceptionally large tides. These are the times when the moon is new and the moon and sun are on the same side of the earth and when the moon is full and the sun and moon are on opposite sides.

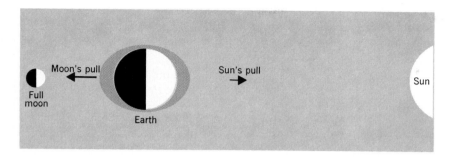

The tides produced on these occasions, called *spring tides,* are 20 percent higher than the average tides.

On two other occasions in each month, the sun and moon are at right angles with respect to the earth. These situations occur when the moon is in the first or third quarter (below).

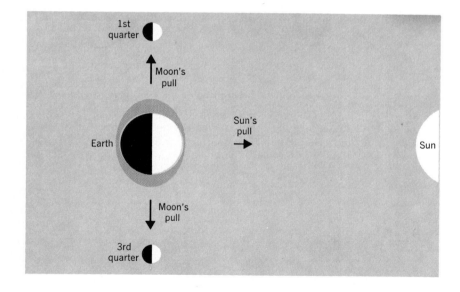

At these times the sun's gravity works to draw away the water from the points on the earth's surface to which it would be pulled by the moon's tidal force. Since the sun's tidal force is not as strong as the moon's, a lunar tide still occurs, but it is lower than it would be if the sun were not pulling away some of the water. The resultant tides, called *neap tides,* are 20 percent lower than the average tide.

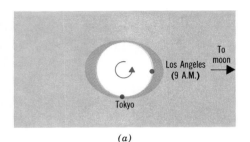

*(a)*

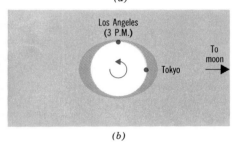

*(b)*

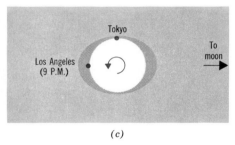

*(c)*

Figure I.22   *Rotation of the earth beneath the tide.*

*The Effect of Tides on the Rotation of the Earth.* During the course of one day, as the earth turns on its axis, the tide remains oriented in the direction of the moon. Thus, the earth rotates underneath the tide, or, from the viewpoint of a person on the earth's surface, the tide moves backward through the surface of the earth. For example, suppose it is high tide in Los Angeles at 9 A.M. on a certain day (Figure I.22a).

Six hours later, Japan will have moved into position under the moon, and it will be high tide in Tokyo (Figure 1.22b).

Los Angeles will not see a high tide again until 9 P.M. in Los Angeles, when it will have rotated into position on the line to the moon, but on the other side (Figure I.22c).

The daily motion of the tides across the oceans and seas, as well as through the solid body of the land, creates a large amount of friction. A huge quantity of energy is dissipated in this friction each day. If the energy could be recovered for useful purposes, it would be sufficient to supply the electrical power requirements of the entire world several times over. The energy is actually dissipated in the turbulence of coastal waters plus a small degree of heating of the rocks in the crust of the earth, and cannot be diverted to constructive work.

However, the tidal friction has another effect of great importance. The tides act as a brake on the spinning earth, slowing down its rotation at a steady rate. In other words, the tides tend to increase the length of the day. Because the earth is so massive, and its inertia is so great, the effect of the tides on the length of the day is extremely small. At the present time, as a result of the tides, each day is one-hundred millionth of a second longer than the preceding day. This small effect projected backward over geological time leads to the conclusion that the day was only 22 hours long, and the year contained 400 days, 300 million years ago in the Devonian era, when our vertebrate ancestors were emerging from the water onto the land. The fossil record contains evidence that there were 400 days in a year in the Devonian era, indicating that tides have, in fact, been slowing down the earth's rotation throughout this long period of earth history.

## THE CELESTIAL SPHERE

The motion of the earth makes it difficult to fix the positions of celestial bodies. In observing the stars, the astronomer as well as the amateur star-gazer is confronted by the problem of the man trying to keep his eye on a distant object as he rides a merry-go-round. The solution adopted by modern astronomers is based on a concept borrowed from the ancients, who thought of the sky as a spherical dome covering the heavens, across which the sun, moon, planets, and stars moved in stately procession. This rigid dome, spanning the vault of the heavens, is called the *celestial sphere.*

Just as we locate the position of an object on the earth by specifying its latitude and longitude, in the same manner astronomers locate the positions[3] of stars in the heavens by marking off lines of latitude and longitude on the celestial sphere. Celestial latitudes are measured in degrees north or south of the celestial equator, defined as the projection of the earth's equator onto the celestial sphere. Latitudes north or south of the equator are given plus or minus signs, respectively. The celestial latitude of a star is called its *declination,* abbreviated Dec.

Celestial longitude, however, cannot be defined in the same way as longitude on the earth, because the earth's lines of longitude rotate with it and sweep across the celestial sphere. Coordinates on the celestial sphere must be fixed in space and not moving with the earth, since they are intended to describe the locations of fixed stars. Astronomers have agreed to select a fixed point on the celestial equator to be used as the starting point for marking off degrees of celestial longitude, just as the Greenwich Meridian is used as the zero-point for measuring degrees of longitude on the surface of the earth. According to their agreement, the chosen place on the celestial equator is marked by the direction of a certain point located in the Constellation Pisces. The zero-point direction is defined more precisely as the direction of the line formed by the intersection between the earth's equatorial plane and the plane of the earth's orbit. This point is called the Vernal Equinox.[4] Because of the precession of the earth's axis of rotation, the chosen direction is not fixed in space but moves slowly through a circle, completing one circuit every 26,000 years (see page I-5). An accurate statement of the coordinates of a celestial body includes a reference to a specific year, indicating that these coordinates are based on the position of the celestial longitude zero-point in that year.

Celestial longitude is marked off in degrees running westward around the celestial equator, ranging from zero degrees at the position of the Vernal Equinox to 360 degrees on the return to this point. With the selection of a zero-point on the celestial equator, we have completed the definition of the coordinate system required to locate stars and other objects on the celestial sphere (Figure I.23).

---

[3] *Position* and *location* are used interchangeably with *direction* in this context. Normally the position of an object means its location in space, including, in the case of stars, not only the direction of the star along the line of sight from the earth, but also its distance from the earth. However, the celestial sphere is assumed to be infinitely far away. Two stars which are at different distances but lie on the same line of sight from the earth will have identical coordinates on the celestial sphere.

[4] The *Vernal* (of the spring) *Equinox* (night equal to day) is the point on the celestial sphere at which the sun is to be found on March 21. Its name derives from the fact that when the sun is at the Vernal Equinox, spring begins in the Northern Hemisphere and the day and night each have 12 hours.

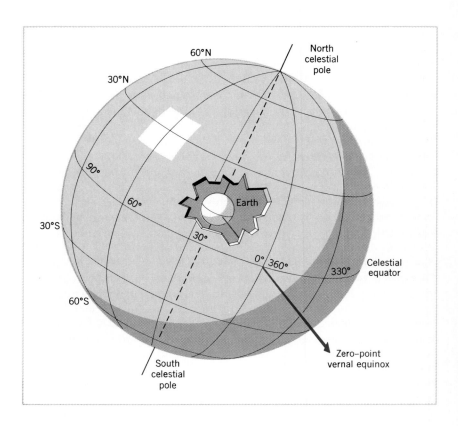

*Figure I.23    The celestial sphere.*

## Sidereal Time: Right Ascension

The stars revolve overhead as the earth turns on its axis. Once in every rotation of the earth, an observer on the ground looks out into the same direction of the sky and sees the same stars in the same positions. For this reason, the period of rotation of the earth is called the *sidereal day* (Latin: *sider* — star). The length of the sidereal day is 23 hours, 56 minutes, and 3 seconds.

A day is usually defined as having precisely 24 hours. Why is the sidereal day — or period of rotation of the earth — shorter by approximately four minutes? The answer is that the 24-hour day is a *solar* day, defined as the interval from noon to noon, that is, the interval from the time the sun is at its highest point in the sky on one day to the time at which it reaches its highest point on the following day. One solar day is longer than the period of rotation of the earth because during the time in which the earth turns on its axis, it also moves along its orbit. As a result, the direction from the earth to the sun changes by a small amount from one day to the next (Figure I.24). It is clear from Figure I.24 that in the period from noon to noon the earth completes slightly more than one turn; consequently, the solar day is slightly longer than the sidereal day.

The difference turns out to be approximately four minutes because the earth completes a full orbit of 360 degrees in 365 days, and thus advances along its orbit through an angle of (360/365) degrees, or nearly one degree, in each day of the year. Consequently, the direction of the sun at noontime shifts by about one degree each day. The earth rotates through 360 degrees on its axis in approximately 24 hours or 1440 minutes, corresponding to a rate of rotation of one degree in four minutes. Thus, four minutes elapse while the earth rotates the extra degree required to complete a solar day.

*Sidereal Time Versus Solar Time.* If you observe the position of a particular star at, say, 9 P.M. on a certain evening, and look for that star at 9 P.M. on the following evening, you will find that it is slightly to the west of the position it had on the previous night, because the earth has rotated four minutes or one degree past a complete turn. However, at 8:56 P.M. on the second night you would find this star at its original position, and again at the same position at 8:52 P.M. on the night after that. If you adjusted your watch to run four minutes fast each day, each star would appear in the same position after an apparent lapse of 24 hours. The watch would then be keeping sidereal time. A sidereal watch—running four minutes fast each day—would be a great convenience to a person who observed the stars a great deal, because all the stars would return to their identical positions in the sky every 24 hours. In sidereal time a star appears on the horizon, rises to its highest point in the sky, and sets, at the same time throughout the year. Astronomical observatories find it useful to have clocks that keep sidereal time, but

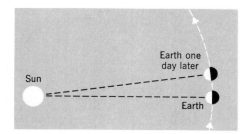

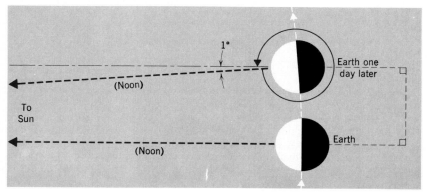

Figure I.24   The daily shift in the apparent position of the sun at noon, as the earth moves in its orbit.

an astronomer will soon get into trouble with his wife if he decides when to go home on the basis of a sidereal clock,[5] because he will be home four minutes earlier every day, and after six months he will expect his dinner at breakfast time. Clocks that keep solar time are preferable for the regulation of practical affairs.

*Right Ascension.* The position of a star on the celestial sphere was defined above in terms of celestial latitude and longitude, each expressed in degrees. Astronomers frequently use a mixed system of coordinates in which celestial latitude is given in degrees north or south of the celestial equator, but celestial longitude is given in hours, minutes, and seconds of sidereal time, calculated from the celestial longitude in degrees by applying a ratio based on the equivalence of 360 degrees to 24 sidereal hours. The celestial longitude of a star expressed in these units, is called its *Right Ascension* (R.A.). The Right Ascension runs eastward around the celestial equator from zero hours at the zero-point on the equator to 24 hours at the same point after completion of a full

Table I.1

**The Twenty Brightest Stars in Order of Decreasing Brightness**

| Name | RA Hours | RA Minutes | DECLINATION Degrees | DECLINATION Minutes | Distance (light-years)[a] |
|------|------|---------|---------|---------|-------------------|
| Sirius | 6 | 42.9 | −16 | 39 | 8.8 |
| Canopus | 6 | 22.8 | −52 | 40 | 98 |
| Arcturus | 14 | 13.4 | +19 | 27 | 36 |
| Alpha Centauri | 14 | 36.2 | −60 | 38 | 4.3 |
| Vega | 18 | 35.2 | +38 | 44 | 26 |
| Capella | 5 | 13.0 | +45 | 57 | 46 |
| Rigel | 5 | 12.1 | − 8 | 15 | 900 |
| Procyon | 7 | 36.7 | + 5 | 21 | 11 |
| Achernar | 1 | 35.9 | −57 | 29 | 150 |
| Hadar | 14 | 0.3 | −60 | 8 | |
| Altair | 19 | 48.3 | + 8 | 44 | 16 |
| Betelgeuse | 5 | 52.5 | + 7 | 24 | 700 |
| Aldebaran | 4 | 33.0 | +16 | 25 | 68 |
| Alpha Crucis | 12 | 23.8 | −62 | 49 | 350 |
| Spica | 13 | 22.6 | −10 | 54 | 230 |
| Antares | 16 | 26.3 | −26 | 19 | 400 |
| Pollux | 7 | 42.3 | +28 | 9 | 35 |
| Fomalhaut | 22 | 54.9 | +29 | 53 | 23 |
| Deneb | 20 | 39.7 | +45 | 6 | 1400 |
| Beta Crucis | 12 | 44.8 | −59 | 25 | 500 |

[a] The light-year = 5.8 trillion miles, the distance traveled by light in one year (see chapter 1).

[5] An observatory sets its sidereal clock to read zero hours when the zero-point of longitude on the celestial equator is ''overhead,'' that is, at its highest point in the sky. Different observatories have sidereal clocks set to different times, just as people live in different time zones. Of course, all sidereal clocks run at the same sidereal rate.

circle. If a star lies 90 degrees east of the zero-point, for example, its Right Ascension is 6 hours. If the star lies 210 degrees east, its Right Ascension is 14 hours. Table I.1 lists the 20 brightest stars in the sky with their coordinates in Declination and Right Ascension, as they would appear in a standard star catalog.

The convenience of Right Ascension is connected with the concept of sidereal time. An astronomer knows that on any night of the year a star he is observing will be "overhead," that is, at its highest point in the sky, when the time on his sidereal clock is the same as the star's Right Ascension.

## THE CHANGING NIGHT SKY

On a clear autumn evening in most locations in the United States, the brightest stars in the sky have the appearance shown below in Figure

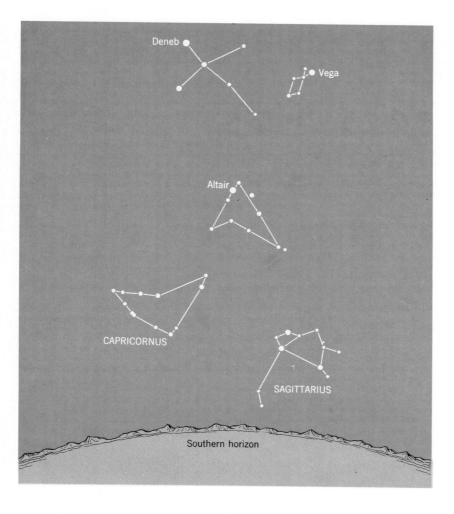

*Figure I.25(a)   The Autumn sky.*

I.25a. Deneb, Vega, and Altair form a conspicuous triangle overhead, and two ensembles of stars nearer the horizon are grouped in the characteristic shapes of the Constellations Capricornus and Sagittarius.[6]

But on a spring night in the same location, most of these stars are gone, and their places are taken by an entirely different group dominated by the Dog Star, Sirius—the brightest star in the sky—and by the Constellations Canis Minor, Gemini, Leo, Orion, and Canis Major[7] (Figure I.25b).

[6] The Goat and the Archer.
[7] Little Dog, Twins, Lion, Hunter, and Great Dog.

*Figure I.25(b)   The Spring sky.*

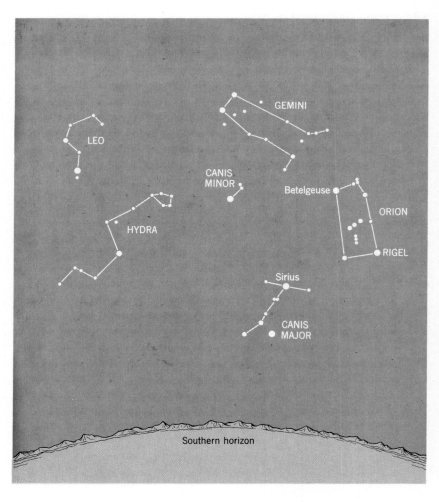

Year after year the same stars and constellations return, each in its proper season. Every winter the Constellation Capricornus appears; every summer Capricornus disappears, and other constellations, such as Leo, appear instead.

The explanation for these changes is connected with the revolution of the earth around the sun. Each night a person looks out into space along a slightly different direction than on the previous night because the earth has moved a small distance along its orbit during the past 24 hours (Figure I.26).

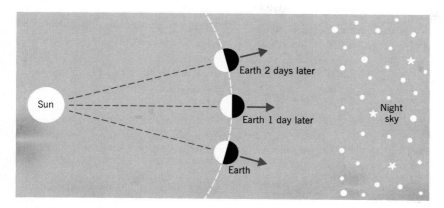

Figure I.26 The changing view of the night sky as the earth moves in its orbit.

The change in direction is only one degree of angle every 24 hours, and its effect on the position of a star is too small to be noticed by the naked eye from one night to the next, but as the earth continues along its orbit during the course of weeks, the change in direction becomes substantial. After six months the earth has moved to the other side of its orbit, and the observer of the night sky, looking out at a different part of the universe, sees a different family of stars.

On page I-22 we noted that a star rises four minutes earlier every night as a consequence of the difference between the sidereal and solar days. The complete alteration in the appearance of the night sky in six months is the result of a large number of four-minute daily changes. Four minutes a day summed over six months adds up to 12 hours, signifying that a star visible in the night sky at one time in the year will be in the day sky, and invisible, six months later.

Figure I.27a shows the earth at the point of its orbit that corresponds to the month of September. At this time, the night side of the earth faces a part of the sky in which the constellations Capricornus and Sagittarius are located, and a person observing the night sky will see these two constellations in that month. Figure I.27b shows the earth in the position of its orbit that corresponds to the month of March. At this time the night side of the earth looks out into the part of the sky containing the Constellation Leo. Therefore, Leo will be visible to a person observing the night sky during March.

## INTERESTING OBJECTS IN THE NIGHT SKY

The changes described above can be verified readily by the untrained observer if he watches the prominent constellations during the course of

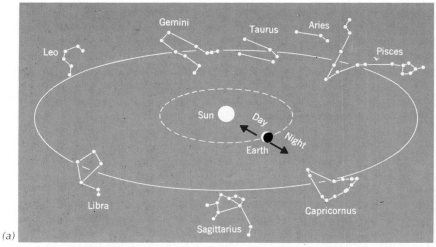

(a)

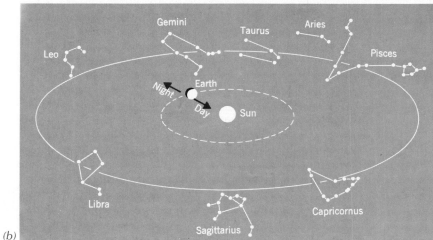

(b)

Figure 1.27(a & b) Changes in the appearance of the sky during the course of the year: (a) September (b) March. The arrows on the night side point toward the constellations that are on the meridian at 9 P.M. in the indicated months.

several months. We assume that the reader is embarking on his study of astronomy at the beginning of the autumn or the spring term, and choose September and February skies as an example. Among the objects visible in these months are some that play an important role in the story of stellar evolution to be unfolded in Part One. These objects, described below, are easy to find provided that the night is clear. The descriptions of their positions assume that the observer is out in the field around 9 P.M. on a September or a February evening.

*September.* As a start, orient yourself with respect to north and south, using the North Star, a compass, or your memory as to where the sun rises and sets. Face south and tilt your head back as far as you can, so that you are looking at the *zenith,* which is the point in the sky directly above your head. You will see a triangle formed by three bright stars (Figure I.28). It is a huge triangle, extending from the zenith nearly halfway to the southern horizon, and it is impossible to miss. The triangle is "upside down," that is, its base is at the top, near the zenith, and the vertex is pointed downward toward the southern horizon. The two stars at the top, forming the base, are very conspicuous because they are the only bright stars near the zenith. These two stars are called Deneb and Vega. The star at the bottom, forming the vertex of the triangle, is called Altair. Altair also is easy to spot because it is a part of a short, straight line of three stars. (Altair is in the middle of the line of three and is brighter than the two stars on either side.)

Of the three stars in the triangle, Vega appears to shine most brightly. Altair is next, and Deneb is the dimmest. The actual brightnesses of the three stars differ radically from their apparent brightnesses, because they lie at widely different distances from us. Deneb is intrinsically the bright-

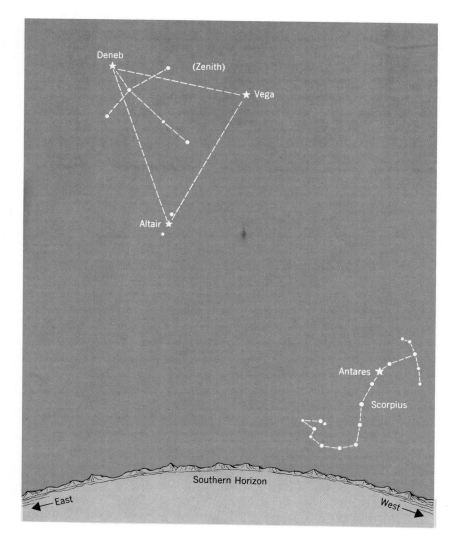

Figure I.28   The summer triangle formed by Deneb, Vega, and Altair.

est of the three stars, 1600 times brighter than Vega and 6400 times brighter than Altair, but it appears to be the dimmest star in the triangle because it is farthest away—1400 light-years from the solar system versus 26 light-years for Vega and 16 light-years for Altair. Because the intensity of light from a star decreases in proportion to the square of its distance from us, the fact that Deneb is 1400/16 = 88 times farther away than Altair means that its brightness is cut down by a factor of $88^2$ or approximately 7700.

Passing through the middle of the triangle of three bright stars is a luminous band, which is the Milky Way. The Milky Way will only be visible if the night is clear and you are away from the lights of large cities. It is formed by the overlapping light from 100 billion or so stars that are packed densely together in the midplane of our Galaxy.

Turning partly to the west from the direction of the triangle, and looking down toward the horizon, you will see the Constellation Scorpius, which will be easily recognizable as a coiled chain of stars resembling a scorpion's tail. It is near the horizon a little bit west of the downward extension of a line joining Deneb and Altair. The brightest star in this constellation is Antares, a distinctly red star. The name Antares is derived from Greek and means "similar to Mars." Because of its red color, ancient astronomers thought Antares was another red planet rivaling Mars.

You can find Antares in another way if you have trouble locating it according to the directions above. Look at Deneb again, and you will see that it marks the top of a cross whose long arm points downward to the southwest through the middle of the huge triangle. If you follow the direction of this arm of the cross with your eye, moving in a straight line, it will bring you to the reddish star, Antares, near the horizon. Figure I.28 shows the cross that runs from Deneb through the triangle toward Antares.

Antares, which lies 400 light-years from the sun, is intrinsically 30,000 times brighter than the sun, about 10 times more massive, and several hundred times larger. It is one of a group of stars called red giants because of their color and size. Originally, Antares was a brilliant blue star but, as with all stars when they grow old and consume their fuel, it increased in brightness and reddened in color as it swelled to its present size. When the sun reaches this stage in its life, in which the hydrogen at its center has been burned and converted to helium, it will also become a red giant, thousands of times brighter than it is today.

Turning to the north, you will see the Big Dipper in the northwest. The bowl of the dipper is less than one-third of the way from the horizon to the zenith, and the handle extends toward the west. The various stars that make up the Big Dipper lie at quite different distances from us. The stars of the Big Dipper provide an example of the fact that the stars making up a constellation may be unrelated to one another and situated at widely different distances from the sun. Usually, the stars in a constellation happen to fall into a closely knit or clearly recognizable pattern only because of the accidental circumstances of the particular line of sight from which we on earth view these stars. Of the seven stars in the Big Dipper, five stars—the middle five of the seven—are all about 75 light-years away from the sun. They form a single group moving through space with approximately the same velocity and direction. The first star of the seven, marking the end of the handle, and the last star, marking the end of the bowl, are completely unrelated to the other five and to one another. The star marking the end of the handle is 109 light-years from us and moving in a direction opposite to that of the middle five, while the star marking the end of the bowl is 250 light-years away and also moving in a direction opposite to that of the middle five stars.

Because of the differences in their directions of movement, the stars of the Big Dipper will not retain the famous form they have today for very long. In 50,000 years the handle will become badly bent, and the bowl will be as flattened as if a child had been banging it on the table. Figure I.29 shows the Big Dipper as it is today, with arrows marking the direction of motion of each of its seven stars, the Big Dipper as it was 50,000 years ago, and the Big Dipper as it will look 50,000 years hence.

If your eyesight is good, you will see that the next to last star in the handle of the Dipper consists of two stars close together forming a so-called *optical double*. One of the two stars shines about twice as brightly as the other. The brighter of the two is called Mizar and the less bright is called Alcor. They are sometimes called the Horse (Mizar) and Rider (Alcor). Actually, the two stars are quite far apart in space, but happen to lie along approximately the same line of sight from the solar system, which makes them seem close together.

To make the situation even more complicated and interesting, the brighter of these two stars—Mizar—is itself a pair of stars, too close together to be resolved except with the aid of a good-sized telescope. They are so close to one another that the force of gravity ties them permanently together. Stars close enough together to be tied to one another by their own forces of gravity are called *binary stars*. The binary in the handle of the Big Dipper is the first that was ever observed. Both stars of the binary are brighter and more massive than the sun, Mizar being altogether three times as massive as the sun. Because of their higher surface temperature, their color is white in comparison to the sun's yellow-white.

One of the most important stars in the sky is the North Star, or Polaris, which happens to lie almost exactly on the earth's axis of rotation, making it a useful star for navigation. To find the North Star, draw a line through the two stars in the Dipper that form the end of the bowl and extend it away from the bowl about five times the distance between these two stars (Figure 1.30).

Because the North Star lies on the earth's axis of rotation, it appears stationary as the earth rotates, but all the stars near it seem to move in circles around the North Star during the course of the night. A photograph of the night sky, taken with a time exposure lasting several hours, shows the paths of the stars circling the North Star. A constellation near the North Star, such as the Big Dipper, seems to turn over during the course of 24 hours as a result of this rotation (Figure I.31). On the other hand, if you turned away from the North Pole and looked to the south, you would find that stars quite far to the south would move along relatively flat paths that

Figure I.29   The Big Dipper (a) today; (b) 50,000 years ago; (c) 50,000 years hence.

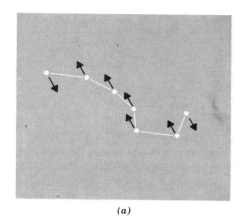

(a)

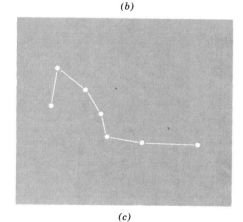

(b)

(c)

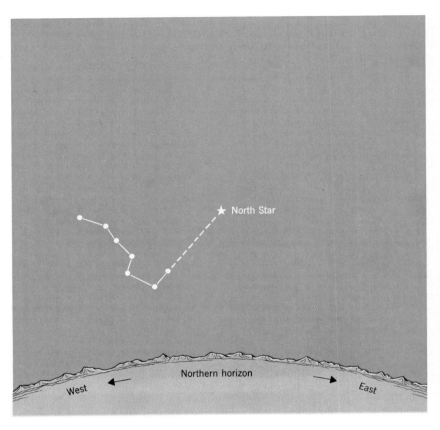

Figure I.30  *Finding the North Star.*

never carried them very high in the sky. These stars, like the stars in the north, are also describing circles around the earth's axis of rotation. However, most of their circular path is blocked from your view by the earth (Figure I.32).

The North Star has an additional interest because it is an example of a group of stars whose light output, or brightness, varies regularly instead of remaining constant for most stars. In the case of Polaris, the brightness increases and decreases by a factor of 9 percent over four days. The variation is produced by a rhythmic breathing, or pulsation, in the size of the star; when it is expanding it shines more brightly and when it is contracting its brightness is at a minimum.

*February.* As before, it is assumed that you are out in the field around 9 P.M. Orient yourself again with respect to north and south, and, facing south, look for a point in the region of the sky approximately midway between the horizon and your zenith. You will see a famous constellation known as Orion, or the Hunter, shown in Figures I.33 and I.34. Orion is the most prominent feature of the winter and spring skies, replacing the summer triangle of the September sky in this respect.

Three close stars in the center of Orion form a straight line running from southeast to northwest. The stars are not exceptionally bright, but the tight line of three is readily recognizable. The three stars are the belt

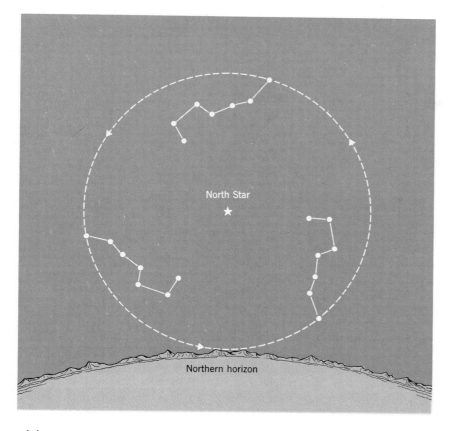

North Star

Northern horizon

Figure I.31   Changes in the position of the Big Dipper during 24 hours.

of the Hunter (Figure I.34). Immediately below the belt lies the sword of the Hunter, and in the middle of the sword is the spectacular Orion Nebula (Color Plate 4). The Orion Nebula is a region filled with a dense concentration of interstellar gases, rich in newborn and young stars. Unfortunately, it is not visible to the naked eye. Figure I.35 shows the position of the nebula in the central region of Orion.

Some distance above the belt stars and slightly to the left, on the right shoulder of the Hunter, is a bright star with a reddish hue, called Betelgeuse.[8] This star is an exceptionally large and luminous red giant, similar to Antares but nearly 10 times brighter. Betelgeuse is located at a distance of 700 light-years from us and is intrinsically 100,000 times brighter than the sun, 20 times more massive, and 500 times larger. If placed at the center of the solar system, this giant star would engulf the earth and extend beyond the orbit of Mars. Betelgeuse, like Antares, was a brilliant blue star in the prime of its life, but as its fuel became exhausted it reddened and expanded into its present form.

To the west and a slight distance downward lies a fairly bright star named Bellatrix, with approximately the same brightness as the stars in

[8] Pronounced Beet-il-jooz.

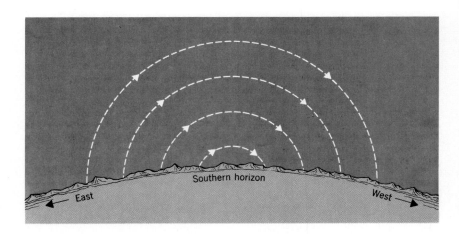

Figure I.32 Star tracks over the southern horizon.

the belt. Bellatrix is located in the left shoulder of the Hunter. Like Betelgeuse, it is considerably hotter than the Sun, with a surface temperature in the neighborhood of 20,000°K, but a blue color in contrast to the red color of Betelgeuse. Bellatrix and similar stars are called blue giants. They are unusually massive stars, still in the prime of their lives, with a plentiful supply of fuel remaining.

A line drawn diagonally downward from Betelgeuse, across the line of the belt stars toward the southwest, will intersect a very bright blue-white star called Rigel, in the left thigh of the Hunter. Rigel, like Bellatrix, is a blue giant in its prime, far more massive, larger and brighter than the sun. A study of Rigel with a telescope reveals it to be a binary—a system of two stars circling about one another under the pull of mutual gravity. The small component, although invisible to the naked eye, is intrinsically far brighter than the sun.

Returning to the belt stars, a line drawn through these three stars and extended downward to the southeast intersects an extremely brilliant star, the brightest star in the heavens, known as Sirius (Figure I.33). Sirius is also called the Dog Star because it is located in the Constellation Canis Major—the Great Dog. Sirius is only 20 times brighter intrinsically than the sun, but it dominates the stellar multitudes in brilliance because it is so close to us, being our fifth closest neighbor. Sirius is also a binary, and of a particularly interesting sort; the brilliant, visible component is circled by a small, invisible companion star, detectable only with large telescopes. The invisible companion is as massive as the sun, but 300 times fainter, and its radius is only 20,000 miles, closer to the size of an earthlike planet than a star. This invisible companion to Sirius belongs to a class of stars known as the white dwarfs. These are dying stars that have exhausted their fuel and have collapsed to a small radius and extraordinary density under the inward force of their great weight. Slowly, with the passage of time, they cool to black dwarfs and disappear from view. It is believed that the sun will follow this course some six billion years from now.

Directly north of Sirius, and at approximately the same elevation above

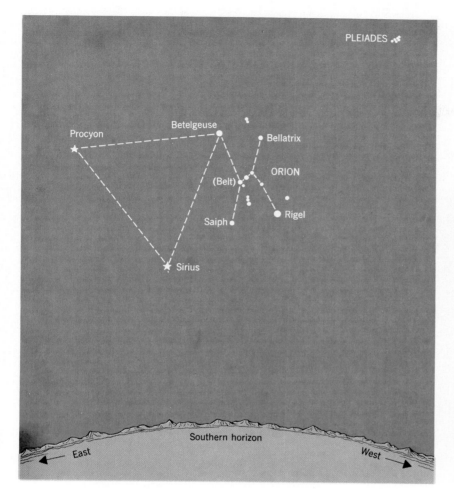

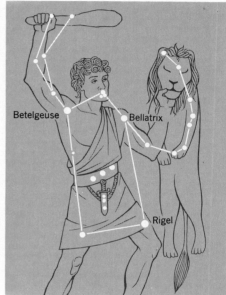

Figure I.34  The Hunter.

the horizon as Betelgeuse, is a fairly bright star called Procyon (Figure I.33). Procyon, Sirius, and Betelgeuse form the nearly equilateral winter triangle in the sky, similar to the summer triangle, but less conspicuous because the February sky contains so many bright stars. Procyon lies at a distance of about 11.5 light-years from the sun and is intrinsically 7.5 times brighter than the sun. It, too, has a companion white dwarf orbiting it.

A line drawn diagonally upward to the northwest from the belt of The Hunter will intersect the small cluster of stars known as the Pleiades (Figure I.36 and Color Plate 9). Six stars can be seen in the Pleiades with the naked eye, but a telescopic study reveals approximately 100 stars within the entire cluster. The stars in the Pleiades cluster are very young, the entire cluster having been formed out of a single concentration of interstellar gas approximately 60 million years ago. They are among the youngest stars visible in the sky.

This brief tour of the autumn and winter skies demonstrates that the

Betelgeuse

Bellatrix

Belt

Nebula    Sword

Rigel

*Figure I.36   A photograph of the Pleiades.*

heavens are populated by stars of many colors, sizes, and ages. The meaning behind the existence of this puzzling variety of stars eluded astronomers until recent years, when rapid progress occurred through the discoveries made in astronomy and other branches of the physical sciences. Some important parts of the story of the stars are now known, and form the subject matter of parts One and Two of this book, but major mysteries remain, and new discoveries are reported by astronomers almost daily. The new developments in astronomy cannot be grasped without an understanding of basic concepts such as the nature of light, the structure of the atom, and the release of nuclear energy. Before the reader embarks on a voyage through these interesting waters, he should first acquire a general perspective on the subject of astronomy by familiarizing himself with the contents of the universe.

**Part One  Stars**

# 1    Contents of the Universe

In recent years the history of man's origins has been extended backward in time from the beginning of life on the earth to the beginning of time in the Universe. The result is a new and more complete story of genesis, which provides fresh perspectives on some of the most profound questions to occupy the mind of man: What am I? How did I get here? What is my relationship to the rest of the Universe?

The ideas involved in this bold inquiry into the origin of the Universe are not a part of everyday thinking. Before the mind can understand them, it must first stretch its concepts of space far beyond their normal limits, and exercise the imagination to vault over the galaxies and span the dimensions of the Universe.

## THE SUN AND ITS NEIGHBORS

The sun, its family of planets with their moons, and a large number of smaller bodies, form the *solar system*. The earth travels around the sun at

*The Andromeda galaxy.*

an average distance of 93 million miles. Between the earth and the sun lie the planets Mercury and Venus. Outside our orbit lie the earthlike planet Mars, the giant planets Jupiter, Saturn, Uranus, and Neptune, and the frozen world of Pluto (see Figure 1.1). Pluto, the outermost planet, travels in an elliptical orbit at distances from the sun ranging between 3 and 4 billion miles. The diameter of the solar system—defined by the farthest sweep of Pluto's orbit—is approximately 8 billion miles.

*Figure 1.1    The solar system.*

Beyond the orbit of Pluto, space contains nothing but a few atoms of hydrogen and occasional comets, until we reach the stars that are the sun's neighbors. These stars are an average of 30 trillion miles away.

An anology will help to clarify the meaning of such enormous distances. Let the sun be the size of an orange; on that scale the earth is a grain of sand circling in orbit around the sun at a distance of 30 feet; Jupiter, eleven times larger than the earth, is a cherry pit revolving at a distance of 200 feet or one city block; Saturn is another cherry pit two blocks from the sun; and Pluto is still another grain of sand at a distance of ten city blocks from the sun. The nearest stars are orange-sized objects more than a thousand miles away.

An orange, a few grains of sand some feet away, and then some cherry pits circling slowly around the orange at a distance of a city block; more than a thousand miles away is another orange, perhaps with a few specks of planetary matter circling around it. That is the void of space.

## The Nearest Stars

The sun's closest neighbor is, according to information available at the present time, the star Alpha Centauri. Alpha Centauri is 24 trillion miles from our solar system, or slightly closer than the average distance from the sun to its neighbors. It is actually a triple star—a family of three stars formed simultaneously out of a single cloud of gas and dust. Ever since their birth, the three stars have circled one another under the attraction of gravity. The largest of the three stars in Alpha Centauri resembles the sun in size and possesses a similar surface temperature and color. The other two are smaller, redder stars. The middle-sized star of the triplet, somewhat smaller than the sun and orange in color, circles around the largest star in a close waltz at a distance of two billion miles. One turn around takes 80 years for this pair. The third member is a very small, faint, red star, a tenth as massive as the sun, which circles the two "close" members of the triplet at a distance of a trillion miles, completing one turn in a million years.

Alpha Centauri is the closest star to us that is bright enough to be visible. However, the sun may have still closer neighbors—very small, dimly luminous stars—too faint to have been detected thus far. There may also be burned-out stars that have exhausted their fuel in the space between the sun and Alpha Centauri. Finally, there may be many bodies the size of planets, too small to glow by their own nuclear energy, in the space around us. All these possibilities await the future exploration of the regions outside the solar system.

The next nearest neighbor of the sun beyond Alpha Centauri is Barnard's Star, 30 trillion miles away. Barnard's Star is smaller than the sun, and almost 300 times fainter. The temperature at the surface of Barnard's Star is 5000 degrees Fahrenheit versus 11,000 degrees Fahrenheit at the surface of the sun, and its color is orange-red rather than yellow. Barnard's Star, unlike Alpha Centauri, is a single star. However, it was discovered recently that two planet-sized objects, both approximately the size of the planet Jupiter, revolve in orbit around it.

Fifty other stars exist within roughly 100 trillion miles of the sun. Some are yellow stars, resembling the sun in size and temperature; a few are larger and brighter than the sun and blue-white in color; most are faint, reddish stars. These stars are our neighbors in space. All have been named, although only a few of the names are familiar. Twenty-six of the 50 belong to multiple stars—doubles or triples. On the average, about one half the stars in the Universe are multiple stars.

## Our Galaxy

The sun and its neighbors are only a few among 100 billion stars that are banded together by gravity in an enormous cluster, called the Galaxy. Most, if not all, the stars in the Universe are held within such clusters. These other clusters also are called galaxies. Our own galaxy, singled out because it contains the sun, is written with a capital "G."

## THE CONTENTS OF THE GALAXY

### The Appearance of the Galaxy

The stars in the Galaxy revolve about its center as the planets revolve about the sun. The sun itself participates in this motion, completing one circuit around the Galaxy in 200 million years. The Galaxy is flattened by its rotating motion into the shape of a disk, whose thickness is roughly one-twentieth of its diameter. The sun is located in the disk about three-fifths of the way out from the center to the edge. A small, spherical clump of stars, called the nucleus of the Galaxy, bulges out of the disk at the center.

The general appearance of the Galaxy is shown very clearly in photographs of three other galaxies that are similar to ours and that happen to be oriented in space so that we see them at several different angles. If you could stand outside the Galaxy and view it edge-on, it would look very much like NGC 4565 (Figure 1.4). Viewed face-on, the Galaxy would look like M74 (Figure 1.2). Viewed obliquely, the Galaxy would look like NGC 7331 (Figure 1.3). The arrow in Figure 1.2 illustrates the position that the sun would have if this were, in fact, a photograph of

*Figure 1.2 The galaxy NGC4565, resembling our Galaxy viewed edge-on.*

Figure 1.3  The galaxy NGC7331, resembling our galaxy in a three-quarter view.

Figure 1.4  The galaxy M74, resembling our Galaxy viewed face-on.

our Galaxy. Our Galaxy, like M74, has spiral arms in which most of the bright stars of the Galaxy are located, including the sun.

When we look into the sky in the plane of the galactic disk, we see so many stars that they are not visible as separate points of light but blend together into a luminous band stretching across the sky (Figure 1.5). The

Figure 1.5   Edge-on view of our Galaxy.

irregular lanes of black running through the center of the Milky Way in Figure 1.5 are caused by the extensive clouds of dust, concentrated in the central plane of the Galaxy, which black out the light of all but the nearest stars.

Additional details can be seen in a montage of many separate but

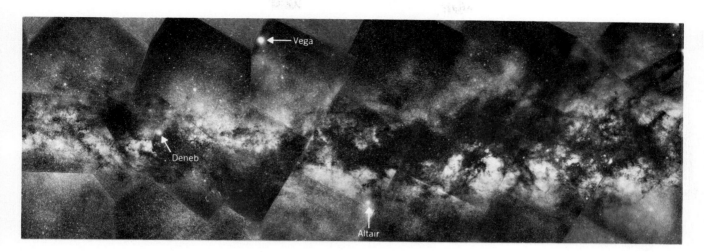

overlapping photographs of the Milky Way (Figure 1.6). Notice the extended black area in the center of the montage. This region, called the Great Rift, is a particularly striking example of the dense clouds of dust that prevent the light of the stars in the Milky Way from reaching us.

If the montage in Figure 1.6 is extended to include the entire Milky Way, it gives a nearly complete view of the Galaxy as seen by an observer located in our solar system. This edge-on view of our Galaxy is shown in Figure 1.7. It presents in an extended detailed view the image of the Galaxy that was captured by a single camera in Figure 1.5. Figure 1.7 also shows the positions of the 7000 brightest stars in the sky, drawn in by an artist. These are the stars that are visible to the naked eye under the best conditions. In addition, we have lettered in the names of several objects of special interest. Three of them are stars—Deneb, Vega, and Altair. The other objects of special interest are background external galaxies. The Andromeda Galaxy is the most distant object visible to the naked eye. The Magellanic Clouds, visible only in the Southern Hemisphere, are two dwarf galaxies anchored permanently to our Galaxy by its gravitational force, each containing only ten billion stars.

An inspection of the Milky Way with even a modest-sized telescope or pair of binoculars reveals the immensity of the number of stars concentrated in this region. In the photograph shown in Figure 1.8, obtained with the enormous light-gathering power of one of the world's large telescopes—the 120-inch instrument at Lick Observatory in California —more than 10,000 stars can be seen, although this photograph shows only a minute portion of the sky.

*Figure 1.6 A montage of Milky Way photographs.*

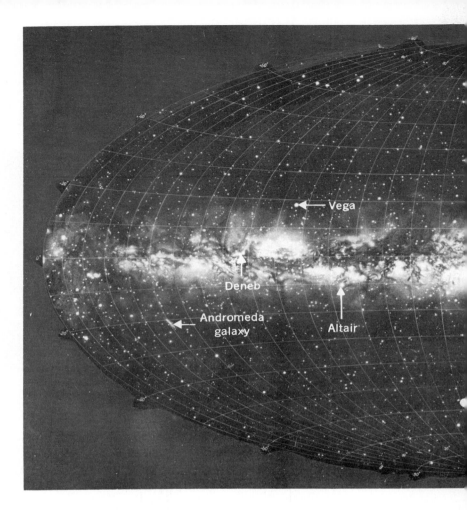

*Figure 1.7 The Milky Way: an edge-on view of our Galaxy.*

The glowing region in the middle of this photograph is called the North American Nebula because of its resemblance to the outline of the North American continent. Notice that very few stars are visible in the dark regions—the "Gulf of Mexico" and "Atlantic Ocean"—that outline the right edge of the North American Nebula. In these regions of the sky, an exceptionally thick concentration of obscuring dust conceals many of the stars in the Milky Way from our sight. The only stars that are visible in this region are the ones that lie on our side of the dust clouds, relatively near to the sun.

Deneb is located just beyond the right edge of the photograph. This fact should enable you to locate this small portion of the Milky Way in the montage of the entire Galaxy in Figure 1.7.

**The Emptiness of the Galaxy**

The vastness of the space between the stars is difficult to comprehend. The same analogy that we used to clarify the meaning of the size of the

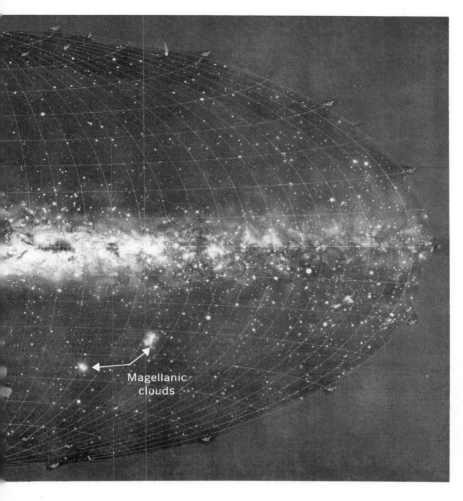

Magellanic clouds

solar system is helpful in attempting to comprehend the emptiness of the Galaxy. Suppose again that the sun is reduced from its million-mile diameter to the size of an orange. The Galaxy, on this scale, is a cluster of 100 billion oranges, each orange separated from its neighbors by an average distance of more than 1000 miles. In the space between, there is nothing but a tenuous distribution of hydrogen atoms and a few molecules and dust grains. That is the emptiness of space in the Galaxy.

## Distances Between Stars; the Light-Year

The stars within the Galaxy are separated from one another by an average distance of 30 trillion miles. To avoid the frequent repetition of such awkwardly large numbers, astronomical distances are usually expressed in units of the light-year, defined as the distance covered in one year by a ray of light traveling 186,000 miles per second. This distance turns out to be approximately 5.8 trillion miles; hence, in these units, the distance from the sun to Alpha Centauri is 4.3 light-years, the average distance between the stars in the Galaxy is 5 light-years, and the diameter of the Galaxy is 100,000 light-years.

Deneb →

Figure 1.8   The Milky Way: 10,000 stars in a small segment.

## THE UNIVERSE

Although the stars within our Galaxy are very thinly scattered, they are, nonetheless, relatively close together in comparison to the space that separates our Galaxy from neighboring galaxies. The distance to the next nearest galaxy comparable in size to ours is 2 million light-years, or 20 times the diameter of our galaxy. It is difficult to imagine the emptiness of intergalactic space. Once outside the Galaxy, we encounter a region of empty stars and nearly empty of dust and stars.

## Neighboring Galaxies

No vacuum ever achieved on earth can match the vacuum of the space outside our Galaxy. But if we go far enough away from the Galaxy, we come to other galaxies, clusters of billions of stars held together, like ours, by the force of gravity. These galaxies are island universes — isolated clusters containing vast numbers of stars and, perhaps, planets — each separated from the others by the void of intergalactic space.

The closest galaxy comparable to the Milky Way galaxy in size is the Andromeda galaxy, which is 2 million light-years from us. This galaxy happens to resemble our own closely in size and shape; it is a disk-shaped spiral of stars, gas, and dust, containing approximately 100 billion stars in all, the entire collection of matter slowly spinning around a central axis like a gigantic pinwheel.

Andromeda is the only major galaxy visible to the naked eye, and it is the most distant object that can be seen without the aid of a telescope. However, it is not conspicuous, despite the fact that its intrinsic brilliance is 100 billion times that of the sun. Because of its enormous distance, Andromeda is barely visible to the naked eye, under the best conditions, as a very faint patch of light.

But if it is photographed with even a modest-sized telescope, the faint patch is seen to have a structure that reminds one of our Galaxy, with a

*Figure 1.9 The Andromeda galaxy, our nearest galactic neighbor comparable in size to the Milky Way.*

brightly glowing center, a distinct impression of spiral arms, and dark lanes presumably formed by obscuring clouds of dust. The photograph of the Andromeda galaxy shown in Figure 1.9 , taken with a 48-inch telescope, indicates these features. The arrows point to two dwarf galaxies very close to the Andromeda, which are similar to the Magellanic Clouds in being captives of the attraction of the mass concentrated in their larger neighbor. The rectangle encloses the area shown in greater detail in Figure 1.10.

If Andromeda is photographed through a larger telescope such as the 100-inch one on Mount Wilson in California, whose light-gathering power and resolution of detail are thousands of times greater than those of the eye, the luminous cloud resolves into billions of individual stars. The photograph of Figure 1.10 taken through the 200-inch telescope, constitutes the area enclosed by a rectangle in the previous photograph. This picture shows many of the separate stars in the upper edge of the Andromeda galaxy. The many stars that appear in the photograph outside the boundaries of the Andromeda galaxy appear to be stars in intergalactic space, but this impression is misleading. Actually, these stars lie in our own Galaxy, but happen to be in the line of sight from the sun to Andromeda.

*Figure 1.10 (opposite page) Detail of the Andromeda galaxy, showing individual stars.*

## The Local Group

Approximately a dozen other galaxies, including the two dwarf galaxies known as the Magellanic Clouds, exist within 3 million light-years of ours. Astronomers call these galaxies the Local Group (Figure 1.11). Of the galaxies in the Local Group, only three—ours, Andromeda, and M33— have the spiral form shown in Figure 1.4. Our remaining galactic neigh-

*Figure 1.11   The Local Group. Only major galaxies are shown.*

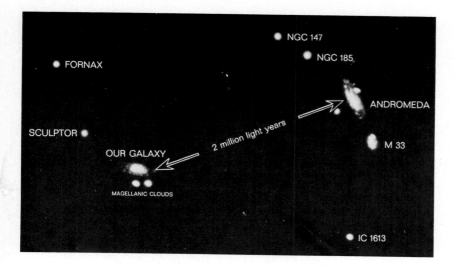

FORNAX

NGC 147

NGC 185

ANDROMEDA

SCULPTOR

OUR GALAXY

2 million light years

M 33

MAGELLANIC CLOUDS

IC 1613

bors are five to ten times smaller in diameter, 100 to 1000 times less populous, and either elliptical or irregular in shape. A photograph of one of the smaller, more irregular galaxies in the Local Group — the large Magellanic Cloud — is shown in Figure 1.12.

Figure 1.12    One of the Magellanic Clouds, a small captive galaxy of the Milky Way.

## Clusters of Galaxies

Enormous though a single galaxy is, it does not constitute the largest collection of matter known in the Universe. Galaxies themselves occur in clusters, held together, once again, by the force of gravitational attraction that each galaxy in the cluster exerts on the others.

Some clusters contain only a few galaxies. An example is the cluster of galaxies in the direction of the constellation Pegasus, shown in the photograph in Figure 1.13. (Astronomers have long used constellations, or groups of stars that seem to form figures in the sky, to impose some order on the huge number of stars in the night sky. Much like signposts, they tell us what part of the heavens we are gazing at.) In Figure 1.13, the spiked objects and the small circular spots are individual stars situated in our own Galaxy along the line of sight to the Pegasus cluster. The large, luminous objects marked by arrows are five galaxies in a small cluster.

*Figure 1.13 Stephan's Quintet: a small cluster of galaxies.*

The Local Group, of which our galaxy is a member, is an aggregate of galaxies including the Milky Way, Andromeda and numerous fainter objects. It is not certain that the Local Group is a clearly defined cluster.

The first group of galaxies outside the Local Group that clearly is a cluster is located in the direction of the constellation Virgo. It contains 2500 galaxies, is located approximately 60 million light-years from the Local Group, and is one of the larger clusters known.

About 300 million light-years from our Galaxy, in the constellation Hercules, is a giant group of galaxies called the Hercules Cluster, which

contains about 10,000 galaxies, each with 10 billion to 100 billion stars. Figure 1.14 shows one small region in the Hercules Cluster, containing approximately 50 galaxies. In this photograph, as in the preceding one, the spiked objects and the small spots of light are stars situated in our Galaxy along the line of sight to the Hercules Cluster; every other object in the photograph is a galaxy of stars.

The Hercules Cluster and other large clusters of galaxies like it are the largest systems of matter definitely known in the Universe. We might expect to find clusters of clusters of galaxies, but no clearcut proof of their existence has been obtained. Perhaps we will find solid evidence for clusters of clusters when astronomical observations improve in accuracy.

Figure 1.14 The Hercules cluster of galaxies.

## THE FORCES THAT HOLD THE UNIVERSE TOGETHER

All objects in the Universe, from the smallest atomic nucleus to the largest galaxy, are held together by only three fundamental forces — a nuclear force, the force of electromagnetism, and the force of gravity.

Most powerful is the *nuclear* force,[1] which binds neutrons and protons together into atomic nuclei. This extremely strong force of attraction pulls the particles of the nucleus together into an exceedingly compact body with a density of one billion tons per cubic inch. Although it is an exceedingly strong force, it has a very short range. The nuclear force will not pull two particles together if they are more than one ten-trillionth of an inch apart.

Next strongest is the *electrical* (electromagnetic) force, which is approximately 100 times weaker than the nuclear force. This force binds electrons to nuclei to form atoms, and it binds atoms together into solid matter. It grows weaker with increasing distance between two particles, although unlike the nuclear force, it does not disappear entirely at any point.

Least powerful is the force of *gravity*. The gravitational force is exceedingly weak, about $10^{36}$ times[2] weaker than the force of electricity. Gravity, like electricity, falls off in strength with increasing distance, but never disappears entirely. In spite of its intrinsic weakness, gravity holds the moon in orbit around the earth, the earth and other planets revolving around the sun, and the sun and other stars clustered together in our Galaxy.

Are there other forces in nature? Are there a fourth force and a fifth

---

[1] Two kinds of nuclear force exist — the one discussed here, and a weaker one, called the Weak Force. The latter plays no substantial role in our discussion and is omitted for simplicity.

[2] $10^{38}$ = 1 followed by 38 zeros. If this number were to have a name it would be 100 trillion, trillion, trillion.

force? Why are there only three forces? No one knows the answers to these questions. It is even more interesting to ask whether there might be connections among the fundamental forces of gravity, electromagnetism, and the nuclear force. Is it possible that there are fewer than three basic forces? Perhaps there is an underlying unity in the Universe that runs even deeper than the unity the physicists have uncovered thus far. Einstein spent the last 25 years of his life in the effort to discover a connection between gravity and electromagnetism which, his instinct told him, must be there. Yet, his instinct, superb though it was in revealing the truths of the special theory of relativity and the general theory of relativity, seems to have failed him in this last endeavor; for he never succeeded in finding a connection between these two forces that was satisfactory to him and to other physicists. Perhaps there is no connection; or perhaps it will require another genius, as great as Einstein, but born in a later age when more is known, to uncover this connection. These are among the unanswered questions of physics today.

**Questions**

1. Compare the emptiness of the atom — that is, the ratio of the size of the electron orbit to the size of the central nucleus — with the emptiness of the solar system.

2. Imagine each disk-shaped galaxy like ours to be the size of a dinner plate. On this scale, what would the distance be from our Galaxy to Andromeda? What would the average distance be between stars in the Galaxy? What would the extent of the observable Universe be, assuming its true radius to be 10 billion light-years?

3. Compare the number of atoms in a glass of water with the number of stars in the observable Universe.

4. List the heirarchy of structure from the smallest subatomic particles to the largest structures in the Universe. After each structure, list the basic force responsible for holding it together.

5. Can you think of a reason why clusters of clusters of galaxies probably do not exist? What characteristic of the gravitational force might explain this phenomenon?

6. Assuming that the Universe is infinite, what intensity of starlight would you expect to observe at the surface of the earth, coming from all the stars in the Universe?

7. It has been suggested that our Universe may be an atom in a larger cosmos, and each atom in our world a universe of its own. What philosophical and scientific arguments can you offer pro and con regarding this suggestion?

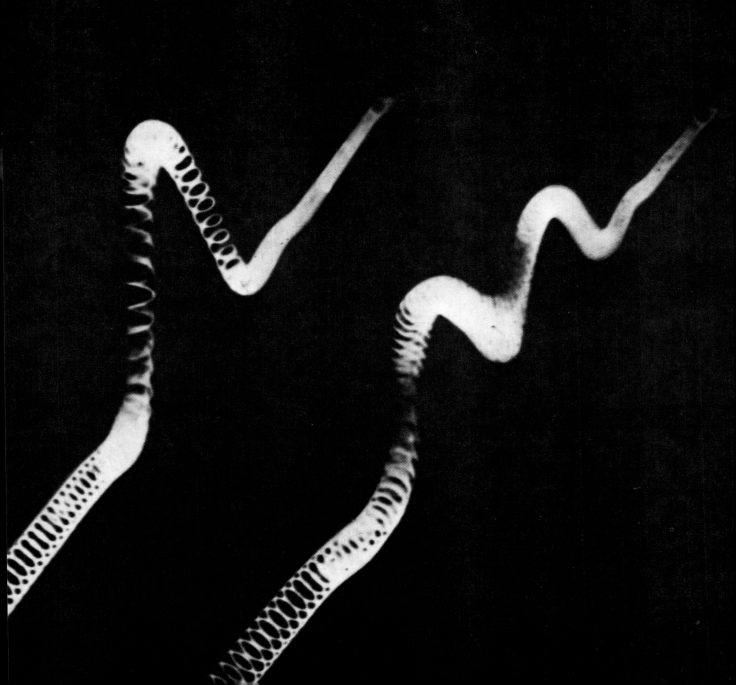

# 2    The Nature of Light

Most of the information we possess about the universe has come to us in the form of light from distant stars and galaxies. The manner in which this light reaches us was once one of the great mysteries of the natural world. Consider Vega, for example, which is a very bright star visible in the autumn sky and located 26 light-years from us. When you look at Vega, your eye responds to light that started on its journey 26 years ago, departing from Vega as a train of waves and moving steadily outward at the rate of 186,000 miles per second. The train of waves covered nearly 20 billion miles every day; year after year these waves traveled across the space between Vega and the sun, until they reached the earth and entered your eye. There they produced chemical changes in your retina that were passed to the brain as the sensation of light.

There is no problem in understanding what happened to the light in the final stage of its journey from Vega to you. The mystery of the phenomenon is connected with the time during which the light was en

*Wave motion.*

route across the space between the stars. Throughout that time the light waves traveled in a nearly perfect vacuum, containing very little matter. But without matter, how can a wave exist? A wave means the repeated vibration of individual particles of matter, in which each particle moves up and down or to and fro, as the crests and troughs of the wave train pass by. The wave is no more or less than this sequence of vibrating particles. Without the particles, there is no wave. This is well known to be the case for sound waves, for example. When a sound wave passes through air, the individual molecules of air move back and forth with each crest and trough of the wave train. If the air is dense so that many molecules are moving back and forth, the sound wave is relatively strong. If the air is less dense so that fewer molecules are vibrating, the sound becomes weaker. If no molecules are present, sound waves cannot exist at all.

But light waves *can* travel through a vacuum. How? What vibrates in the emptiness of space when a train of light waves passes through?

## LIGHT AS AN ELECTRIC VIBRATION

The answer was discovered one hundred years ago by James Maxwell, an English physicist. Vibrations of electric force can pass through a vacuum without any particles to carry them. Light is one particular kind of electric vibration. When a star radiates light into space, it is actually sending out trains of electric vibrations. No particles are involved.

What is meant by a vibrating electric force? It sounds like a peculiar notion. Actually, vibrations of electric force can be created very easily. Consider an electrically charged particle, such as an electron, and imagine that it is at rest initially. The electron will exert an electric force on any other charged particle near it. It is the center of an electrical sphere of influence stretching out into space in all directions. The electron is surrounded by a *field* of electric force (Figure 2.1).

Suppose that another charged particle, such as a proton, for example, is located at some distance from the electron. Since opposite electrical charges attract, a force will pull the two charged particles toward each other. This force of attraction is like a rubber band, drawing the two particles together.

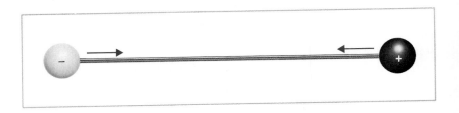

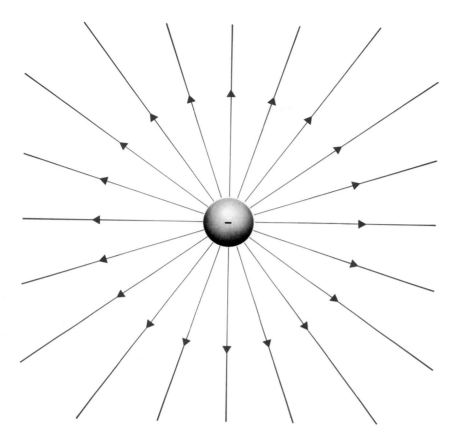

Figure 2.1 The field of force around an electron.

Suppose that the two particles are held fast and prevented from approaching one another. Now jiggle the electron up and down very quickly.

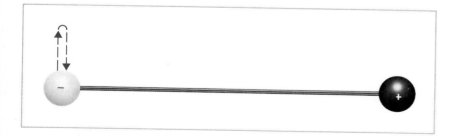

The quick change in the position of the electron produces a corresponding change in the electric force felt by the other particle. The key question is, when does the other particle feel the change in electric force? Does it feel it instantly? Or does it take some time for the effect of the change in the electron's position to reach this particle?

If the particles really were connected by a rubber band, a wave would move across the band, taking some time to travel from one particle to the other.

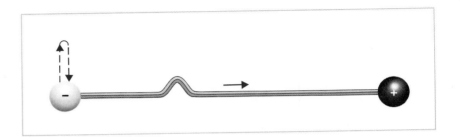

That is what happens with the electric force between two charged particles. If one particle moves, the other doesn't feel the motion until some time later. The particles respond just as if the force between them were in fact a stretched rubber band. An electrical wave generated by the motion of the electron moves rapidly along the imaginary "rubber band"; when it reaches the proton at the other end, the proton jumps up and down.

How fast do such electric waves move? We know their speed from the mathematical investigations of James Clerk Maxwell, the nineteenth century English physicist. In 1865, Maxwell was studying electricity and predicted, on the basis of complex mathematical reasoning, that waves of electric force should travel through space at a speed of 186,000 miles per second.

Maxwell's prediction for the speed of electric waves is extremely high. It means that if two charged particles are separated by, say, a distance of 10 feet, and one particle is jiggled sharply, the other particle feels the resultant electric disturbance only a 100-millionth of a second later.

Maxwell was not thinking about light at all when he made this calculation, but as soon as he arrived at his result he realized that the speed of these electric waves was very close to the speed of light, which had been measured many years earlier, and was known also to be approximately 186,000 miles per second. Some people thought this was a coincidence, but Maxwell was certain that it was not. He concluded that light waves were waves of electric force.

Returning to the example of the two electric charges, let us suppose now that the electron vibrates up and down continuously. A train of waves of electric force will move out into space from the vibrating electron, traveling at 186,000 miles per second. When the train of electric waves reaches the proton, it will cause the proton to move up and down in response to the motion of the electron. A signal has been transmitted from the electron to the proton by means of the electric vibrations traveling through the space between them.

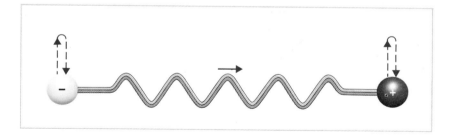

The situation is the same as if you and a friend were holding on to an elastic cord, and you jiggled your end of the cord at a steady rate. A train of waves would travel down the cord toward your friend. The vibrations would exert a force on his hand when they reached it, causing his hand to vibrate at the same frequency as yours (Figure 2.2).

Figure 2.2 *Waves generated by a vibrating object.*

Now we can see how light signals are transmitted from one point to another. The electrons in the outer layers of a star send out electric waves into space. When these waves reach some other material, such as the retina of your eye, they cause the electrons in the atoms of that substance to vibrate also, at the same frequency. In the case of the retina, the vibrations produce chemical changes, which create a signal that is passed along to the brain. That is how you receive the light from a star.

THE NATURE OF LIGHT

# THE ELECTROMAGNETIC SPECTRUM

A fundamental law of magnetism tells us that a changing electric force will produce a magnetic field. If the electric force varies regularly, as in a train of electric waves, the accompanying magnetic field will also vary regularly. As a consequence, the train of electric waves set up by a jiggling electric charge is accompanied by a train of magnetic waves that move through space at the same speed, that is, the speed of light. The two kinds of waves are tied together inseparably; one cannot exist without the other. For this reason, light is called an electromagnetic wave, or electromagnetic radiation, rather than simply an electric wave.

Light is not the only kind of electromagnetic radiation. Gamma rays, x-rays, ultraviolet radiation, infrared rays, radar, television signals, and radio waves are also types of electromagnetic radiation. Together with light, they form the *electromagnetic spectrum*.

All electromagnetic waves travel at the same speed, which is the speed of light. This is $186,283 \pm 0.6$ miles/second or $299,793 \pm 1$ km/sec in a vacuum and less in material substances. The different parts of the electromagnetic spectrum differ from one another only with respect to the frequency of the electromagnetic waves.

The retina of the human eye is sensitive to electromagnetic waves with frequencies between $4.3 \times 10^{14}$ vibrations per second (usually written as cycles per second and abbreviated cps) and $7.5 \times 10^{14}$ cps. Hence this band of frequencies is called the *visible* region of the electromagnetic spectrum. When electromagnetic waves with a frequency of $7.5 \times 10^{14}$ cps strike the retina of the eye, the signal sent to the brain registers this light as "blue-violet" in color. If electromagnetic waves with a somewhat lower frequency strike the retina as, for example, waves whose frequency is $6.5 \times 10^{14}$ cps, the eye signals to the brain that "blue" light has arrived. If the frequency of the light striking the retina decreases further, the eye sends to the brain, in succession, signals indicating the sensation of "blue-green," "green," "yellow," "orange," and "red" light. Light whose frequency is $4.2 \times 10^{14}$ cps provides the sensation of a "deep red" color (Color Plate 2).

If the electromagnetic waves have a frequency higher than $7.5 \times 10^{14}$ cps, the eye does not respond to them. Such waves, lying beyond the violet edge of the spectrum, are called *ultraviolet* light. If the waves have a frequency lower than $4.3 \times 10^{14}$ cps, the eye again does not respond to them. These waves, whose frequency is lower than the lowest frequency of visible light at the red end of the spectrum, are called *infrared light* or *infrared radiation*.

Why does the eye respond to the band of frequencies between $7.5 \times 10^{14}$ cps and $4.3 \times 10^{14}$ cps, and to no others? The answer is that this band of frequencies can pass freely through the gases that make up the earth's atmosphere without being absorbed, whereas other electromagnetic waves coming from the sun, with frequencies higher than $7.5 \times 10^{14}$ cps or lower than $4.3 \times 10^{14}$ cps, are heavily absorbed in their passage through the atmosphere.[1] The human eye has evolved in response to the need for seeing objects on the earth's surface with the aid of sunlight. If it were sensitive to frequencies far outside the visible region, that is, far outside the band of electromagnetic radiation that gets through the atmosphere, there would be little difference between night and day, and your eyes would be of relatively little value to you. An eye with a peak sensitivity in the middle of the band of frequencies that passes through the atmosphere is the most useful kind of eye that a person can have. On another planet, with a different atmosphere that is, perhaps, strongly absorbing in what we call the visible region, but transparent in the infrared region, evolution might generate creatures with infrared-sensitive eyes; or, on still another planet, there might be creatures with ultraviolet-sensitive eyes, depending on the electromagnetic wavelength for which the atmosphere of the planet is transparent.

The *x-ray* region of the electromagnetic spectrum consists of waves with frequencies ranging from $3 \times 10^{16}$ cps to $5 \times 10^{18}$ cps. This region lies next to the ultraviolet region in the spectrum. The *gamma ray* region of the spectrum has the highest frequencies of all, ranging upward from $3 \times 10^{19}$ cps.

At the other end of the spectrum, *microwaves* or *radar* cover the frequency ranges from $3 \times 10^{10}$ cps to $3 \times 10^{12}$ cps, just beyond the infrared portion of the spectrum. Beyond the radar region is the *television* and *FM* band of frequencies extending from $3 \times 10^{8}$ cps to $3 \times 10^{5}$ cps. *Radio* waves have the lowest frequencies in the spectrum, extending downward from $3 \times 10^{5}$ cps.

Electromagnetic waves are often described in terms of their wavelength rather than their frequency. An example will demonstrate the conversion from wavelength to frequency. Suppose an electron vibrates one million times per second, creating an electromagnetic wave with a frequency of $10^{6}$ cps. That is, in one second a million waves are sent out. Since the velocity of light is 186,000 miles per second, this train of one million waves has moved 186,000 miles in that second. Clearly the length of each wave is $186,000/1,000,000 = 0.186$ miles $= 1230$ feet.

This is an example of a general formula connecting frequency and wavelength:

$$\text{wavelength} = \frac{\text{speed of wave}}{\text{frequency}}$$

The formula holds for any type of wave and is not restricted to light waves.

Figure 2.3 shows the complete electromagnetic spectrum with frequencies and wavelengths indicated. The names given to the various parts of the spectrum also are shown.

Column two of the figure indicates how transparent the earth's atmos-

---

[1] Long radio waves also pass through the atmosphere, but they are considerably weaker than visible light.

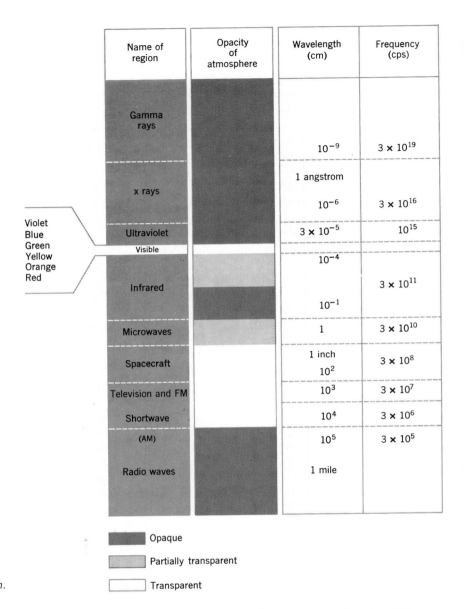

| Name of region | Opacity of atmosphere | Wavelength (cm) | Frequency (cps) |
|---|---|---|---|
| Gamma rays | | | |
| | | $10^{-9}$ | $3 \times 10^{19}$ |
| x rays | | 1 angstrom | |
| | | $10^{-6}$ | $3 \times 10^{16}$ |
| Ultraviolet | | $3 \times 10^{-5}$ | $10^{15}$ |
| Visible | | $10^{-4}$ | |
| Infrared | | | $3 \times 10^{11}$ |
| | | $10^{-1}$ | |
| Microwaves | | 1 | $3 \times 10^{10}$ |
| Spacecraft | | 1 inch | $3 \times 10^{8}$ |
| | | $10^{2}$ | |
| Television and FM | | $10^{3}$ | $3 \times 10^{7}$ |
| Shortwave | | $10^{4}$ | $3 \times 10^{6}$ |
| (AM) | | $10^{5}$ | $3 \times 10^{5}$ |
| Radio waves | | 1 mile | |

Violet
Blue
Green
Yellow
Orange
Red

■ Opaque

▨ Partially transparent

☐ Transparent

*Figure 2.3 The electromagnetic spectrum.*

phere is to electromagnetic radiation of various wavelengths. This part of the diagram indicates the types of electromagnetic waves that are useful as tools in astronomy. It shows that the atmosphere blocks radiation from outside the earth everywhere except in the visible region, a part of the infrared, and a part of the radio region. These bands of electromagnetic radiation are three windows of the atmosphere through which earthbound astronomers can look out at the Universe.

Until quite recently, astronomers thought that the visible window was the only one that they had, but in recent years several additional windows in the sky have opened up to the astronomer. The first of these is the band of wavelengths in the radio region extending from a few millimeters up to 100 meters. This is a true window, with very little atmospheric absorption of radio waves over that large range of wavelengths. The second recently acquired astronomical window is in the infrared region, extending from the long-wave end of the visible spectrum in the deep red at a wavelength of approximately 7000 angstroms* to approximately one millimeter, on the lower border of the radio region. The infrared window, unlike the radio window, is murky or opaque in many places, but a few transparent bands have been found in this region, and they have already yielded vital information that will be discussed in later chapters.

In the future, astronomers look forward to conducting their observations above the earth's atmosphere, either from telescopes circling in orbit around the earth, or from observatories on the moon. The age of astronomy in space, which has already commenced with the launching of the orbiting astronomical observatory, will place a wealth of previously unavailable materials in their hands.

## ELECTROMAGNETIC RADIATION FROM A HEATED OBJECT

A star, or any other hot object, radiates electromagnetic waves stretching across the entire spectrum from infinitely short wavelengths to infinitely long ones. However, the waves emitted at various wavelengths do not have equal intensities. The intensity is always weak at very short and very long wavelengths, and it is strongest at some wavelength in between. What determines the wavelength at which the intensity of the radiation is strongest?

A moment's reflection will tell you the answer. Suppose that you place a bar of iron in a furnace and steadily raise its temperature. At first, the iron feels hot and sends out heat but does not glow visibly; at this point the peak of its radiation is in the infrared region. As the temperature of the iron increases, it begins to emit radiation in the visible region, first turning dull red in color, then cherry-red, then yellow, and finally, at the highest furnace temperature, becoming white-hot.

This simple thought-experiment—which has its counterpart in many experiences of everyday life involving kitchen stoves, soldering irons, and so on—shows that the *temperature* of an object determines the wavelength at which it radiates most of its energy. As the temperature of the object increases, the radiation emitted by it moves toward shorter wavelengths—from the infrared to the red, then to the yellow, to the blue, and eventually, if the temperature mounts high enough, presumably past the blue and the violet into the ultraviolet.

As a parenthetical question, you may ask why an object becomes white-hot at furnace temperatures. This seems difficult to understand because white is not one of the colors of the spectrum. The answer is that when an object is heated to the point where its peak intensity is in the visible spectrum—in the yellow, let us say—it will also emit

*One angstrom ($\overset{\circ}{A}$) = $10^{-8}$ cm.

radiation at surrounding wavelengths in the blue, green, and red. That is, it will radiate a mixture of all wavelengths in the visible region. Such a mixture of all colors is recorded by the eye and brain as the sensation of *white* light.

There is a formula connecting the temperature of a glowing object and the wavelength at which it radiates with the greatest intensity. This formula, known as Wien's law, is

$$T = \frac{2.89 \times 10^7}{\lambda_{max}}$$

where $T$ is the temperature in degrees Kelvin and $\lambda_{max}$ is the wavelength at which the intensity has its peak. (The wavelength is expressed in angstroms.) Notice that the formula agrees with your everyday experience on the way in which the color of a heated object changes as its temperature increases: the hotter the object, the shorter the wavelength of its most intense radiation.

## THE DOPPLER SHIFT

Imagine yourself on an accelerating spacecraft moving rapidly away from the solar system en route to Barnard's Star in a search for earthlike planets orbiting one of the sun's near neighbors. The ship is equipped with engines powerful enough to accelerate it to half the speed of light so that the round trip can be completed in the lifetime of the crew. As the ship accelerates to its cruising speed, you look back occasionally at the sun, now reduced to one star among many in the black sky. To your surprise you find that the sun seems to be cooling down. Its color changes gradually from yellow-white to orange, and finally to red.

Looking ahead to Barnard's Star, a relatively cool, reddish star as seen from the earth, you find that this star seems to be getting hotter. Its color changes from red to orange, yellow, and finally to yellow-white.

When the ship reaches Barnard's Star and decelerates preparatory to dropping into orbit, a backward glance at the sun reveals that it seems to be growing hotter and recovering its normal yellow-white color, while Barnard's Star appears to be cooling rapidly to its familiar reddish hue.

The flight is imaginary and may never take place, but the description of the changing star colors is based on a concrete physical fact: if you are moving toward or away from a heated object, you see an apparent change in the wavelength of the light emitted by it. If you are moving away from a star, the radiation that you see from that star will seem to be shifted toward longer wavelengths, that is, toward the red end of the spectrum. If you are moving toward the star, the radiation will seem to be shifted toward shorter wavelengths, that is, toward the blue end of the spectrum.

Similar changes occur when the source of the radiation is moving instead of the observer. If a star is moving away from you, its radiation seems to be shifted toward the red end of the spectrum; if it is moving toward you, its radiation seems to be shifted toward the blue end of the spectrum.

These cases are examples of a phenomenon discovered in 1842 by an Austrian physicist named Christian Doppler and called the *Doppler effect* or the *Doppler shift*. This effect can occur in any kind of wave motion, and it is not limited to electromagnetic waves. The effect arises whenever the observer and the source of the waves are moving toward or away from one another. It does not matter if one is moving and the other is at rest, or if both are moving; the observed Doppler shift is the same.

The explanation of the Doppler shift is not complicated. We will use electromagnetic waves as an example. Suppose an oscillating electric charge—the source of the waves—is sending waves into space in the direction of the observer (Figure 2.4a).

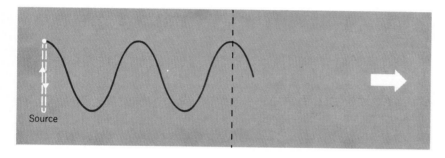

*Figure 2.4a   A source of light generates an electromagnetic wave.*

Every time the charge completes one oscillation, one more wave is added to the outgoing wave train (Figure 2.4b).

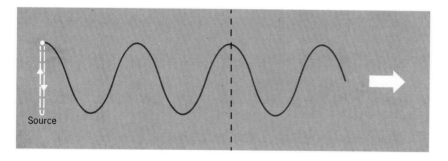

*Figure 2.4b   Another vibration of the source adds a wave to the train.*

Suppose that the source commences to move backward in the direction away from the observer. New waves continue to be added to the train, one after the other, as the source oscillates up and down. Each of these waves moves out into the medium at the same speed as before. However, during the time in which the oscillating charge creates one wave, the charge itself moves backward, so that the length of the wave is stretched (Figure 2.5).

The amount of the increase in the wavelength depends on the velocity of the source and the velocity of the wave train. If the source moves

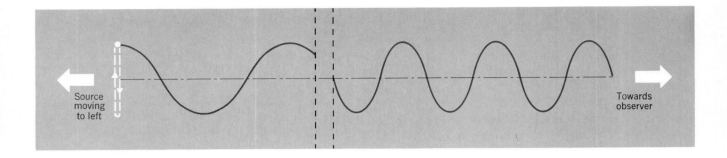

Source moving to left

Towards observer

*Figure 2.5   The stretching of the waves by the receding motion of the source.*

slowly in comparison to the wave train, each wave will be stretched by a correspondingly small amount. If the source moves backward nearly as fast as the wave train moves forward, the waves will be stretched enormously. These ideas can be expressed in a formula giving the change in wavelength $\Delta\lambda$ produced by a motion of the source at a velocity $v$, when the speed of the waves is $V$ and their original, unstretched wavelength is $\lambda$:

$$\frac{\Delta\lambda}{\lambda} = \frac{v}{V}.$$

The same type of reasoning applies when the source moves toward the observer instead of away from him. In this case, the length of a wave is decreased because the oscillating charge moves forward during the time in which the wave is produced. Again the amount of the effect depends on the speed of the source and the speed of the wave train: the larger the speed of the source in comparison to the speed of the waves, the greater the effect. The same formula can be used for this case as for the case of backward motion, if we agree to attach a plus sign to the velocity of the source when the source is moving away from the observer and a minus sign to it when it is moving toward the observer. This rule is reasonable because the distance from the source to the observer increases in the first case and decreases in the second.

This formula also applies to the case in which the source is fixed and the observer is moving, or the case in which both are moving. The formula can be used in any of these cases if $v$ is defined as the net speed at which the source and the observer approach or recede from one another.[2]

Let us now return to the imaginary spacecraft speeding from the sun to Barnard's Star at one-half the speed of light. According to observations, the radiation from these stars is peaked at a wavelength of 4700Å for the sun, and 10,000Å for Barnard's Star. Using the above formula, we can calculate the shift in wavelength when the velocity $v$ of the spacecraft is

[2] This formula is not precise because it does not include effects of relativity theory. The exact formula derived from relativity theory is $\lambda$ (new) $= \lambda$ (old) $\sqrt{(1 \pm v/C) / (1 \mp v/C)}$, where the upper sign is valid for a receding source and the lower sign for an approaching source. When $v/C$ is small the approximate and exact formulas are in close agreement.

equal to one-half the speed of light or 93,000 miles per second.

Sun: $\dfrac{\Delta\lambda}{\lambda} = \dfrac{93{,}000\ \text{mps}}{186{,}000\ \text{mps}} = \dfrac{1}{2}$

$$\Delta\lambda = \lambda \times \frac{1}{2} = 4700 \times \frac{1}{2} = 2350\overset{\circ}{A}$$

Thus, the shifted wavelength corresponding to the sun's radiation at 4700Å is now

$$\lambda\ (\text{new}) = \lambda\ (\text{old}) + \Delta\lambda = 4700\overset{\circ}{A} + 2350\overset{\circ}{A} = 7050\overset{\circ}{A}$$

This wavelength lies at the far red end of the visible spectrum.[3] The eye would see the sun as a deep red star if the observer were traveling away from it at this speed.

Barnard's Star: $\dfrac{\Delta\lambda}{\lambda} = \dfrac{93{,}000\ \text{mps}}{186{,}000\ \text{mps}} = \dfrac{1}{2}$

$$\Delta\lambda = \lambda \times \frac{1}{2} = 10{,}000\ \overset{\circ}{A} \times \frac{1}{2} = 5000\overset{\circ}{A}$$

Thus, the shifted wavelength corresponding to 10,000Å is

$$\lambda\ (\text{new}) = \lambda\ (\text{old}) - \Delta\lambda = 10{,}000\overset{\circ}{A} - 5000\overset{\circ}{A}$$

$$= 5000\overset{\circ}{A}$$

This wavelength lies in the middle of the visible spectrum. The eye would see Barnard's Star as a yellow-white star if the observer were traveling toward it at the same speed.


**The Cosmological Red Shift**

The Doppler shift is connected with one of the most remarkable discoveries in modern astronomy. Around 1913 the American astronomer V. M. Slipher discovered that light from distant objects is shifted in wavelength toward the red end of the spectrum in a substantial amount. The effect discovered by Slipher is known today as the *cosmological red shift.* The Doppler effect is the only satisfactory way to explain the red shift. Applying the law for the Doppler shift to Slipher's discovery, we see that every other galaxy must be moving away from us. In the case of the most distant galaxies the red shift is very large, corresponding to a velocity of recession close to the speed of light. In a later chapter we will come back to the red shift in the light from galaxies and discuss its implications for the birth of the Universe.


Questions

1. Invent your own definition of a wave. Assume that your task is to describe a wave to a person who has never seen one.

2. If two physicists with perfect reflexes attempt to measure the speed of

[3] The relativistic formula (footnote 2) yields 8150Å. Below, when approaching Barnard's Star, the relativistic formula yields 5800Å.

light by standing on hills 2 miles apart with flashlights and synchronized watches, what precision is needed in their watches to yield a 10 percent accuracy in the measurement?

3. What would be the wavelength of the radiation produced by an electron vibrating 1000 times per second?

4. A radar signal is beamed toward Mars and the echo is received 7 minutes later. How far is Mars from the earth on this occasion?

5. What portions of the electromagnetic spectrum would you select for viewing from an orbiting astronomical observatory? Why?

6. The sun's temperature is 5800°K. At what wavelength does it radiate the maximum energy? To what color does this correspond? Barnard's Star is red-orange in color. What temperature would you predict for its surface?

7. The eye is insensitive to wavelengths longer than 7000Å. To what temperature must an iron be heated before it begins to glow visibly?

8. When the electromagnetic spectrum emitted by a distant galaxy is analyzed, it is found that the characteristic patterns of radiation emitted by familiar elements are present, but with their wavelengths shifted toward the red by 15 percent. Is this galaxy moving away from or toward the earth? At what speed?

9. Describe a world in which the speed of light is 100 feet per second.

# 3    The Tools of the Astronomer

Electromagnetic radiation is the principal link between the earth and the stars. This radiation passes through "windows" in our atmosphere to the surface of the earth, where it is collected by the telescope and analyzed into its component wavelengths by the spectroscope. These instruments—the telescope and the spectroscope—are the essential tools of the astronomer.

## THE OPTICAL TELESCOPE

The telescope was probably invented sometime between 1600 and 1608. Opticians in Germany and Holland were designing telescopes in 1608, and by 1609 some of the Dutch instruments had found their way to Italy, where they aroused the interest of a professor of mathematics at Padua named Galileo Galilei. Galileo either copied the known design of the instrument or figured out its principle himself; in either case he built one, grinding his own lenses for that purpose. The military value of the in-

*The Mt. Palomar Observatory photographed in moonlight.*

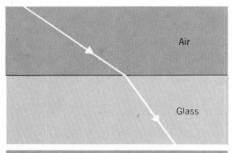

*Figure 3.1a   The bending of a ray of light by glass.*

Air

Glass

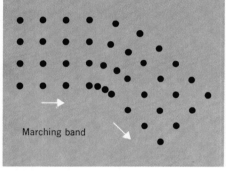

Marching band

*Figure 3.1b   A marching band turns right.*

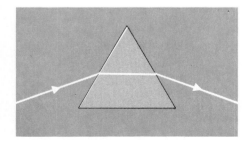

*Figure 3.2   A light ray is bent toward the base of a prism.*

vention struck him first, and he made a gift of his instrument to the powerful government of the city of Venice, explaining that it could protect the city against seaborne invasions by bringing into view "sails and shipping . . . two hours before they were seen with the naked eye." The senate of Venice immediately doubled Galileo's salary.

Next Galileo raised his telescope to the heavens. He examined the moon first and, although his instrument had a magnification of only 30 powers, he could see that the surface of the moon, far from being of a polished smoothness, as had been thought, was "uneven, full of hollows and protuberances, just like the surface of the earth itself." Then he looked at the sun and found its bright surface marred by dark, ugly blotches—the sunspots. He looked at Jupiter and found that it was a disk, like the moon but smaller, instead of the point of light that the naked eye saw. He also detected four small stars, invisible to the naked eye, that moved around Jupiter as the earth moves around the sun, or the moon around the earth. These were the four largest of Jupiter's twelve moons; today they are known as the Galilean satellites of that planet.

Finally, Galileo discovered that he could see far more stars through the telescope than were visible to the naked eye. A person with sharp eyesight can count 5000 stars in the sky, if he has the patience for the task. Galileo, with his primitive telescope, increased the number to 50,000.

**The Principle of the Telescope**

The telescope revolutionized astronomy. How does this miraculous instrument work? The answer requires a detour through the basic ideas of optics and lenses. The working of the telescope depends on the effect produced by glass on rays of light passing through it. Light entering a piece of glass at an angle is bent downward as it passes through the glass. This property of glass, called *refraction,* is caused by the fact that light travels more slowly in glass than air. As a beam of light enters a piece of glass traveling at an angle to the surface, the part of the beam that hits the glass first slows down, but the remainder of the beam continues to move at full velocity, close to 186,000 miles per second. As a result, the beam turns downward toward the glass (Figure 3.1a).

A band marching down a football field changes its direction in the same way. If the file of marchers on one side of the band—say the right side—takes shorter steps than the marchers on the left side, the band will wheel around the slow-moving marchers and turn to the right (Figure 3.1b).

Suppose we apply these ideas to a beam of light entering a glass prism at an angle. The light is deflected downward as it enters the glass, and deflected a second time as it emerges from the other side. We can describe its complete path by saying that it is bent toward the base of the prism.

Now place two prisms together, base to base, and trace rays from a

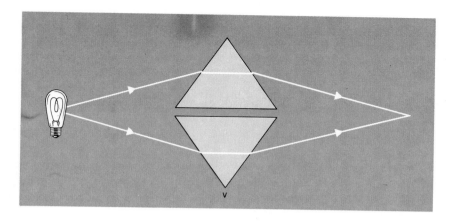

light bulb, located at the left. All rays of light passing through the upper prism are bent downward and rays of light passing through the lower prism are bent upward. As a result, the rays converge on the other side of the double prism. If we place a piece of paper at the right, we will see on it a bright spot formed by the converging rays of light. The two prisms have *focused* the light from the bulb. They are a crude, imperfect lens.

Unfortunately, not all the rays converge to the same point. The rays passing through the prisms near the apex come together at a greater distance than the rays passing through near the base.

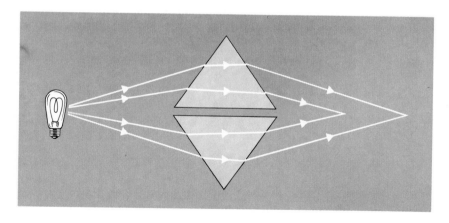

That is, the focusing is blurred. To improve it, the angle of the prisms must be changed near the base.

Now the rays converge to one place. If a sheet of paper is placed there, a small, bright spot of light should appear on it. We have constructed a focusing device out of the prisms.

The focusing effect can be improved by filling in the cracks between the prisms and rounding off their outer surfaces in such a way that all rays of light from the source at the left will converge to a single point on the right. Now the combination of prisms has been converted to a *lens*.

Beyond the point at which the rays of light converge, they diverge

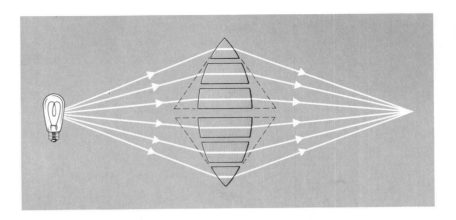

again in the same manner as they diverged from the original object. If you place your eye beyond this point, and look toward the lens, the diverging rays of light enter the eye and fall on the retina, conveying a message to

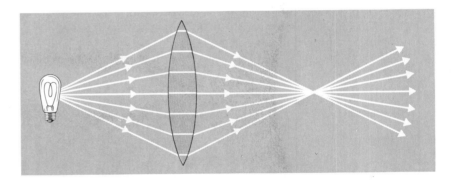

the brain that they are emanating from a copy of the original object located at the new point of divergence.

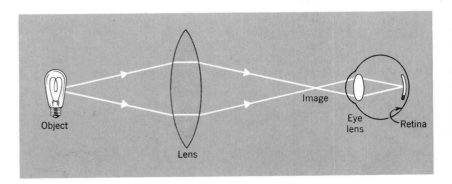

No object is actually there, but we say that an *image* of the object has been formed by the lens at this point.

Now we are ready to apply our understanding of lenses and magnification to the telescope. The telescope consists of two lenses, the objective and the ocular. The light from a distant object enters the objective lens first. This is the larger of the two lenses. The objective lens forms an image of the distant object, just as in the preceding diagrams. The purpose of the ocular lens is to serve as a magnifying glass. The person using the telescope puts his eye close to the ocular, and uses it to magnify the image formed by the objective. Because the distant object's image is magnified the object appears to be much closer than it actually is (Figure 3.3).

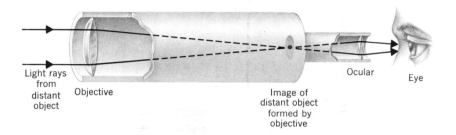

Light rays from distant object    Objective    Image of distant object formed by objective    Ocular    Eye

Figure 3.3  The principle of the telescope. Light rays from a distant object enter the objective lens at left. The objective lens forms an image of the distant object at the other end of the telescope, near the ocular. The ocular acts as a magnifying lens. The eye looks through the ocular and sees the image magnified, making the distant object seem closer than it really is.

The telescope is an ingenious instrument and an immense improvement over the naked eye; but it has its limitations, as all instruments do. The most important limit is on the sharpness of the image. This limitation stems from two causes: *diffraction* and *atmospheric blurring*.

## Diffraction

The magnifying power of a telescope may be increased by increasing the power of the ocular lens; but the objective lens determines the sharpness of the picture. As the image formed with the telescope becomes more magnified, it also becomes increasingly blurred, so that no additional details can be seen, although the image is larger. As a rough rule of thumb, the maximum useful magnification is approximately 50 times the diameter of the objective in inches. Greater magnifications than this may be impressive but do not reveal any new information.

Why does the size of the objective lens limit the useful magnification of a telescope?

The answer lies in the nature of light itself, which sets a fundamental limitation on the magnifying powers of a telescope. A beam of light traveling freely through space moves in a straight line. However, when the beam passes close to the edge of an object, the part of the beam nearest to the edge bends and travels in a slightly different direction. This effect, known as diffraction, spoils the convergence of light rays passing through a lens and blurs the image produced by the lens in a way that cannot be removed, no matter how much expense and trouble are put into the design of the lens.

The reason for diffraction becomes clear if we remember that light moves in a series of waves, like sound waves, or waves on the ocean. As the waves go past the edge of an obstacle, some spill over the side, just as sound waves spill over the side in passing the corner of a building. If you are standing around the corner from a friend and he shouts to you, you can hear him even if there are no buildings or objects to reflect the sound to you, because the sound wave bends around the corner (see Figure 3.4a).

The degree of bending of light waves is a million times smaller than for sound waves. In fact, the amount by which light bends in passing an obstacle is so minute that usually the naked eye cannot detect it. Only carefully arranged instruments can disclose it. For this reason, most people believe that light always travels in straight lines (see Figure 3.4b).

*Figure 3.4(a) Bending of sound at the edge of an obstacle.*

*Figure 3.4(b) Bending of light at the edge of an obstacle.*

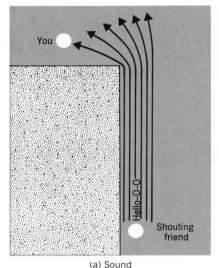
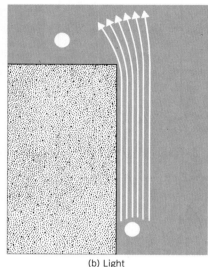

(a) Sound          (b) Light

*Figure 3.5 Bending of light at the edge of a lens.*

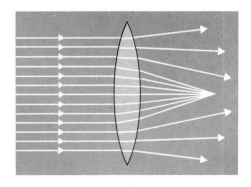

Let us see what diffraction does to the clarity of the image formed by a convex lens. Suppose that the lens is mounted in a tube, as it would be if it were part of a telescope. The outside of the tube blocks off all parts of the light beam except those parts passing through the lens itself. Thus, in this case, the edge of the lens itself provides the obstacle that the light beam must pass on its way to the eye. The parts of the beam of light that pass through the objective lens near its center are undisturbed by the edge of the lens. After leaving the lens, they converge to a single point. However, the parts of the beam passing through the lens near its edge are bent away from their course by the diffraction effect, so that these parts of the beam do not converge to the same place as the parts of the beam that pass through the center. Instead, they converge farther out (Figure 3.5).

If a screen is placed where the main beam of the lens should be focused, it will show a blurred patch of light with a bright spot in the center (Figure 3.6).

The bright, central spot is produced by the converging action of the lens on the main beam. The surrounding blurred patch results from diffraction of light waves at the edge of the lens.

When an astronomer takes a photograph of a star through a large telescope, he often overexposes the central spot and surrounding patch of light, so that they appear as a uniformly bright disk. This bright area is called the *diffraction disk*. If two stars lie close together, their diffraction disks may overlap to such an extent that they appear as a single star on photographs taken through a telescope.

Suppose that we look through a telescope at an object with many individual details, such as the moon. Every point on the moon becomes a spot of light in the image through the focusing action of the objective lens. However, the spots of light are not sharply defined points. Because of the diffraction effect, each point on the image is actually a blurred disk of light that overlaps neighboring points. For this reason, the eye cannot see any details of the image that are closer together than the width of the disc of light produced by diffraction.

An inexpensive telescope usually has an objective lens with a diameter of about 2 inches. The blurring effect of diffraction on a lens of this size prevents the observer from seeing details on the moon that are less than 10 *miles* in size. A hole on the moon big enough to swallow up the city of Paris would be invisible through a 2-inch telescope.

For large lenses, the relative effect of the edge is less than for smaller lenses, and the diffraction is correspondingly less important. With a telescope containing a lens 12 inches in diameter, for example, features on the moon as small as 1.5 miles across can be seen. A good 12-inch telescope costs a thousand dollars or more, and it is the largest telescope ordinarily used by anyone except a professional astronomer.

Instruments used in astronomical research are both larger and considerably more expensive. The 40-inch telescope at Yerkes Observatory, for example, cost approximately $300,000 (Figure 3.7). The money for this telescope was donated to the University of Chicago in 1892 by Charles Yerkes, owner of the Chicago streetcar system. Pictures of the moon obtainable with the Yerkes telescope are vastly superior to the ones taken by smaller instruments, because of the decreased diffraction effect.

Figure 3.8a and b shows the region of the crater Clavius on the moon taken with telescopes of two different sizes. Figure 3.8a was taken through a telescope of 10-inch diameter. Figure 3.8b on the right was taken with the 36-inch telescope at the Lick Observatory in California. Examination of the photographs shows many small craters visible in the Lick photograph, especially on the floor of the large crater Clavius near the top of the photograph, that cannot be seen in the photograph made through the 10-inch telescope.

## Atmospheric Blurring

When engineers and scientists working for the National Aeronautics and Space Administration began to design a spacecraft for the manned

Figure 3.6   Blurring of the image of a point of light by the bending of light rays at the edge of the lens.

landing on the moon, they required information even more detailed than that provided by a 40-inch telescope. Jagged rocks a few feet long at the landing site could damage the landing gear, or soft dust might swallow up the entire landing craft and astronauts. These conditions had to be known beforehand so that the astronauts and the mission could be protected from disaster. According to the theory of diffraction, NASA should

Figure 3.8 Clavius crater region of the moon; (a) through a 10-inch telescope, (b) through the 36-inch Lick telescope.

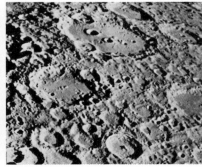

have been able to secure needle-sharp lunar photographs simply by using a telescope with a very large objective lens. With such a telescope we might also be able to settle the question of the canals on Mars and look for other signs of life on that planet.

Unfortunately, there is a limit to the amount of detail that can be seen from the earth, no matter how large a telescope is used. Rays of light from the moon or other objects in the heavens are bent slightly by the earth's atmosphere as they travel through it on their way to the telescope. This bending of the rays in the atmosphere has nothing to do with diffraction. It stems from the fact that light travels at a slightly slower speed in air than in a vacuum. In a vacuum it travels at 186,283 miles per second, while in air at ground level, it travels at 186,235 miles per second. When a ray of light from a star enters the earth's atmosphere at an oblique angle, it bends because of the slightly slower speed of travel in the atmosphere, just as a ray of light is bent when it enters a piece of glass (Figure 3.9a and b). The bending of rays of light in the atmosphere would have no effect on the clarity of the star image formed by the telescope, *if the air were perfectly quiet*; it would only shift the apparent position of a star by a small amount. The atmosphere, however, is continuously in motion and, as a result, the position of the image dances about continually, many times a second. It moves rapidly on the retina of the astronomer's eye if he looks through the telescope, or on the photographic film, if he uses a camera instead. If the astronomer takes a photograph through the telescope, the dancing motion will blur the photographic image.

## Good and Bad "Seeing"

A clear, cold night, when the stars are twinkling brightly, is a bad night for the astronomer, for he knows that when the stars twinkle most, the image in his telescope is moving rapidly and all photographs will be very blurred. Astronomers refer to the conditions of the atmosphere as the "seeing." If the atmosphere is quiet on a particular night and the stars shine steadily, the "seeing" is said to be good that night. When the stars twinkle, the "seeing" is bad.

When an astronomer photographs the telescopic image of a star he can reduce the blurring if he makes the exposure time of the film short enough to freeze the star or other astronomical object into one position. Astronomers, however, take most of their photographs with long exposures so that they can capture as much light as possible from faint, distant objects. Even the moon, which is the brightest astronomical object next to the sun, requires an exposure time of as much as a second when photographed through a large telescope. As a result, photographs of the moon taken through large telescopes are badly blurred by the atmospheric jiggling of the image.

The effect of the atmosphere on lunar photographs is evident in the contrast between a photograph of the rim of Alphonsus crater, taken with

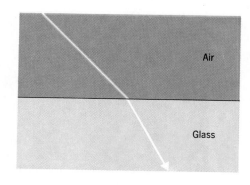

*Figure 3.9(a) Light bending as it enters glass.*

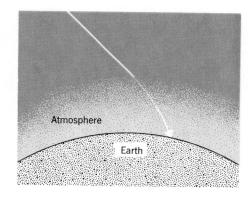

*Figure 3.9(b) Light bending in the atmosphere.*

the 100-inch telescope at the Mount Wilson Observatory in California (Figure 3.10a) and a photograph of the same region taken from the Ranger spacecraft as it hurtled toward the moon's surface (Figure 3.10b). The earth's atmosphere produced all the blurring and loss of detail evident in the Mount Wilson photograph, the effect of the diffraction being negligible when the telescope diameter is as large as 100 inches.

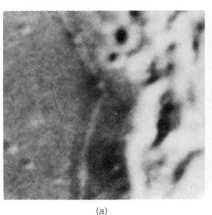 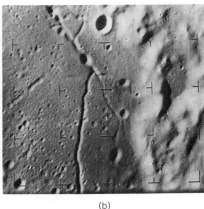

(a)              (b)

*Figure 3.10(a)  A section of the floor and surrounding rampart of Alphonsus crater, photographed by the 100-inch telescope at Mount Wilson. No feature smaller than one mile in diameter can be seen.*

*Figure 3.10(b)  The same area, photographed from a height of 115 miles above the moon by a camera on the Ranger 9 spacecraft. Surface features as small as 300 feet in diameter may be seen.*

The amount of atmospheric blurring varies from night to night, but on the average it is such as to erase any lunar features in the photographs if the feature is less than one-half mile in size. The diffraction effect in a 24-inch telescope produces the same amount of blurring in lunar photographs: about one-half mile. If the diameter of the telescope is made larger than 24 inches, this diffraction effect will be decreased, but no change in telescope design can lessen the problem posed by the earth's atmosphere. That is why we will never know, by photographing Mars from telescopes on the earth, whether any forms of life exist on that planet; that is also why it was impossible to prepare for the lunar landing by taking photographs of the moon from the earth, but was necessary instead to send a spacecraft above the earth's atmosphere to take better pictures.

**Superiority of the Eye**

One increase in the sharpness of earthbound observations is possible. The eye can see more details through a telescope than the photographic plate can record. As the telescope image dances about because of atmospheric motions, the eye moves very rapidly, following the erratic movements of the image. In this way, the eye eliminates some of the blur produced by the atmosphere. Astronomers find that the eye can observe about three times as much detail as can be recorded with a photographic plate. For this reason, the best views of the moon and Mars from earth have always been obtained by the human observer rather than the camera. A trained astronomer observes details on the moon's surface as small as 500 to 1000 feet across, when the seeing is good. When the planetary astronomer, Giovanni Virginio Schiaparelli, asserted that he saw canals on Mars in 1877 at the very limits of his visual acuity, it was impossible to test his claim by taking photographs of Mars, because the photographs surely would not reveal details that the eye could barely see.

To obtain the maximum amount of detail on the moon or Mars, the observer must use his eye rather than the camera, and he must also use a telescope large enough to produce a diffraction disk that is small in comparison to the blur produced by the atmosphere under the best seeing conditions. Telescopes with diameters of approximately 40 inches satisfy this requirement. Still larger telescopes produce a slight improvement in the sharpness of the image because of the smaller diffraction effects, but the gain in clarity does not offset the vastly increased cost of a big telescope. A 200-inch telescope costs roughly 30 times as much as a 40-inch telescope. The diffraction disk is considerably smaller, but the improvement in sharpness of the image is hardly noticeable unless the atmospheric conditions are exceptionally good. With average "seeing," photographs of the moon taken through a 200-inch telescope are no better than those taken through the 36-inch Lick telescope. Figure 3.11 shows Clavius photographed through the Lick telescope (a) and the Palomar telescope (b).

(a)          (b)

*Figure 3.11 Clavius crater; (a) photographed through the 36-inch Lick telescope, (b) photographed through the 200-inch Mount Palomar telescope.*

## Light-Gathering Power

Why are telescopes ever made in sizes larger than 40 inches? There must be a reason why Andrew Carnegie, a hardheaded steel magnate, gave millions of dollars to the work on the Mount Wilson 100-inch telescope, and why the Rockefeller Foundation gave six million dollars for the construction of the 200-inch telescope on Mount Palomar in California.

The answer is that professional astronomers do not use their telescopes solely to see details on the surface of the moon, or to search for evidence of life on Mars. High magnification is not their only goal; they are also interested in searching out the secrets of the universe at great distances from the earth, and what they need for that purpose is light-gathering power.

A telescope of large diameter gathers more light than a small telescope and can therefore reveal stars that are too faint or far away to be seen with a smaller instrument. With the eye we can see stars about 2000 light-years away. A 12-inch telescope collects 2500 times as much light as the naked eye and enables us to see out to the edge of our Galaxy, about 100,000 light-years distant. We can also see other galaxies outside ours with a 12-inch telescope, some of them as far away as 100 million light-years. The 200-inch telescope on Mount Palomar collects two million times as much light as the eye, and enables us to see ordinary galaxies as far away as 2 billion light-years.

With this incentive, astronomers built telescopes with lenses of larger and larger diameters during the last century until, in 1913, a telescope was constructed with an objective lens that had a diameter of 40 inches. Since that time, no telescope has been built with a larger lens. It is impossible, with all the refinements of twentieth century technology, to make a lens larger than 40 inches in diameter, except at prohibitive expense. To mention only one of the difficulties, a large lens sags in the middle under its own weight, destroying the shape into which it was ground for a perfect image.

## REFLECTING TELESCOPES

Telescopes with a diameter larger than 40 inches have been built to achieve greater light-gathering power but, to build them, astronomers have been forced to use a different principle of telescope construction, one that dates back to the time of Isaac Newton. When Newton was a student at Cambridge, and about 23 years old, he had an ingenious idea for making a telescope without going to the trouble of grinding a glass objective lens. Newton was not the first to have this idea; others had thought of it years before him, but Newton was the first to make it work.

### The Newtonian Reflector

Newton reasoned that the important aspect of a telescope lens was simply its ability to converge rays of light. A concave spherical mirror has the same property, thought Newton; it will reflect rays of light from a distant object in such a way as to bring them together near a single point (Figure 3.12).

After passing through this point, the rays of light will diverge and, to the eye placed near the mirror, it will seem that they are coming from the point of convergence. All that will then be necessary is to place a

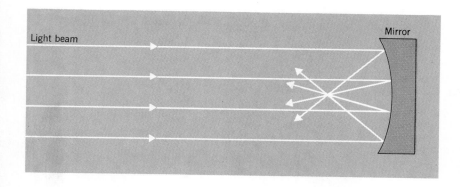

Figure 3.12 Concave mirror converging light.

magnifying lens in front of the eye so that the image can be enlarged for examination, and we will have a telescope. That is, we will have a device for forming an enlarged image of distant objects.

There is one problem with this reflecting mirror telescope. Clearly, Newton's head would have been in the way when he tried to examine the image. To solve this difficulty, he placed a small mirror at an angle of 45 degrees, as shown in Figure 3.13 so that the rays of light were reflected to the side. Only a small part of the original light, the part blocked off by the relatively small mirror placed inside, was lost.

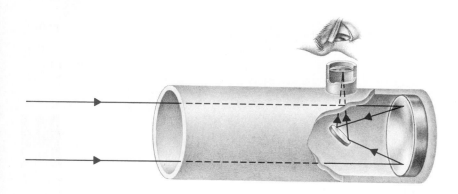

Figure 3.13 The Newtonian reflecting telescope.

A magnifying lens placed at the side as an ocular then permitted the observer to look at the magnified image, as in an ordinary telescope.

The first reflecting telescope Newton built had a mirror only 2 inches in diameter. Being so small, it offered no advantage over the refracting telescope, or telescope with an objective lens, as a means of collecting more light (Figure 3.14 a and b). Newton built it for an entirely different reason; he built it to get rid of troublesome color fringing that plagued the makers of the simple telescopes based on lenses. A lens, like a prism, tends to break up white light into its component colors. A mirror reflects all colors of light at the same angle and, therefore, all the colors contained in the beam of light are brought to a focus at the same point. That is, a mirror has no color fringing (Figure 3.15).

Figure 3.14 Newton's first telescope; (a) side view, (b) front view.

Of course, when the image formed by the mirror is viewed through a magnifying lens, some color fringing is produced by that lens, but if the color fringing caused by the objective is eliminated, the clarity of the final image is much improved.

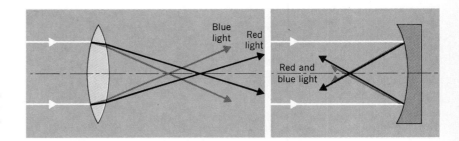

Blue light  Red light

Red and blue light

Figure 3.15 Red light and blue light are focused at different points by a lens (left); a mirror (right) focuses all colors at the same point.

## The Cassegrain Reflector

Another version of the reflecting mirror telescope, in some ways even more ingenious than Newton's, was invented by the French astronomer Cassegrain in 1672. Cassegrain proposed to cut a small hole in the center of the concave mirror. He placed a small mirror opposite the main one, near the point of convergence, so as to reflect the rays of light as they converged. However, Cassegrain's small mirror was oriented so that it reflected the converging rays straight down the axis of the telescope and through the hole in the main mirror (Figure 3.16). The small mirror had to be *convex* in shape, so as to spread the rays of light enough to bring them to a focus outside the tube. With a magnifying lens placed at the hole in the main mirror, the observer could now view the magnified image in comfort.

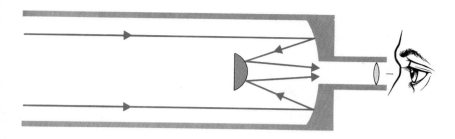

*Figure 3.16   The Cassegrainian telescope.*

The Cassegrain reflector offered a great practical advantage over the Newtonian reflector because the astronomer was stationed at the bottom end of the telescope tube, and could stay on the ground in safety and comfort while he studied the sky. In using a large-sized Newtonian reflector, on the other hand, the astronomer had to perch precariously on a bosun's chair or scaffolding as much as 40 feet above the ground, at considerable danger.

Newton sneered at Cassegrain's invention as a proposal without merit and succeeded in discouraging Cassegrain, who sank back into obscurity and never built a model based on his design. "Its advantages are none," said Newton to the Royal Society. It was only much later, after Newton's death, that astronomers came to appreciate the virtues of this clever design.

## Mirrors: Metal versus Glass

In spite of the advantages of the reflecting telescope over the refracting instrument, the reflector did not come into wide use until the twentieth century. Prior to that time, telescope mirrors were cast out of metal alloys, usually in mixtures of copper, zinc, and tin that could be polished fairly well but tarnished rapidly on exposure to the air, seriously reducing the brightness of the image. In 1850, a technique was developed for depositing a thin film of brightly reflecting silver on glass. This technical development tipped the balance in favor of reflecting telescopes from that point on. It was then possible to make the telescope mirror out of glass, which could be ground accurately to a concave surface and coated with a thin film of silver or some other brightly reflecting metal to give it a high polish. In principle, when the metal coating became tarnished, the coating could be applied again, so that the mirror had an indefinitely long lifetime.

However, silver-coated glass mirrors suffered from a serious defect. Glass expands and contracts with changes in temperature. If the telescope mirror was ground to the right shape at room temperature and used at night at a lower temperature, it would not have the correct shape during its actual use because of the contraction of the glass. As a result, the telescope image would be blurred and distorted.

In addition, the silver coating tarnished rapidly and had to be repolished inconveniently often. In repolishing there was always the danger that the shape of the mirror would be affected, ruining the focusing properties of the telescope and clarity of the image.

The road to success in making large reflecting telescopes opened up with the development of the technology for casting mirrors out of Pyrex glass, about 50 years ago. Pyrex glass expands and contracts very little in comparison with ordinary glass and can be counted on to preserve the precisely concave shape given to it during the grinding and polishing process. Today, mirrors are made of Pyrex glass or of fused quartz, a hard glasslike substance, more expensive than Pyrex and difficult to manufacture, but even better than Pyrex at keeping its size and shape at different temperatures.

At about the same time that the Pyrex mirrors came into use, a physics professor at the Johns Hopkins University developed a practical way of coating the mirror with aluminum rather than silver. The aluminum mirror reflected more light, did not tarnish noticeably, and was hard

*Figure 3.17 The 200-inch mirror completed for testing before an aluminum coating is applied. The ribbed structure, designed to give strength with minimum weight, shows through the surface.*

enough to be washed with soap and water, an operation that would entirely wipe away a thin silver film.

The 200-inch telescope on Mount Palomar has a mirror made out of Pyrex glass covered by aluminum. The mirror is in the shape of a parabola, a shape superior because all the light rays are reflected to a single point rather than near a point, as with a spherical mirror. The grinding and polishing of the Mount Palomar mirror required 11 years, including a 4-year interruption during the Second World War. Five tons of glass were removed by polishing during that 11-year period. When the polishing was completed, the mirror surface was true to the desired parabolic form within one-millionth of an inch over every point of the 30,000 square inches on its surface (Figure 3.17).

## RADIO TELESCOPES

The same kind of curved reflector that collects light waves can also be used to collect and focus radio waves. If the curved reflector is pointed at the sky and used to study radio signals coming to us from various directions in space, it is called a radio telescope. Although the radio telescope was invented only 40 years ago, and although astronomers have actively employed it for less than 20 years, it has already added a large body of information to our knowledge of the Universe.

An electrical engineer named Karl Jansky was responsible for the birth of radio astronomy. In 1931 Jansky, while working on a practical problem in telephone communications and not thinking about astronomy at all, stumbled across a phenomenon that led to the opening of a second window in the sky. Radio telephone conversations between London and New York were frequently obliterated by static and hissing noises, and Jansky had the assignment of finding out where the static came from so that it could be eliminated. He found that when obvious causes of static, such as thunderstorms, were eliminated, one type of noise still remained. This was a hissing noise that did not seem to be connected with any earthly phenomenon. Probing into the nature of this mysterious hiss, he found that it came from a fixed direction in space, which later turned out to be the center of our Galaxy. Jansky had stumbled on radio waves from space. His accidental discovery opened up the new field of radio astronomy. The radio window soon joined the visible window as a second channel through which information passed to the earth from the heavens.

### Advantages of the Radio Telescope

One of the basic reasons for the importance of radio telescopes is the fact that a radio wave can pass freely through a cloud of dust that would completely absorb light rays. Astronomers use them to see through the dust in space in the same way that air traffic controllers use radar waves to see through clouds.[1] With his telescope, the radio astronomer has been able to chart the structure of our own Galaxy, in spite of the fact that most of the Galaxy is completely obscured by huge clouds of dust that lie in the space between the stars. By plotting the intensity of radio waves coming to us from various parts of our Galaxy, astronomers have learned that the Galaxy has the form of a spiral, similar to the shape of our neighbor, the Andromeda galaxy. They have also made fundamental discoveries regarding other galaxies—discoveries that could never have been made by studying the visible light emitted from these galaxies.

Another reason for the importance of radio astronomy is the fact that we can see farther into space with radio telescopes than we can with optical telescopes. That is, radio telescopes can detect galaxies—by means of the radio waves they emit—that lie at least twice as far away as the faintest and most distant galaxies that an astronomer can see with an optical telescope. In this sense, radio telescopes have doubled the radius of the observable universe.

It seems surprising at first that radio telescopes have a greater range than optical telescopes, because distant galaxies emit far less energy in the form of radio waves than they do in the form of visible light. One obvious reason for their greater range is the fact that they have a larger diameter than optical telescopes and, therefore, a larger collecting area. Another important reason is the fact that when a galaxy is far away, and relatively faint, the rays of visible light from it are blanketed by a general background of illumination in the night sky, which is produced by the so-called night glow—a kind of phosphorescence of air molecules in the upper atmosphere—and by the zodiacal light—a soft and very faint glow of light that results from sunlight hitting particles of dust in the solar system and bouncing back to the dark side of the earth. Usually the background radiation is too faint to be seen by the eye; you may look up into the sky on a moonless night and find it completely black; but, nonetheless, your eye is being bombarded by a small amount of light at all times, at an intensity too low for your retina to detect it, but sufficient to obscure the light from the faintest stars.

[1] Airport radars send out a radar beam into space, which bounces off the aircraft and returns an echo to the airport, indicating the plane's distance. This technique is called "active" radar. Radio astronomers usually "listen" passively to radio signals from distant stars and galaxies. These objects lie too far away for a radio signal to be sent out to them and bounced back. The only exceptions to this passive role of the radio astronomer are provided by the objects in our own solar system—the moon, the sun, and the planets—which are close enough to provide an echo of detectable strength if a radar or radio signal is transmitted to them. Extensive studies of the moon and planets have been carried out by this technique, which has grown into a new branch of astronomy called "radar astronomy."

In the radio region, however, the sky harbors practically no radiation at all. With respect to radio waves, the sky is always blacker than the blackest nighttime sky in the visible region. For this reason, the radio signal from a very distant galaxy will emerge out of the general sky background and be easily detectable, even though the intensity of the radio signal is far lower than the intensity of the light beam from this galaxy.

It is interesting to note that with respect to radio waves the *daytime* sky is as black as the *nighttime* sky. Thus radio astronomers, unlike optical astronomers, can see galaxies as well during the day as they can at night. They can even see them on a cloudy day or night since radio waves pass freely through clouds, although at the shorter wavelengths in the radio region, the clouds do present some problem because they absorb and reflect small amounts of radio energy, as well as emitting some radio waves themselves.

## Comparison with Optical Telescopes

The radio telescope differs from the optical telescope in two important respects. First, it uses an antenna plus a radio receiver as a receiving device, instead of the eye or a photographic film. It would do no good to "look" through a radio telescope, because the eye is completely insensitive to radio waves. Photographic film is also insensitive to radio waves; hence, you cannot use a camera to record the "images" that a radio telescope focuses. The only device that reacts sensitively to radio waves is an antenna—an electrically conducting wire along which electrons can move freely. To understand how the antenna works, remember that a radio wave is a kind of electromagnetic wave. As the train of electric waves moves past the antenna, it exerts a changing electric force on the electrons in the antenna. The electric force reverses its direction repeatedly as the waves move by. In response to this continually reversing force, the electrons surge up and down the length of the antenna. The antenna is connected to a radio receiver, which detects the tiny electrical changes produced by the surging electron currents in the antenna and magnifies their strength a million times or more so that they can be easily measured and recorded (Figure 3.18a).

The receiver is connected to a recording device, which may be a pen moving across a slowly revolving drum of paper. As the radio telescope scans the sky, it may happen to point in the direction of a strong source of radio waves such as a radio galaxy, which is the equivalent of a bright object for an optical telescope. The radio receiver will detect a sharp increase in the strength of the signal, and the pen will record a peak on the drum of paper. An example is shown in Figure 3.18b, which is a drawing of a recording made by a radio telescope pointed at a galaxy.

If the signal from the radio receiver were connected to a loudspeaker

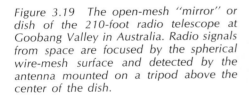

Figure 3.18(a) *Drawing of a radio telescope.*

Figure 3.18(b) *Recording of a signal from a radio galaxy received by a radio telescope.*

Antenna

Reflector

Receiver

Recorder

*(a)*

*(b)*

instead of to a recording pen, you would hear the distant galaxy as a crescendo of static emerging out of the general background of hissing noises, as the axis of the telescope swept past the direction of the radio galaxy. The effect would be similar to that produced on your automobile radio when you drive under a high-voltage wire.

The second difference between radio and optical telescopes concerns the surface of the "mirror." The mirror in a radio telescope does not have to be smooth and brightly polished, as is necessary for the mirror of a high-quality optical telescope. In fact, the "mirror" of a radio telescope can have holes in it, or even be constructed out of wire mesh instead of solid glass or metal, and work just as well as if it were a beautifully finished and perfectly reflecting spherical mirror. Figure 3.19

Figure 3.19 *The open-mesh "mirror" or dish of the 210-foot radio telescope at Goobang Valley in Australia. Radio signals from space are focused by the spherical wire-mesh surface and detected by the antenna mounted on a tripod above the center of the dish.*

shows the wire mesh 210-foot telescope in Australia.

How can a "mirror" with holes in it reflect waves? The explanation is that an electromagnetic wave cannot "feel" any hole, irregularity, or roughness that is smaller than its wavelength in size. Most radio astronomy is done at wavelengths ranging from about one-half inch to a few yards. For use at these wavelengths, the reflecting surface need only be "true" to a spherical or parabolic surface to within a fraction of an inch or so. The surface itself can resemble a gravelly or pebbly pavement, but as long as the bumps, hollows, and holes are no more than a fraction of an inch in size, the telescope mirror will reflect radio waves just as well as if it were a finely polished piece of metal. In fact, the reflecting surface of a radio telescope is usually so rough that it is not called a mirror at all. Instead, it is called a dish.

*Figure 3.20   The 250-foot radio telescope at Jodrell Bank.*

Figure 3.20 shows the radio telescope at Jodrell Bank in England. For many years this telescope was the largest dish in the world that was *fully steerable,* with a construction enabling it to point at any direction in the sky. It has recently been supplanted as the world's largest fully steerable telescope by a 300-foot dish in Bonn, Germany. The surface of the Jodrell Bank dish was originally constructed out of an open wire mesh, but later the mesh was covered with metal to improve the accuracy of the telescope at shorter wavelengths. This improvement made the telescope heavier, but permitted it to be used with wavelengths as short as 4 inches. The original mesh covering limited the use of the telescope to wavelengths of 10 inches and greater.

Much of the expense of a large optical telescope comes from the cost of casting the mirror and polishing its surface to an accuracy of a fraction of a wavelength in the optical region. Because the required accuracy is about one-millionth of an inch, the degree of polishing required is extensive, and the amount of skilled labor and the costs are correspondingly high. That is one of the reasons why the largest optical telescope in use thus far has a diameter of only 200 inches or 16.6 feet. The dish of a radio telescope can be made much larger at reasonable expense. Radio telescopes have been constructed with diameters ranging up to 300 feet. For special purposes, in which the telescope does not have to be capable of pointing at any direction in the sky, the diameter can be as much as 1000 feet.

The largest radio telescope in the world—1000 feet in diameter—is located in Puerto Rico, in the hills near Arecibo on the western end of the island. It is actually a fixed dish built into a natural bowl in the hills (Figure. 3.21a). However, the direction from which it receives signals, that is, the direction in which it "looks," can be varied by as much as 15 degrees on either side of vertical. This flexibility comes from suspending the antenna-receiver from cables strung above the bowl, which permits the antenna-receiver to travel across the valley like a cable car (Figure 3.21b).

*Figure 3.21 The thousand-foot dish at Arecibo, Puerto Rico. The dish was made by placing wire mesh over a natural bowl in the mountains.*
*(a) The bowl is oriented vertically upward and is immovable, but receives signals from many directions as a result of the earth's rotation.*

(a)

(b) Additional flexibility in direction is obtained by shifting the antenna, which is mounted on a trolley suspended from cables 500 feet above the valley floor.

The direction in which the telescope points also varies during the course of the day with the rotation of the earth. The combined variation is sufficiently great for the instrument to be "pointed" at all the planets in the solar system as well as at a number of galaxies and nebulas.

## Signals from Intelligent Beings?

Occasionally confusion arises concerning the nature of the radio signals that come to us from outer space. The term "radio signal" seems to have the meaning of a voice communication, or a stream of dots and dashes sent out by intelligent beings. However, the signals that radio astronomers record are not of this kind. In fact, no astronomer has yet received a signal that he feels indicates the presence of intelligent life in space. The radio signals picked up from stars and galaxies resemble a steady noise of static. They are rapidly fluctuating radio waves containing all frequencies jumbled together in a chaotic fashion, and they can be explained entirely by natural causes, without resorting to a theory of transmitters constructed by intelligent beings.

Although the radio signals received from space show no signs of having been produced by intelligent life, many people have wondered whether a few signals produced by intelligent beings might be mixed into the jumbled static of the natural radio waves. Thus far, radio astronomers have made only one serious effort to listen for such artificial signals, ignoring the natural waves that are their normal interest. The project was carried on in 1957 at the National Radio Astronomy Observatory at Green Bank,

West Virginia, using the 85-foot dish antenna. The astronomers detected nothing that could not be connected with passing trucks or other manifestations of terrestrial life, and they abandoned their effort after one month, being anxious to get back to their regular research projects. In the future astronomers may listen again for a longer time.


## NEW WINDOWS ON SPACE

### Infrared Telescopes

The science of infrared astronomy was born only a few short years ago, although the existence of a window in the sky in the infrared region had been known for a long time. The problem plaguing earlier would-be infrared astronomers was that there was no good way of detecting infrared waves. Photographic plates, which are the heart of modern optical astronomy, are only sensitive to light of visible and shorter wavelengths, and do not react at all to infrared waves. Radio receivers, which are the heart of radio astronomy, also do not respond to infrared waves. But in 1963 a physicist named Frank Low, working on a problem not related to astronomy, built a device that turned out to be an extremely good detector of infrared radiation, one thousand times more sensitive than any infrared detector that had been built up to that time. Low's secret was that he cooled his detector to the temperature of liquid helium, a few degrees above absolute zero, making it extremely responsive to even the faintest trace of heat radiation that fell on it. Once the supersensitive infrared detector was built, Low immediately saw its applications to astronomy, although, like Jansky, he had not been thinking about new astronomical windows in the sky when he began his project. Placed at the focus of an ordinary reflecting telescope, in place of the normal photographic plate, the liquid helium-cooled detector provided an excellent way of collecting the very faint infrared radiation from distant stars and galaxies. In the first year of his work with the new detector, Low made several remarkable discoveries, which are described in subsequent chapters. His work marked the birth of infrared astronomy.

The type of sensitive infrared detector first designed by Low has come into extensive use among astronomers. Because so many exciting discoveries have been made with infrared telescopes working on the ground, the astronomers engaged in this branch of research have been strongly motivated to send their infrared telescopes up into the atmosphere in balloons and high-altitude aircraft, so that some of the murkiness that obscures the infrared windows on the ground will be cleared up. This murkiness is almost entirely the result of water vapor in the atmosphere, which is confined to low altitudes. An aircraft flying at 40,000 feet has only half as much water above it as exists above a telescope situated on the ground, and a balloon, flying at altitudes up to 100,000 feet, has only 1 percent as much water vapor above it as is found at ground level.

The balloon flights have been particularly fruitful, the amount of water vapor remaining above balloon altitudes being so small that certain regions of the infrared spectrum which are completely blacked out from the ground—such as the region between one-tenth of a millimeter and 1 millimeter—become accessible. Figure 3.22 shows the launching of a large helium-filled balloon, more than ten stories high when fully inflated, from which there is suspended a gondola containing a telescope with a liquid-helium infrared detector mounted at its focus.

This balloon-borne infrared telescope was responsible for the mapping of the Galaxy at a wavelength of one-tenth of a millimeter. The discovery of intense infrared emission in the plane of the Milky Way Galaxy was made in the course of a balloon flight similar to the one whose beginning is shown in this photograph.

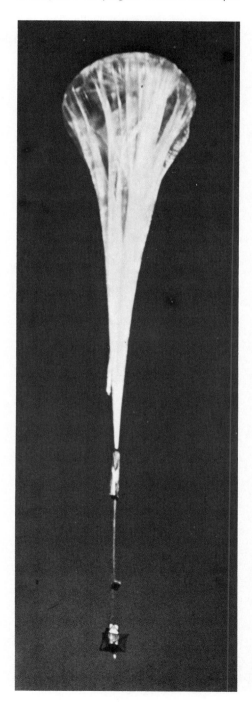

*Figure 3.22 A balloon ascends into the stratosphere carrying an infrared telescope.*

## Telescopes in Space

With the advent of the space age, the entire electromagnetic spectrum has opened up to the astronomer. From the first astronomical use of the telescope by Galileo in 1609 until the discovery of radio waves coming from space in 1931, telescope observations in the visible band of wavelengths occupied the center of the stage in all astronomical investigations. After the birth of radio astronomy in 1931, radio telescopes shared the burden of exploring the universe with optical telescopes. Now many other regions of the electromagnetic spectrum are becoming accessible to the astronomer, in addition to the radio and optical wavelengths. Most of the newly accessible wavelengths are strongly absorbed by the atmosphere and could not be seen until the opening of the spage age, when astronomers were able to send their telescopes into orbit. At relatively long wavelengths in the infrared region, some radiation penetrates to the ground where it can be observed with telescopes, or to the stratosphere where it can be observed from balloons. However, the very short-wavelength regions of the spectrum—extending from the far ultraviolet into the x-ray region and then into the gamma-ray region—can only be observed from platforms in space, because the atmosphere blanks out all these types of radiation below an altitude of 60 miles.

The potential of the short-wavelength end of the electromagnetic spectrum has just begun to be tapped with the launching of the Orbiting Astronomical Observatory in 1968 for a preliminary survey of the sky in the ultraviolet, the launch of the Small Astronomical Satellite in 1971 carrying an x-ray telescope, and the launching of the first crude version of a gamma-ray telescope. Although the information collected thus far by these orbiting telescopes does not compare in importance with the results of ground observations, their future importance warrants a brief discussion of the new instruments and their findings.

*Ultraviolet Astronomy.* The ultraviolet region has always tantalized astronomers because it lies on the edge of the visible region, and some of its radiation—that in the *near-ultraviolet* just beyond the visible band of violet wavelengths—penetrates to the ground in sufficient amounts to be observed to a limited degree. It has always seemed to astronomers that if the curtain could be rolled up to reveal the full ultraviolet region, their knowledge of the universe would be enormously enhanced.

A satellite containing instruments designed solely for ultraviolet astronomy was launched in 1968 and provided the first comprehensive survey of the sky at wavelengths in the far ultraviolet—entirely inaccessible from the ground—extending from 2650Å down to approximately 1000Å. The ultraviolet-telescope satellite, called the Orbiting Astronomical Observatory, or OAO, was the largest and most complex scientific satellite ever launched to date. Figure 3.23 shows the OAO. It weighed 4400 pounds, was 10 feet long and 7 feet in diameter, and contained seven telescopes looking out one end and four telescopes looking out the other. The largest telescope had a mirror with a diameter of 16 inches. The mirrors were made of beryllium instead of quartz or glass in order to minimize their weight.

Solar cells mounted on paddles 10 feet in length extended out from the sides of the main satellite body like stubby wings, collecting sunlight and converting it into electrical power to operate the satellite's instruments as well as its radio transmitter and receiver.

Astronomers controlling the satellite from the ground could send up

*Figure 3.23 The Orbiting Astronomical Observatory (OAO) photographed during a pre-launch test.*

radio commands, causing the satellite to swivel about in the sky so as to look at any chosen star or galaxy.

The telescope was equipped with photoelectric cells constructed to search for and lock into certain guide-stars. Spinning flywheels located inside the satellite body were used to shift the orientation of the satellite in response to ground commands, working on an action-reaction principle; that is, as the flywheel turned one way, the satellite turned the other. With the aid of these devices, the body of the satellite, and the satellite's telescopes, could be pointed to a chosen star with an accuracy of 1 minute of arc, or one-sixtieth of a degree. This pointing accuracy corresponds to hitting a twenty-five-cent coin at a distance of 100 yards.

In place of the photographic plates normally used to capture the image of a star in a ground-based telescope, the OAO telescopes employ photoelectric cells.

Among the initial results from the OAO are the first observations of planets, galaxies, and quasar 3C273, the brightest object observed in the heavens thus far (see Chapter 10). The results show that the Andromeda galaxy is abnormally luminous at wavelengths shorter than 2700Å, a result that may be difficult to explain if it is true for all galaxies.

*X-Ray Astronomy.* The next-shortest region of the spectrum beyond the ultraviolet is the x-ray band of wavelengths, extending from roughly 100Å down to 1Å. Astronomical observations of the sky in this band of wavelengths have been made from rockets launched into nearly vertical trajectories. During each flight the rocket stays above an altitude of 60 miles for approximately 10 minutes during which period the x-ray observations are carried out. Although the accumulated total of x-ray observations made from rockets in this way has amounted to only a few hours, the rocket observations have revealed that a number of points in the sky are copious x-ray emitters.

Approximately 40 separate sources of x rays were discovered during the rocket flights, nearly all lying in our Galaxy, and all charcterized by an enormous amount of x-ray power. On the average, each of these sources sent out to space, in the form of x rays alone, about 1000 times the total energy radiated by the sun at all wavelengths. The most powerful x-ray source discovered by the rockets was Scorpius X-1, so named because it was located in the direction of the constellation Scorpius. An inconspicuous, faint star, hardly detectable by optical or radio astronomers, is located at the position of this x-ray source. No explanation has been provided for the strong emission of x rays by the faint, starlike object.

These discoveries have given astronomers a strong motivation for carrying out a complete survey of the sky in the x-ray region, using an x-ray telescope mounted in a satellite from which observations can be conducted for weeks or months instead of hours. The first x-ray satellite was launched in 1970 from a floating platform on the equator off the coast of Kenya in the Indian Ocean. It was named Uhuru, meaning "freedom" in Swahili. The Uhuru satellite weighs 315 pounds and is 8 feet long.

As the satellite circles the earth in orbit, it rotates once every 12 minutes, scanning the sky to obtain a complete survey of x-ray sources. The direction of its telescopes is controlled by an electromagnet mounted in the body of the satellite, which can be magnetized by a radio command from the ground, causing the electromagnet to act as a compass needle in the earth's magnetic field. The tendency of the electromagnet to align itself with the earth's magnetic field provides a force that is used to control the satellite orientation. Using this device, the satellite can be focused on strong x-ray sources with a precision of 1 minute of arc.

Several important discoveries were made with the Uhuru satellite immediately after its launch. An x-ray pulsar, emitting precisely timed x-ray pulses at the rate of about 15 per second, was discovered in the direction of the constellation Cygnus (see Chapter 7 for a discussion of pulsars). From the rapid pulse rate of this pulsar, theoretical astronomers could deduce that it must be relatively young. Presumably, it is the compressed core of a star that exploded into a supernova a fairly short time ago on the scale of stellar lifetimes. Using radio telescopes, astronomers have discovered a relatively young pulsar, emitting approximately 30 flashes a second at the center of the expanding cloud of gas known as the Crab Nebula, which is known to be the remains of a star that exploded in A.D. 1054. However, no visible remains of an exploded star can be found in the neighborhood of the x-ray pulsar in Cygnus. Do these discoveries indicate that pulsars can be formed in other ways than by the compressive forces of a supernova collapse and explosion? No one knows at the time of this writing. The discovery of the puzzling x-ray pulsar in Cygnus has strengthened the interest of astronomers in these new regions of the electromagnetic spectrum.

### Gravitational Astronomy

In Chapter 2 we described how light and other kinds of electromagnetic radiation can be explained as vibrations of the electric force traveling through space. In 1916, Einstein predicted that vibrations of gravitational force could also travel through space. These vibrations would be *gravity waves* or *gravitational radiation*. To understand how gravity waves can be created, let us consider a situation involving two particles of matter, analogous to the situation involving charges of electricity that was described in Chapter 2.[2] Suppose that one particle is initially at rest. It will exert a gravitational force on any other particle of matter near it. Suppose that a second particle of matter is located at some distance from the first

---

[2] The theory of gravitational waves shows that the motion of massive objects must be somewhat more complicated than in the electromagnetic case, in order to generate gravitational radiation. However, the physical process in creating gravitational waves is essentially the one described.

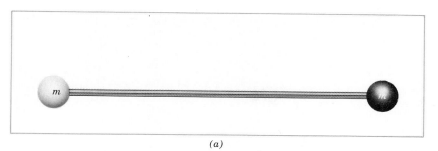

(a)

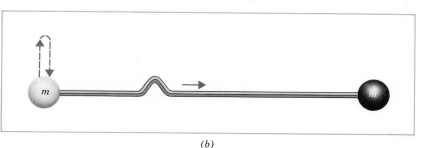

(b)

*Figure 3.24a  Gravitational attraction between two masses.*

*Figure 3.24b  Movement of a mass generates a gravitational pulse.*

one. The gravitational attraction between the two particles will pull them toward each other. As in the case of the electrical attraction, this gravitational attraction is like a rubber band drawing the two particles together (Figure 3.24a).

Suppose now that the two particles are prevented from approaching one another, and that the particle on the left is jiggled up and down very quickly. The jiggling motion has the same effect as though a kink had been produced in the rubber band stretched between the two particles. This kink travels down the rubber band toward the particle on the right (Figure 3.24b).

When the kink in the rubber band reaches the second particle, it causes that particle to move up and down rapidly, responding to the original jiggling motion of the first particle. In precise analogy to the situation described for electric forces, the second particle has experienced a gravitational pulse that was transmitted across space from the first particle. If the particle on the left vibrates up and down repeatedly, a train of gravitational pulses will move out into space from it. This train of pulses is a gravitational wave. When it reaches the second particle, the gravitational wave will cause that particle to move up and down repeatedly.

How fast does a train of gravitational waves move through space? According to Einstein, gravity waves travel at the speed of light. Whether or not they actually travel at that speed is not known yet, but if Einstein's theory of relativity is valid, we can be certain that they do not travel any faster. Since there is a large amount of evidence in support of the relativity theory, it is considered certain by astronomers and physicists that gravity waves travel no faster than the speed of light.

Figure 3.25a   The Skylab Space Station.

Figure 3.25b   Physicist Owen Garriot setting up instruments near the cluster of solar telescopes mounted outside the Skylab.

*The Weber Experiment.* Although the existence of gravity waves has been assumed by theorists for a long time, the detection of these waves has always been considered extremely difficult, if not impossible, because of the weakness of the gravitational force. This force is $10^{36}$ times weaker than the electromagnetic force that gives rise to light waves. If electromagnetic forces were as weak as gravity forces, our entire Galaxy would shine with the brightness of a 10-watt bulb.

A physicist named Joseph Weber was the first person to claim success in detecting gravity waves. Although astronomy was not his major interest originally, he was intrigued by the extremely difficult experimental problem that the search for these waves presented, and he began to build detectors for them. In 1969, after 10 years of effort, he announced that he had discovered gravity radiation. Several research teams in the United States, the United Kingdom, and the USSR are attempting to confirm Weber's results.

*Astronomy from Skylab.* A large instrumented space station called Skylab (Figure 3.25a) was launched in 1973 with astronomy as one of its principal scientific objectives. Three scientist-astronaut crews were sent into orbit for periods of one to several months to conduct observations with telescopes, cameras and other instruments (Figure 3.25b). The astronomical instruments were designed mainly for the study of the sun. Solar flares and other disturbances on the sun's surface were ob-

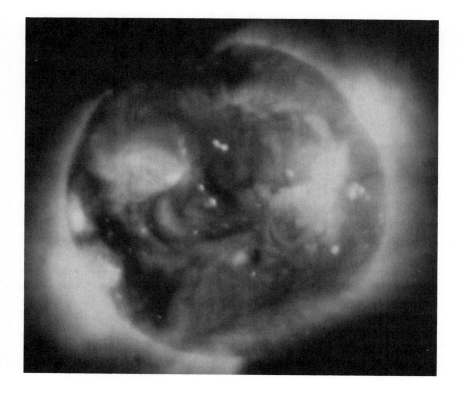

*Figure 3.26 An x-ray photograph of the sun taken from the Skylab on May 28, 1973.*

served during the missions. One of the most interesting results was a series of photographs of the sun taken with an x-ray telescope. Figure 3.26 is an example. The x-ray photographs show intense x-ray emission from the disturbed areas on the sun's surface. They also show curves and loops that seem to follow magnetic lines of force above the sun's surface. The x-ray emitting regions are connected with the presence of strong, localized magnetic fields on the sun's surface (Page 330), from which lines of magnetic force loop upward into the corona or outer atmosphere of the sun.

## THE SPECTROSCOPE

We will see in the next chapter that astronomers acquire most of their information about the stars by breaking up starlight into separate wavelengths and measuring the intensity of light at each wavelength. Each element in the star radiates out a characteristic pattern of wavelengths. By separating the light into its individual wavelengths, we can detect these patterns, determine which elements are in the star, how much of each the star contains, the temperature of the star, how fast it is moving through space, and a great deal of other important information about it.

The eye cannot be used for this purpose because it is unable to resolve a ray of light into its individual wavelengths. To separate the wavelengths in a beam of light, the eye must be aided by an instrument called the *spectroscope*. Spectroscopes separate the various wavelengths in a beam of light from one another so that each wavelength can be examined individually. If the separate wavelengths are photographed, instead of being examined by eye, the instrument is called a *spectrograph*.

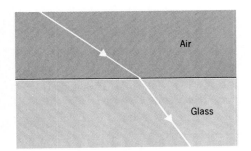

Figure 3.27 *The bending of a ray of light as it enters glass.*

### Prism Spectroscopes

The commonest type of spectroscope makes use of a property of light traveling through glass. As we pointed out in our explanation of the telescope, a ray of light is bent as it enters a piece of glass because light travels more slowly in glass than in air (Figure 3.27). The speed of light in glass depends on its wavelengths, the speed of blue light in ordinary glass being approximately 121,000 miles per second, whereas the speed of red light is 122,000 miles per second. Because of the difference in speeds, blue light is bent by a larger angle than red light on entering a piece of glass obliquely.

When a ray of light composed by many wavelengths enters a piece of glass, the light breaks up into a number of different beams, each beam consisting of one wavelength (Figure 3.28). The beams travel through the

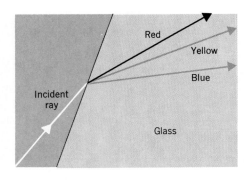

Figure 3.28 *The separation of a ray of light into component colors as it enters the glass.*

glass in different directions according to their wavelengths. In other words, the glass separates the wavelengths in the ray of light. When the light emerges from the glass into the air again, it should be possible for the eye to see each of these wavelengths.

If the glass is cut in the shape of a prism, the rays will be bent twice — once when they enter the prism, and a second time when they leave it. On entering and leaving the prism, they will bend by angles that depend on their wavelengths. Each time, the blue wavelengths will bend more than the red. Thus, using a glass in the shape of a prism roughly doubles the spread between the blue and the red wavelengths in the beam. This is the principle of the prism spectroscope or spectrograph. If the ray of light contains two wavelengths that are very close together, they will still overlap after passing through the prism. In order to separate close-lying wavelengths, the light is usually passed through a narrow slit before it enters the prism. After leaving the prism, the light will consist of several narrow beams, each beam being an image of the slit in the one color or wavelength characteristic of the substance emitting the light. The eye or the camera sees the images of the slit as narrow lines of light. These are called *spectral lines* (color plate 1).

The set of spectral lines makes up the spectrum of the light that is being analyzed. We will see that each element's atoms radiate their own characteristic pattern of wavelengths. These wavelengths, which pass through a spectroscope and are analyzed into a line spectrum, are a means of identifying the presence of that element. If the light from a star contains the spectral lines of a familiar chemical element, we know that that element exists in the star.

Figure 3.29 shows the components of a complete spectroscope. The

*Figure 3.29 The principle of the prism spectroscope.*

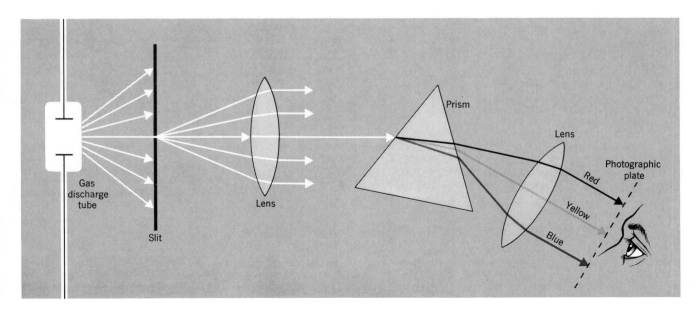

gas discharge tube shown at the left is a typical light source. The end of the spectroscope facing the light source is sealed off from light except for a narrow slit, which admits a small amount of light from the source. The light from the slit passes through the prism after first passing through a lens placed between the slit and the prism. The lens makes the beams diverging from the slit run parallel. The parallel beam enters the prism and emerges from the other side as a number of distinct beams, one for every wavelength in the original light from the gas discharge tube. The beams travel in different directions according to their wavelengths. The light then passes through a second lens that focuses its components into a number of images of the slit. There is one slit image or spectral line for each wavelength in the original light.

If the instrument is used as a spectroscope, the eye is placed beyond the second lens and sees many slit images, one for each wavelength in the light source. If it is used as a spectrograph, photographic film is placed at the focus of the second lens to record the slit images. The entire series of slit images makes up the spectrum of the light emitted from the source.

The size and appearance of a typical spectrograph can be seen in Figure 3.30, which shows a prism spectrograph attached to the end of the 72-inch reflecting telescope of the Dominion Observatory in Canada.

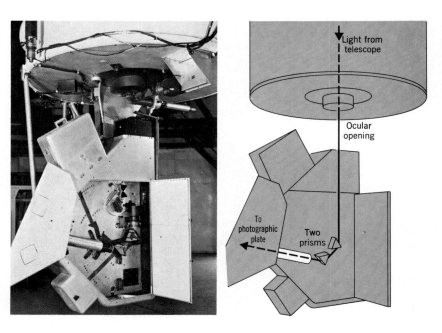

*Figure 3.30 A prism spectrograph attached to a large reflecting telescope. The spectrograph contains two prisms arranged in tandem to increase the separation of the various wavelengths in the light beam. The diagram at right shows the path of the ray of light from the ocular opening through the two prisms to the photographic plate on which the spectrum will be recorded. (The ocular is removed when the spectrograph is attached, because its function is taken over by one of the lenses in the spectrograph.)*

The telescope is in a vertical position in this photograph. The mounting for the ocular appears at the top of the photograph with the main tube of the telescope extending upward from it. The Cassegrainian design is used here, with the light coming downward from above through a hole in the center of the main mirror.

## The Importance of the Spectroscope

A big telescope is a very impressive object; and even a small telescope has great appeal to all, because one can actually look through its eyepiece and see for oneself many of the interesting objects in the heavens. The spectroscope, on the other hand, is no more than an adjunct to the main instrument, mounted on the end of the telescope as a minor attachment. Although it is capable of producing a pretty pattern of colors, normally one does not look through it, but instead a photographic film or plate is placed at the exit end of the instrument to record the distribution of wavelengths as an uninteresting pattern of gray and black lines. Yet if this seemingly modest instrument is removed, the value of the telescope is enormously reduced; the telescope captures the appearance of celestial bodies, but the spectroscope analyzes their nature.

## The Grating Spectroscope

The glass prism is not the only device that can spread light into a spectrum. A device called a diffraction grating can also be used. Astronomers prefer the diffraction grating because it spreads light into a broader spectrum than the prism and can, therefore, separate very closely spaced lines.

In addition, the grating works equally well at wavelengths in the infrared, visible, and ultraviolet regions of the electromagnetic spectrum whereas the use of the glass prism is limited to visible wavelengths only, since glass is opaque to infrared and ultraviolet radiation. This is the most important advantage of the grating over the prism, since the infrared and ultraviolet regions of the spectrum have supplied much valuable information about stars and galaxies that astronomers could never have gathered from visible radiation alone.

The diffraction grating is based on the property of light waves known as diffraction. As we said earlier, diffraction is the bending of a light ray that passes close to an obstacle. Suppose that we consider a black screen in which there is a narrow opening—a slit. Assume that a beam of light falls on this screen. Assume also that this light beam contains light of only one wavelength. If the width of the slit is larger than the wavelength of light, most of the beam goes through the slit unchanged in direction. At the edges of the beam, however, a pronounced spreading, or change in direction of the light occurs because of the closeness of the edge of the slit (Figure 3.31).

The important property of diffraction, with respect to its possible use in a spectroscope, lies in the fact that the amount of the spreading depends on the wavelength: The longer the wavelength, the greater the angle through which the light is bent as it passes the edge of the slit.

A single slit, however, does not yet provide an adequate basis for constructing a spectroscope. It turns out that if we use only one slit, or a small number of slits, the spectral lines produced by the diffraction effect are

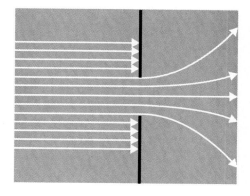

*Figure 3.31 The bending of light at the edges of an opening.*

fuzzy and ill-defined, and the separation according to wavelength is not very pronounced; but if a large number of exceedingly narrow, precisely parallel, and finely spaced slits are used, the diffraction effect leads to spectral lines that are sharp, bright, and very widely spaced according to their wavelength. In other words, it makes an excellent tool for producing a spectrum.

Such a device—a large number of narrow, closely spaced slits in a screen—is called a diffraction grating.

The diffraction gratings actually used in spectroscopes differ in one respect from the one we have described. From a practical viewpoint, it would be very difficult to construct a screen with thousands of narrow slits in it, and so instead the grating is made by scratching many narrow, parallel lines on a smooth, hard surface (Figure 3.32). Usually the lines are scratched on a metal surface with a diamond point. Each scratch scatters the original beam and produces an outgoing wavelet of light, just as if it were a narrow slit through which light was passing from the other side.

When the wavelets from these many scratches are added up, they have exactly the same effect as the addition of the waves from a large number of closely spaced slits. However, the ruled scratches have the advantage that they can be much more finely spaced than slits in a piece of glass or metal. Standard diffraction gratings are ruled with about 15,000 lines per inch, and astronomers use gratings ruled with as many as 30,000 lines per inch. Figure 3.33 shows a ruling engine, specially designed for making diffraction gratings, in the process of ruling lines on a glass disk. The lines

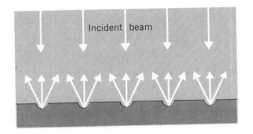

Figure 3.32 Principle of a diffraction grating based on reflection.

Figure 3.33 Lines being ruled on a reflection-type grating.

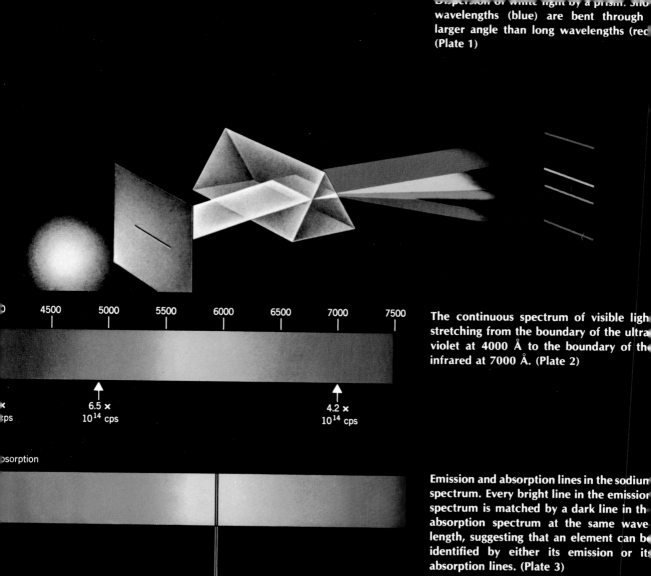

Dispersion of white light by a prism. Short wavelengths (blue) are bent through larger angle than long wavelengths (red) (Plate 1)

The continuous spectrum of visible light stretching from the boundary of the ultra violet at 4000 Å to the boundary of the infrared at 7000 Å. (Plate 2)

4500    5000    5500    6000    6500    7000    7500

6.5 ×
$10^{14}$ cps

4.2 ×
$10^{14}$ cps

Emission and absorption lines in the sodium spectrum. Every bright line in the emission spectrum is matched by a dark line in the absorption spectrum at the same wave length, suggesting that an element can be identified by either its emission or its absorption lines. (Plate 3)

absorption

emission

5890 Å      5896 Å

**YOUNG STARS:** The Orion Nebula — a
region of intense star formation in the
sword of the Hunter. Hot, newly formed
stars in the region emit most of their
energy in the ultraviolet. The ultraviolet
radiation is absorbed by hydrogen gas
surrounding the stars and reradiated as
visible light. (Plate 4)

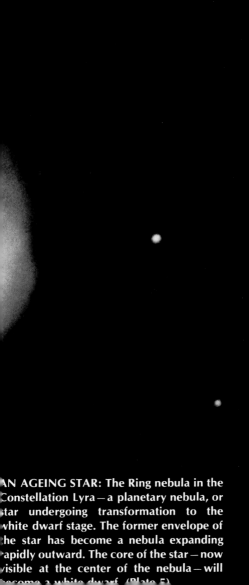

AN AGEING STAR: The Ring nebula in the
Constellation Lyra — a planetary nebula, or
star undergoing transformation to the
white dwarf stage. The former envelope of
the star has become a nebula expanding
rapidly outward. The core of the star — now
visible at the center of the nebula — will
become a white dwarf. (Plate 5)

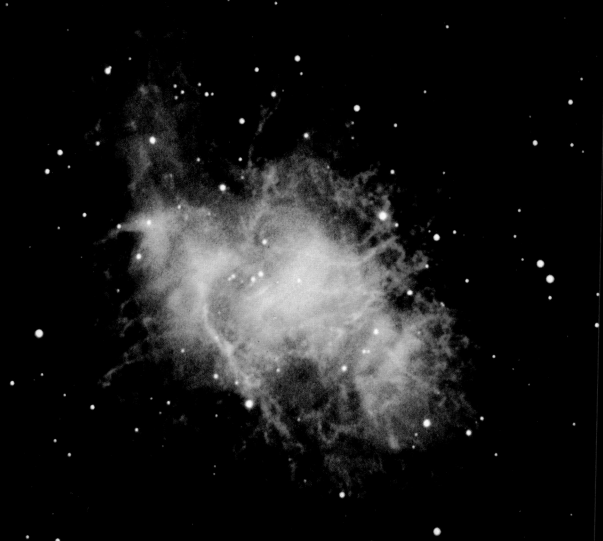

A DYING STAR: The Crab Nebula—a remnant of the supernova explosion of A.D. 1054. The compressed core of the supernova is a pulsar barely visible in the photograph as a faint star in the center of the nebula. (Plate 6)

THE MILKY WAY: The Trifid Nebula, a luminous region similar to the Orion Nebula, seemingly divided into three parts by dark lanes of obscuring dust in the Milky Way. (Plate 7)

THE MILKY WAY: The nebula in the Constellation Serpens, showing star formation in the Galactic disk or Milky Way. The small dark regions are believed to be protostars — pockets of condensed gas on the way to forming new stars. (Plate 8)

CLUSTERS: The Pleiades Cluster — a young galactic cluster containing about 100 stars, formed 60 million years ago. Each of the hottest stars in the cluster is surrounded by the blue glow of its own light reflected from the surrounding interstellar dust. (Plate 9)

CLUSTERS: M13 — a globular cluster containing approximately 300 thousand stars, about 10 billion years old. (Plate 10)

GALAXIES: M82 – an exploding galaxy in the Constellation Ursa Major, about 13 million light-years away. (Plate 11)

GALAXIES: M-51 – the Whirlpool Galaxy, showing a blue color in the spiral arms produced by many newly formed hot stars (blue giants). The white nucleus contains older stars. The irregular galaxy at the left may be a satellite galaxy. (Plate 12)

**THE EARTH:** Shown in gibbous phase, photographed from an Apollo spacecraft. Parts of Africa, Europe, and the Middle East are visible. (Plate 13)

MARS: Photographed from the Earth, showing the dark "seas," lighter "continents," and white polar caps. (Plate 14)

SATURN: (left) The dark band at the equator is the shadow of the rings. (Plate 15)
JUPITER: (above) The Red Spot at left,
30,000 miles across, does not appear to be
anchored over a fixed point on the planet.
Its existence has not been satisfactorily
explained. (Plate 16)

THE SURFACE OF THE MOON: Showing Mt. Hadley in the background and the Swamp of Decay in the foreground, photographed during the Apollo 15 mission in 1971. (Plate 17)

are scratched by a diamond point controlled by a screw that has been machined to the highest possible precision. The lines are ruled very slowly, the making of one grating requiring several months. Once the grating is ruled, however, it is possible to obtain many replicas from it of surprisingly high precision by pouring a solution of collodion or gelatin onto the surface and stripping it off after it has hardened. Gratings made in this way for student use can be obtained for as little as 10 cents.

Diffraction gratings stretch the visible range of wavelengths out into a spectrum spanning an angle of as much as 60 degrees. In contrast, the most effective glass prisms spread the visible spectrum over an angle of only 20 degrees. The ability of a grating or a prism to spread out the spectrum is called its dispersive power. The high dispersive power of the diffraction grating enables the astronomer or the laboratory spectroscopist to separate different spectral lines that lie extremely close together. This accuracy is essential when the astronomer seeks to identify the elements occurring in stars by comparing their spectral lines with spectral lines of known elements measured in the laboratory.

## Questions

1. Describe how a converging lens produces an image. Gradually move a small magnifying lens away from this page. Describe the changes you see. Can you explain them?

2. Describe the role of each of the two lenses in a simple refracting telescope.

3. The lens at right is called a diverging lens. Predict how parallel rays of light would bend as they pass through this lens. Draw the paths of the rays. Can you invent a telescope by combining a diverging lens with a converging lens?

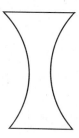

4. What factors blur the sharpness of the image produced by a telescope? How can you reduce each of them?

5. A two-inch telescope can distinguish features on the moon larger than 10 miles in size. What is the smallest sunspot that can be seen with this instrument?

6. Discuss the relative advantages and disadvantages of refracting and reflecting telescopes.

7. Discuss the relative advantages and disadvantages of optical reflecting and radio telescopes. Why can a radio telescope be used for a larger portion of the day than an optical telescope? The radio astronomer is plagued by terrestrial radio noise, or static. List the possible sources of static.

8. List the parts of a complete prism spectrograph and state the purpose of each.

9. Discuss the relative advantages and disadvantages of an observatory in earth orbit and an observatory located on the moon, from both the practical and the pure research viewpoints.

# 4    The Message of Starlight

During the early years of the nineteenth century the telescope and the spectroscope evolved rapidly, and by the midpoint of the century they had reached a relatively advanced state of development. With these instruments in hand, the astronomer possessed the essential apparatus necessary for an inquiry into the properties of stars. Yet in 1850 he knew no more about the true nature of the stars than astronomers had known in Newton's time. The instruments were available, but new theoretical concepts were needed. The new concepts came out of laboratory investigations into the nature of the atom.

## THE ATOM

Astronomy began to grind to a halt in the middle of the nineteenth century. The optical telescope had reached a high state of perfection, and nobody yet dreamed of other windows in the sky. Tens of thousands of stars had been catalogued and put away on the shelf. Two new planets

*Prominent hydrogen lines in a sequence of stellar spectra.*

unknown to the ancients had been discovered, but a thorough search had failed to reveal signs of any others. Astronomers seemed to know everything that could be learned about the stars as points of light. They did not know what the stars are made of, or why they shine, but this knowledge seemed to be forever denied to man because of the enormous distances that separated our solar system from other stars. Astronomy—the oldest science—was ready to enter the graveyard of dead subjects. It was revitalized by a basic discovery in physics.

The discovery was that a hot atom radiates light at a series of wavelengths peculiar to that element. The chemical composition of a star could therefore be determined by analyzing the light coming from that star. This is the message of starlight. It reveals to us what the stars are made of, and how they were born, evolve, and die.

A German physics professor named Kirchhoff, working with the chemist Bunsen, who invented the Bunsen burner, was the first to realize that flaming objects send out a coded message about themselves to all who have the wisdom to read it. Kirchhoff discovered how to break the code, and with his discovery astronomy was reborn. Let us begin the study of modern astronomy by entering the laboratory of the atomic physicist to study the unique properties of the atom.

## The Difference Between the World of the Atom and the Everyday World

The electrons in an atom circle around the nucleus, attracted to it by the electrical force, in the same way that satellites circle in their orbits around the earth under the attraction of the gravitational force. However, the orbiting electron and the orbiting satellite turn out to behave very differently. When we launch a satellite into an orbit, the orbit can be at any distance from the earth that we choose, depending on the power of the rocket. If a certain amount of rocket power will get a satellite into a given orbit, an increase in that rocket power will put it into a higher orbit. Any orbit is possible for the satellite if we put enough power into launching it. This is a simple fact that agrees with our everyday experience on the earth: If you throw a ball up into the air, it will rise a certain distance; if you throw it up a little harder, it will rise a little higher.

But the familiar laws of nature, as we know them in the everyday world, do not apply to the world of the atom. A completely different set of laws governs that world. According to these laws, an electron can only circle the nucleus in certain allowed orbits of definite radius. In the hydrogen atom, for example, the smallest orbit possible to the electron has a radius of 0.53Å. (The angstrom, equal to $10^{-8}$ cm, is a convenient unit for expressing distances between atoms; its symbol is Å.) No orbits smaller than 0.53Å are allowed in the hydrogen atom. That is, no hydrogen atom has ever been found in which the electron's orbit is smaller than 0.53Å.

Moving outward, the next allowed orbit in the hydrogen atom has a radius of 2.12Å. No hydrogen atom exists in the world in which the electron orbit has a radius between 0.53Å and 2.12Å. The rules of the atomic world forbid these intermediate radii. Furthermore, the rules are very precise; the orbit radius has to be *exactly* 0.53Å or 2.12Å. An orbit with a very slightly different radius—0.52Å, say, or 2.10Å—never occurs.

The limitation on possible sizes of orbits is not the only peculiarity of the atom. The *number* of electrons in each orbit is also limited. When dealing with orbiting satellites, we can place as many satellites in an orbit of a given radius as we wish; if a satellite is placed in an orbit with an altitude of 110 miles, no law of nature prevents us from placing a second satellite in an orbit with precisely the same altitude. But in the world of the atom, the number of electrons in a given orbit is limited by a complicated set of rules, which have been worked out by atomic physicists over many years and are a part of the laws of the atom. If these rules specify that a given orbit may contain no more than two electrons, for example, we can be sure that we will never find an atom in which there are more than two electrons in that particular orbit.

*Electron Shells.* The allowed orbits for the electrons in an atom are called *shells*. The laws of the atom specify the radii of the electron shells in each atom, and the maximum number of electrons that may occupy each shell.

The carbon atom, for example, has precisely two electrons in a shell with a radius of 0.2Å, and four electrons circling in a shell outside the inner group, at a radius of 0.5Å. According to astronomers, this is true not only of all the carbon atoms examined on the earth but also of every carbon atom in the universe.

Our experience with everyday events provides no precedent for such peculiar laws. These laws become important only when we investigate the properties of an object as small as an atom or smaller; on a larger scale of distances, such as an inch, a foot, or a mile, they do not apply. When the Danish physicist, Niels Bohr, proposed a special set of laws for the world of the atom around 1910, many other physicists objected, but within a few years it became clear that the properties of the atom could not be explained in any way except by the special rules proposed by Bohr, and now these laws of atomic physics are universally accepted. However, they are still just as hard to understand in terms of everyday experience as they were 60 years ago.

## Electron Shells in the Atom

Every type of atom has its own special set of electron orbits or shells; the radii of these shells are peculiar to each element. The number of shells increases from small atoms to bigger ones. However, the maximum number of electrons in each shell remains the same for all elements. The innermost of the allowed orbits, called the first shell, can hold up to two electrons and no more. It is this shell that contains two electrons in the carbon atom. The second allowed orbit is called the second shell. Eight electrons are the maximum number permitted in the second shell. In carbon, as we noted, the second shell contains only four electrons. When four electrons are placed in the second shell of the carbon atom, they make up, with the two electrons in the first shell, a total of six electrons. Six electrons exactly cancel the positive charges on the six protons in the carbon nucleus. Thus an additional electron passing by feels no electrical attraction toward the carbon atom and cannot be drawn into an orbit in the second shell, even though there is a place for it there.

We have now laid the foundations for building up the entire periodic table of the elements in terms of the number of electrons in the various shells. We start with hydrogen, which contains one electron in the first shell (Figure 4.1a).

Next is helium which contains two electrons in the first shell (Figure 4.1b).

Beyond helium is lithium, with three electrons. In the world of the atom, it is not possible to place a third electron in the inner shell. Therefore, the third electron must go into the second shell (Figure 4.2).

After lithium comes beryllium, with a total of four electrons. Once more, two go into the first shell, and the next two must go into the second shell. Beryllium is followed by boron with five electrons, of which two must be located in the first shell and three must be placed in the second. Then comes carbon, which we have already discussed. Carbon is followed by atoms with successively five, six, seven, and eight electrons in the second shell. These atoms are, respectively, nitrogen, oxygen,

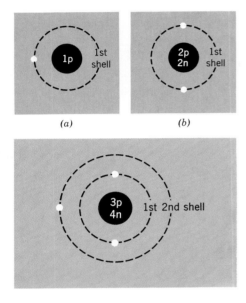

Figure 4.1 (a) and (b)  Atoms of hydrogen and helium $p$ = proton, $n$ = neutron, $e$ = electron).

Figure 4.2  The lithium atom.

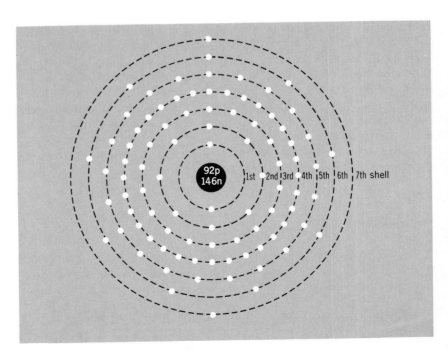

Figure 4.3  An atom of uranium.

fluorine, and neon. At neon, the process of filling the second shell stops because the neon atom has eight electrons in this shell, and eight is the maximum number allowed by the laws of the atom.

By the time we come to the elements near the end of the periodic table, for example, uranium, we find the first four shells fully occupied, the fifth shell nearly full, and the sixth and seventh shells partly occupied (Figure 4.3).

Notice, as we proceed in building up the table of the elements, that it is always the charge of the nucleus, or the number of protons in the nucleus, that sets a limit on the number of electrons we can place in the shells. The number of protons in the nucleus determines the electron structure of the atom.


## GROUND STATES AND EXCITED STATES OF THE ATOM

We have been describing the orbits of the electrons in the atom when it is undisturbed. This undisturbed state is called the ground state of the atom. If another particle collides with the atom, it may be forced out of its ground state. For example, in a gas, the atoms continually collide with one another; or, as another example, the atom may be hit by a fast-moving atomic projectile in a nuclear physics experiment. Occasionally, an electron will absorb enough energy in such a collision to break the bond that holds it in its orbit. This excited electron will then jump outward to a new orbit, in which it circles the nucleus at a greater distance than it did before. An atom in which an electron has absorbed energy and jumped into a larger orbit is said to be in an excited state.

If the force of the collision is very great, the electron may leave the atom entirely. An atom that has lost an electron is said to be ionized; the part of the atom left behind by the departing electron is called an atomic ion, or simply an ion. If the atom has lost one electron, it is sometimes referred to as being singly ionized; if it has lost two electrons, it is doubly ionized; and so on.

Usually it is the electrons of the outermost shell that are affected by collisions. The electrons of the inner shells are more tightly bound to the nucleus and less readily disturbed by outside forces; also, the electrons in the shells surrounding them act as an electrical screen against disturbances.


### An Example: Excited States of the Hydrogen Atom

In the ground state of the hydrogen atom the radius of the electron's orbit is 0.53Å. Because of the special nature of the laws of the atom, the electron in the hydrogen atom cannot circle in an orbit smaller than 0.53Å; orbits larger than 0.53Å also are forbidden, until we come to an

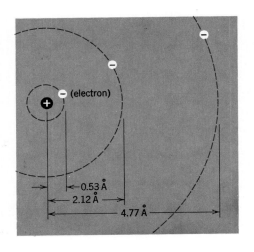

Figure 4.4 The first three allowed orbits and orbital radii for the hydrogen atom.

orbit with a radius of 2.12Å. This is the first orbit above the ground state that is allowed for the electron in the hydrogen atom; it is the first excited state of the hydrogen atom. All orbits with radii greater than 2.12Å are also forbidden, until an orbit radius of 4.77Å is reached. This is again an allowed orbit. It is the second excited state of the hydrogen atom. A third excited state and higher excited states are possible, each one corresponding to a larger orbit than the last. It develops that an infinite number of excited states are possible for the hydrogen atom (and for all other atoms as well), with orbit radii stretching out to infinity.[1]

Figure 4.4 shows the first three allowed orbits (the ground state plus the first two excited states) of the electron in the hydrogen atom.

### Emission of Light from an Excited Atom

Suppose an atom has been excited. What happens to this excited atom? Does the electron stay in the excited orbit, circling there indefinitely? If the orbiting electron were similar to an orbiting satellite, the answer would be yes, it could stay there almost indefinitely. However, the laws of the atom give a very different answer. When an electron has been kicked upward to an excited orbit, it circles in that orbit for a definite amount of time, characteristic of that particular orbit and that particular atom, and then collapses back down to the original orbit in a sudden transition. Typically, the electron stays in an excited orbit for one one-hundred-millionth of a second before collapsing. At the same time that the electron goes back to the ground state orbit, the atom emits a flash of light, or a

Figure 4.5 An excited hydrogen atom collapses to the ground state, emitting a photon.

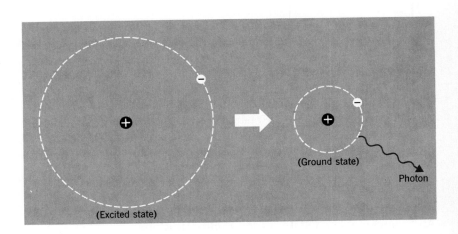

[1] From the laws of the atom, theorists have derived a surprisingly simple rule for the radii of the allowed orbits of electrons in the hydrogen atom: the allowed orbits have radii given by the formula: RADIUS = $n^2 \times 0.53$Å, where $n$ is any integer from 1 to infinity.

photon (Figure 4.5). The energy of the photon is equal to the difference between the energy of the electron in the ground state orbit and its energy in the excited orbit.

## Exciting an Atom by Absorption of Light

Collisions are not the only means by which the electrons in the atom can be propelled from the ground state to a higher state. Suppose a beam of light shines through a gas composed of atoms. This beam of light is equivalent to a hail of photons. If the energy of the photons in the beam is precisely equal to the difference between the energy of the electron in its ground state orbit and the energy of the electron in one of its excited orbits, a photon can be absorbed by an atom in the gas, kicking the electron in the atom from its ground state to an excited orbit (Figure 4.6). The electron remains in the excited orbit for about one-hundred-millionth of a second and then collapses down to the ground state, emitting a photon of the same energy and the same wavelength as the absorbed photon.

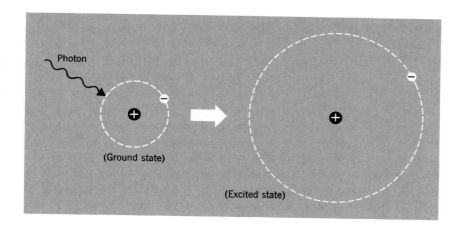

*Figure 4.6  A hydrogen atom is raised to an excited state by the absorption of a photon.*

## ATOMIC SPECTRA

The size of an electron's orbit, and the amount of energy required to get the electron from its ground state to various excited states, depend on the strength of the force holding the electron to the nucleus. This force depends, in turn, on the number of protons in the nucleus. That is, it depends on the nature of the element involved. Thus, each element has its own set of electron orbits, which are the same for every atom of that element, but they are substantially different from the electron orbits of every other element. This set of facts—the uniqueness of electron orbits in an element—is the reason that underlies the use of atomic spectral lines as the means of identifying an element. Many branches of modern science depend on this property of the atom.

## The Spectrum of a Heated Gas

Suppose we have a glass container filled with a pure gas consisting of atoms of one kind, such as hydrogen or helium. If the temperature of the gas is low, nearly all the atoms will be in the ground state. If the gas is now heated, collisions will become more violent, and an increasing fraction of the atoms will be raised to excited states. Collapsing from these excited states to the ground state, these atoms will emit photons with wavelengths characteristic of the kind of atom that makes up the gas in the container. Of course, photons of many different energies and wavelengths will be emitted by the atoms of the heated gas, because each atom possesses many excited states, and transitions can occur between any one of these excited states and the ground state, or from one excited state to any other excited state. As a result of the emission of all these photons with various energies and wavelengths, the heated gas will glow with a light composed of many different colors.

If the temperature of the gas is only moderately high, collisions will occur with only a moderate degree of force, and it is possible that no more than the first excited state will be reached. In this case, the light emitted by the heated gas will then consist of a single color or wavelength, corresponding to the wavelength of the photon emitted in the transition from the first excited state to the ground state. If this light is examined through a spectroscope, a single bright line will be seen at the wavelength or color of the emitted photon. As the temperature of the gas is increased, collisions grow more violent, and the atoms are raised to higher and higher states of excitation. To the single line that appeared in the spectroscope when the temperature of the gas was moderate, there will be added more and more lines until, at very high temperatures, the spectroscope will show a great number of lines representing all possible transitions between allowed states of the atoms in the gas. This set of lines is called the spectrum of this kind of atom.

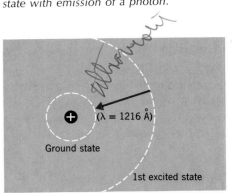

Figure 4.7  *Transition of a hydrogen atom from the first excited state to the ground state with emission of a photon.*

(λ = 1216 Å)

Ground state

1st excited state

## An Example: The Spectrum of the Hydrogen Atom

Consider the hydrogen atom once again as an example. When a hydrogen atom undergoes a transition from its first excited state to the ground state, a photon is emitted with a wavelength of 1216Å, deep in the ultraviolet portion of the spectrum (Figure 4.7).

A line at this wavelength should be the first to appear if hydrogen gas is heated to a high temperature and examined through a spectroscope. If the temperature is raised still higher, or the hydrogen atoms are excited in some other way, as for example, by passing an electric discharge through the gas, some of the atoms will be raised to the second excited state of hydrogen. A hydrogen atom raised to the second excited state can collapse down to the ground state in either of two ways: it can proceed directly from the second excited state to the ground state, or it can undergo a transition to the first excited state as an intermediate step, and from there go to the ground state. In the first case, a single photon is emitted at a wavelength of 1026Å, corresponding to the energy difference between the second excited state and the ground state. In the second case, two photons are emitted, the first with a wavelength of 6562Å, in the red region of the spectrum, and the second with a wavelength of 1216Å, in the ultraviolet region. The arrows in Figure 4.8 show three transitions between the first and second excited states and the ground state. Each transition produces a photon whose wavelength is shown in parentheses. The sum of the energies of the 1216Å photon and the 6563Å photon is equal to the energy of the single photon emitted in the direct transition to the ground state.

As higher states of excitation are reached, the number of alternative ways in which the atom can return to the ground state also increases, and the series of lines emitted by the gas becomes very complex. However, we can introduce a degree of order into this complicated series of lines of different wavelengths by separating the lines into the following groups.

In the first group, place all the lines produced when the hydrogen atom undergoes a transition from any excited state to the *ground* state; this series of lines, of which the first one has a wavelength of 1216Å, is known as the Lyman series.

In the second group, place all the lines produced when a hydrogen atom undergoes a transition from any excited state to the *first excited* state; this series, in which the first line has a wavelength of 6563Å, is known as the Balmer series.

In the third group, place all lines that are produced when the atom undergoes a transition from a higher excited state down to the *second excited* state; this series, in which the first line has a wavelength of 18,756Å, is known as the Paschen series.

Figure 4.9 shows the transitions, indicated by arrows, that give rise to the various lines in the above series.

Figure 4.10 shows the first twelve lines in the Balmer series of the hydrogen atom.

Figure 4.9 brings home the fact that each group, or series of lines, represents a transition to one particular final state of the hydrogen atom. This interpretation of the lines emitted by the hydrogen atom was first put forward by the great Danish physicist, Niels Bohr, about 60 years ago, at the beginning of the twentieth century. Prior to that time, no one had any idea of the meaning of the spectra emitted by atoms. In spite of the failure of physicists to make any sense out of these singular sequences of spectral lines, an enormous amount of work went into measuring the wavelengths of the lines and classifying them, partly for their usefulness

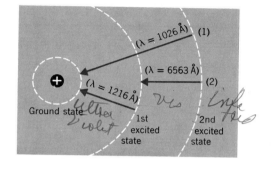

Figure 4.8   Transitions from the first and second excited states of the hydrogen atom to the ground state.

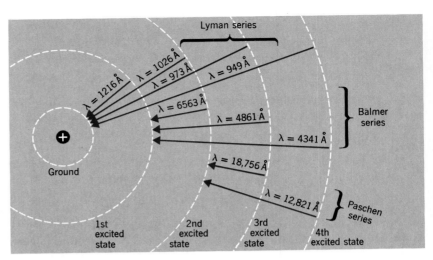

Figure 4.9 Formation of the Lyman, Balmer, and Paschen series of spectral lines in the hydrogen atom.

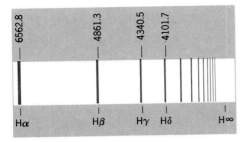

Figure 4.10 The Balmer series.

as the fingerprints of atoms, and partly in the hope of learning something about the structure of the atom, which was at that time an entirely mysterious object.

### Emission Spectra

The sequence of lines emitted by atoms of a given kind, when heated to a high temperature, is called the emission spectrum, or sometimes the bright-line spectrum of that atom. An enormous amount of effort has been invested in spectroscopic examination of the light emitted by atoms. The wavelengths of the lines in the spectra of all the elements have been carefully measured, tabulated, and published. The wavelengths of most lines have been measured with an accuracy of a fraction of an angstrom; and, in the case of certain lines, the precision of the wavelength measurement is one ten-thousandth of an angstrom. The wavelengths of the lines in the spectrum of an element are as unique a characteristic of that element as your fingerprints are of your identity. The published tables of wavelengths of the lines in the spectra of the elements are equivalent to the fingerprint files of the FBI.

### Absorption Spectra

Suppose we have a source of light that emits radiation of all visible wavelengths, such as, for example, a tungsten lamp. If the light from this lamp is examined through a spectroscope, we will see the full range of colors in the visible spectrum, starting with violet at the short-wavelength end, and shading imperceptibly into indigo, blue, green, yellow, orange, and finally to red in the long-wavelength end of the visible region. Now let us place a glass container, filled with a gas of unidentified atoms, between the spectroscope and the light source, so that the light from the source must pass through the gas on its way to the spectroscope (Figure 4.12).

The original light contains photons of all wavelengths and energies; among these photons are some whose energy is precisely equal to the difference between the energies of the ground state and one of the excited states of the type of atom that fills the container. Such a photon can be absorbed by an atom in the container, raising the atom from its ground state to an excited state. This absorption process removes the photon of that particular wavelength from the beam of light. The excited atom collapses quickly to its ground state again, emitting another photon of precisely the same energy and wavelength as the photon it had just previously absorbed. However, the new photon need not be emitted *in the same direction* in which the old one was traveling; in fact, it is usually emitted in a different direction (Figure 4.11).

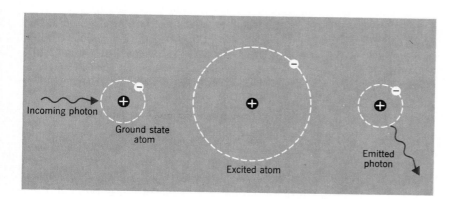

*Figure 4.11   Atom absorbing photon and emitting it in a different direction.*

Incoming photon

Ground state atom

Excited atom

Emitted photon

Thus, the reemitted photons do not enter the spectroscope because they are not traveling in the right direction to do so. They leave the container at the sides, the top, or the bottom, and they are removed from the beam as far as an observer looking through the spectroscope can determine (Figure 4.12). Therefore, the continuous spectrum of the light source will, as seen through the spectroscope, be marked by dark lines at certain wavelengths. The wavelengths that are "dark" are those whose photons have been removed by absorption in passing through the gas.

But what wavelengths are these? They are the wavelengths that correspond to the difference between the energy of the ground state and the energy of an excited state for the atoms of the gas. It is precisely these wavelengths at which light would be emitted if the gas were heated to a high temperature, so as to produce a bright-line or emission spectrum. If we photograph the continuous spectrum in the above experiment, marked by dark lines at certain wavelengths as we have described, and place it alongside a photograph of the emission spectrum of the atoms of the gas

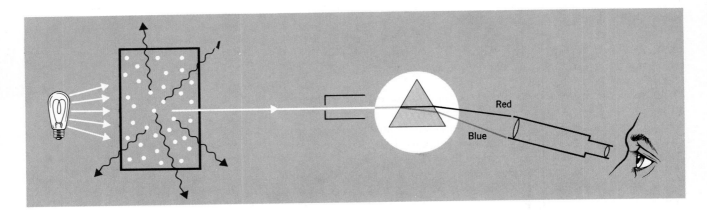

Figure 4.12 The removal of selected wavelengths from a beam of light passing through a gas.

in the container, we will find that every dark line in the continuous spectrum is matched by a bright line in the emission spectrum of the atom.

As an example, the sodium atom has two very intense emission lines located close to one another in the yellow region of the spectrum, at 5890Å and 5896Å. The states of the sodium atom that produce these lines are easily excited when the sodium atoms are placed in a flame; it is because of this that the color of a Bunsen burner flame turns yellow if salt is thrown on it. Color plate 3 shows the two yellow lines in the emission spectrum of sodium, as seen through a spectroscope that is pointed at a flame to which sodium has been added. Above the emission spectrum of sodium is an illustration of an absorption spectrum of sodium, produced when a glass container of sodium vapor is placed between the spectroscope and a light source. The two dark lines in the yellow region of the spectrum occur at precisely the same wavelengths as the two bright lines. Either pair of lines indicates that sodium is present.

That is, the dark lines in the continuous spectrum, produced when a beam of light passes through the gas and some of the photons are absorbed from this beam, are the fingerprint of the type of atom that makes up that gas, in just the same sense that the emission spectrum produced by that atom when heated also is its fingerprint. The dark lines in the continuous spectrum of the gas are called its *absorption spectrum*. Either the absorption spectrum or the emission spectrum can be used to identify an element.

## STELLAR SPECTRA

Stars are flaming balls of gas whose surfaces are at high temperatures ranging up to tens of thousands of degrees. Violent collisions occur at these temperatures, raising many atoms to excited states, and producing a rich emission spectrum with a great number of lines. These lines may be so many in number as to nearly fill in the spectrum of colors. Moreover, because of the high density of the gas in the star, collisions are very frequent, so that an electron in an excited state often is disturbed by a second collision before it has a chance to collapse down to its ground state orbit. Such disturbances blur the sharpness of the transition to the ground state, and spread each line in the spectrum out into a broad band of color. Neighboring lines in the spectrum overlap one another as a result, so that by the time the light leaves the surface of the star the separate spectral lines have been blurred into a *continuous spectrum,* that is, a rainbow of light including all wavelengths.

Stars have atmospheres just as planets do. The atmosphere of a star lies above its surface, and consists of the same elements that are contained in the body of the star, but at very low density. The light radiated from the surface of a star must pass through this atmosphere on its way out. This light is, as we have noted, a continuous spectrum. The atoms in the star's atmosphere will absorb some of the wavelengths in the continuous spectrum of radiation coming up from the surface beneath, just as, in the laboratory experiment described above, the atoms of gas in the glass container absorbed some of the wavelengths in the beam of light passing through them. As a result, when the spectrum of the star is examined in a spectroscope by an astronomer on the earth, he will find that it is crossed by dark lines that are the absorption lines of the atoms composing the star's atmosphere. These atoms have left their fingerprints on the star's light. This is the way in which astronomers determine the materials out of which stars are made: they collect the starlight in telescopes, spread it out into its component wavelengths, and photograph the resultant absorption spectrum.

## The Sun

The first star to be studied through a spectroscope was the sun, which the British astronomer William Hyde Wollaston examined in 1802. Wollaston found that the sun emitted a continuous spectrum that was interrupted by a series of dark lines . In 1814 a German telescope maker named Joseph von Fraunhofer repeated the observations, using a diffraction grating instead of a prism. He counted 754 dark lines, which scientists now call the Fraunhofer lines. Fraunhofer noticed that some of the dark lines fell in the same positions—that is, had the same wavelengths—as the bright lines in the spectra of certain elements studied in his laboratory. For example, he found that many dark lines in the sun's spectrum lay at the same wavelengths at which iron emits bright spectral lines when heated and vaporized. Figure 4.13 shows a part of the sun's spectrum and the spectrum of excited iron atoms. The agreement between the dark lines in the sun's spectrum and the bright lines emitted by iron is too close to result from mere coincidence. Fraunhofer realized this fact, but he was unable to supply the reason for the apparent coincidence. Had he done so, he would have founded the modern science of astrophysics.

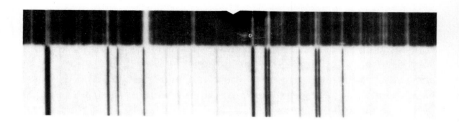

Figure 4.13 Solar spectrum compared with laboratory spectrum of iron, indicating presence of iron in the sun.

Fraunhofer's dark lines remained unexplained for nearly half a century, until in 1859 when another German scientist, named Gustav Robert Kirchhoff, realized their meaning. Kirchhoff came to his realization more or less by accident, as is often the case with great discoveries, not that he deserves less credit for his discovery on that account. Accidental discoveries often make no impact on the world, because their discoverers lack the insight and the touch of genius necessary to appreciate the larger significance of their own work. Kirchhoff possessed the insight needed to convert his experimental result into a great addition to knowledge.

Kirchhoff's discovery was that sodium vapor placed between a spectroscope and a continuous light source — he used sunlight — produced two dark lines at the same wavelengths at which heated sodium vapor produced two bright lines. Kirchhoff thought that he was going to *intensify* the normally bright yellow lines of sodium by inserting a flask of sodium vapor into the path of his light beam; he was not expecting to produce dark lines when he performed the experiment. This is the sense in which we can say that he came on his discovery by accident. But, once he had produced the dark lines, he saw their meaning, which was that the atoms of sodium vapor had *absorbed* the characteristic wavelengths of the sodium atom from the beam of light passing through them. Also, he immediately saw that if this explanation was correct, he possessed the key to determining the composition of the sun and other stars. If sodium vapor could absorb its characteristic wavelengths, so could the elements in the atmosphere of the sun. The Fraunhofer lines, then, were proof of the existence of certain elements in the sun.

By this experiment, and his interpretation of it, Kirchhoff became the father of astrophysics. His work ushered in the modern age of astronomy.

Scientists have observed thousands of dark absorption lines in the sun's spectrum, and by comparing their wavelengths with the wavelengths of bright lines emitted by elements in laboratory experiments on the earth, they have detected the presence of 67 elements in the sun. The particularly fine photograph of the sun's spectrum shown in Figure 4.14 stretches across the full range of the visible spectrum, from the violet end at 3900Å to the red end at 6900Å. The wavelengths and colors are noted above the spectrum, and some dark lines are labeled below the spectrum with the name of the element responsible. Approximately 1000 lines are visible in this photograph.

Figure 4.15 shows the solar astronomer R. S. Richardson with a complete spectrum of the sun that he has spread out on the floor in a strip of photographs 40 feet long. He needs this degree of detail to study the thousands of lines that are crowded into the sun's absorption spectrum.

Figure 4.14 Fraunhofer lines in the solar spectrum.

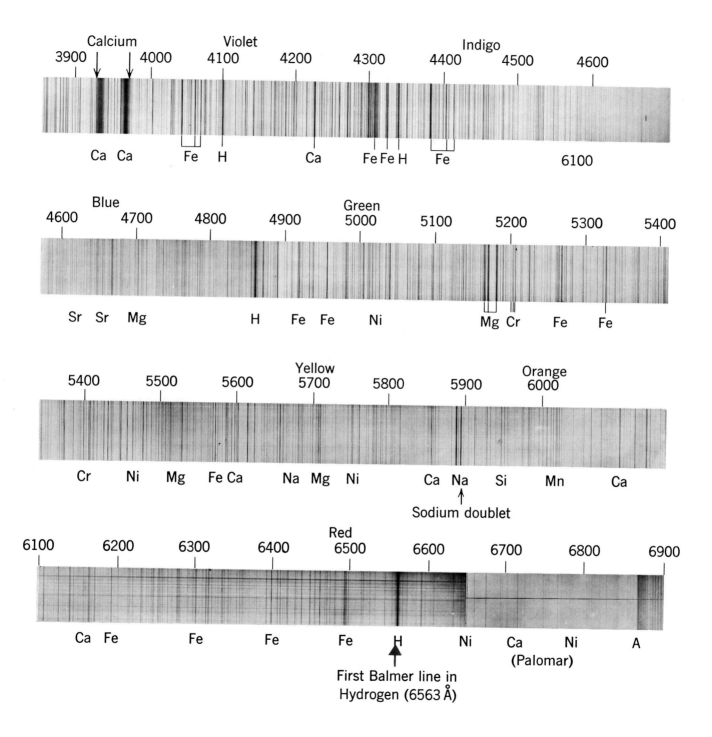

Calcium

3900     4000     Violet 4100     4200     4300     4400 Indigo     4500     4600

Ca   Ca    Fe   H     Ca    Fe Fe H   Fe     6100

Blue

4600    4700    4800    4900   Green 5000    5100    5200    5300    5400

Sr   Sr   Mg     H    Fe   Fe    Ni     Mg Cr    Fe     Fe

5400    5500    5600   Yellow 5700    5800    5900   Orange 6000

Cr    Ni    Mg    Fe Ca    Na Mg   Ni     Ca Na    Si    Mn     Ca

Sodium doublet

6100    6200    6300    6400   Red 6500    6600    6700    6800    6900

Ca   Fe     Fe     Fe     Fe    H     Ni    Ca    Ni     A
                                                           (Palomar)

First Balmer line in
Hydrogen (6563 Å)

Figure 4.15   A detailed solar spectrum 40 feet long.

## Other Stars

When you compare the absorption spectrum of the sun with the spectra of other stars, you find that some stars have sunlike spectra but the spectra of other stars look quite different. For example, consider the blue-violet region, running from 3900Å to 4600Å. In this region, the sun's spectrum shows two lines for hydrogen (at 4101Å and 4340Å) plus hundreds of lines produced by iron and other elements (Figure 4.16a). The spectrum of the star Vega shows the same two hydrogen lines in the blue-violet (Figure 4.16b), but they are much blacker and more intense than the sun's spectrum and, apart from the two hydrogen lines, the spectrum of Vega shows very little else. The many lines that are present in the sun's spectrum in the blue-violet region are absent.

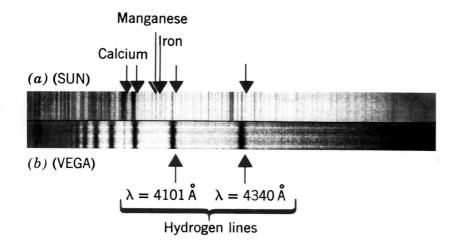

Figure 4.16 *The absorption spectra of the sun (a) and Vega (b).*

At first, you might conclude that Vega has a much higher percentage of hydrogen than the sun, because its spectrum shows stronger absorption lines for that element. You might also think that Vega has no iron, because the lines of this element are missing from its spectrum, and that the sun, whose spectrum is crowded with iron lines, must have a rich abundance of this metal.

Actually, the differences in the spectra of the stars are deceptive. All stars are made of a similar mixture of materials. There are some differences in composition from one star to another, but these differences are far less than the variations in the spectra of stars would seem to indicate.

To understand why the spectrum of one star can differ greatly from the spectrum of another, even though their compositions are the same or very nearly the same, consider the two hydrogen lines in the spectra of the sun and Vega (Figure 4.16). One of these lines is produced when the electron in the hydrogen atom absorbs a photon and is kicked upward from the first excited orbit to the fourth excited orbit. The other line is produced when the electron is kicked upward from the first excited orbit to the fifth excited orbit (Figure 4.17). In both cases, the atom must be in the first excited state above the ground state to produce this line. Now suppose that we are dealing with a star whose surface temperature is relatively low. In this star, because of the low temperature, nearly all hydrogen atoms at the surface will be in the ground state. Few atoms will be in excited states. Therefore, in the spectrum of such a star the Lyman series of hydrogen lines in the ultraviolet, corresponding to transitions from the ground state to excited states, will appear prominently. However, the Balmer series, including the two lines referred to above, will be relatively weak, because the Balmer lines can be formed only when the hydrogen atom is initially in the first excited state, and at the low temperature that we are assuming for this star very few atoms will be in any excited state.

Now consider a star whose surface temperature is somewhat higher. A greater number of hydrogen atoms on the surface of this star will be in excited states than in the case of the cooler star; therefore, transitions

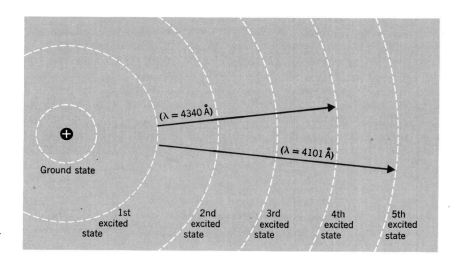

$(\lambda = 4340 \, \text{Å})$

$(\lambda = 4101 \, \text{Å})$

Ground state

| 1st excited state | 2nd excited state | 3rd excited state | 4th excited state | 5th excited state |

Figure 4.17 Transitions producing hydrogen lines in the solar spectrum.

from the first excited state upward to higher states will occur frequently. The spectrum of this star should show the hydrogen lines at 4101Å and 4340Å with greater intensity than the cooler star.

Suppose we consider a star of still higher temperature. The fraction of its hydrogen atoms in the first excited state will be still greater, and the intensity of the absorption lines corresponding to the Balmer series will also be greater.

Looking at the spectra of these stars of increasing temperature, one after the other, we will see that the hydrogen absorption lines grow stronger from one star to the next. We will be tempted to conclude that the amount of hydrogen increases from star to star. But the conclusion would be false. The variations in spectra are caused by differences in the surface *temperatures* of the stars, and not by differences in the *amount* of hydrogen.

Figure 4.18 shows this effect clearly. The figure presents the spectra of seven different stars, from Vega to the sun, placed under one another and lined up so that the wavelengths match from spectrum to spectrum. The positions of the third and fourth lines in the Balmer series of hydrogen are marked. The surface temperature of each star is indicated next to its spectrum at the right. We see that the intensity of the hydrogen lines increases from star to star as the surface temperatures go up from 5800°K to 9600°K.

The high temperature of Vega also explains the absence of the iron lines from its spectrum, mentioned on page 95. These lines, which appear prominently in the solar spectrum, are produced by neutral iron atoms, possessing their full complement of electrons; but, because of the relatively high temperature at Vega's surface, nearly all iron atoms in Vega are *ionized;* that is, they have lost at least one electron in collisions. An ionized iron atom also possesses an absorption spectrum, but it lies in the ultraviolet rather than the visible region.

Suppose a star is even hotter than Vega. Will the hydrogen lines be

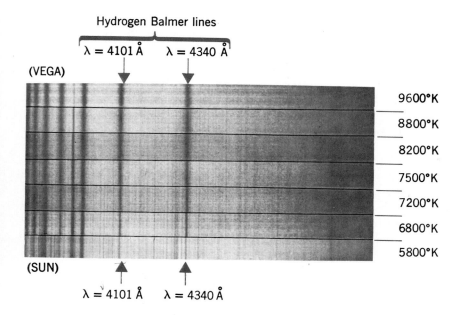

Figure 4.18 *Effect of temperature on the intensity of hydrogen lines.*

still stronger than they are in Vega's spectrum? It would seem so according to our explanation. However, another factor entering into the case of extremely hot stars cuts down the intensity of the hydrogen lines. At an extremely high temperature most of the hydrogen atoms are stripped of their electrons—that is, ionized—by the violent collisions that take place in the hot gas. An ionized hydrogen atom is a proton, which cannot absorb light. Thus, in extremely hot stars, the hydrogen lines become fainter again.

You can see this effect clearly in Figure 4.19. The figure shows the spectrum of a star like Vega and the spectra of five other stars, once again all lined up under one another so that the wavelength scales match, and arranged in a sequence of surface temperatures with the Vegalike star on the bottom and the hotter stars above it. The sequence of surface temperatures marked at the right runs from 9600°K at the bottom to approximately 35,000°K at the top. The hydrogen lines are discernibly

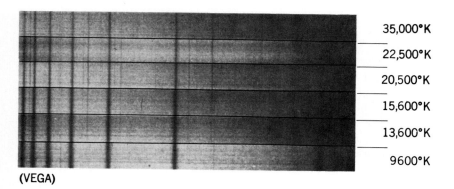

Figure 4.19 *Effect of increasing temperature on the intensity of hydrogen lines.*

fainter in the hotter stars at the top of this figure. In very hot stars, with surface temperatures in the neighborhood of 50,000°K, the hydrogen lines are hardly visible, because only a negligible fraction of the hydrogen atoms in these stars retain their electrons.

## Classification of Stellar Spectra

Suppose we arrange all the stars in the sky in a sequence, with the hottest stars at the top of the list and the coolest at the bottom. What will their spectra look like? Astronomers use precisely this scheme to classify the complicated varieties of stellar spectra.

*O Stars.* Astronomers have grouped together all stars whose spectra show the lines of singly ionized helium. The presence of ionized helium lines in a star's spectrum is a direct sign that the temperature of the star is at least 30,000°K. The reason is that the energy required to ionize helium—to strip an electron from the helium atom—is greater than for any other kind of atom. For this reason, only the very hottest stars contain large numbers of ionized helium atoms.

The group of stars with ionized helium lines in their spectra are called "O" stars or "O-type" stars. The hottest O-type stars ever observed have temperatures in the neighborhood of 80,000°K. Thus, the O-type stars may be considered to have temperatures ranging from 30,000°K to 80,000°K.

The spectra of O-type stars are relatively clean and uncluttered, and very unlike the spectrum of the sun. These spectra show lines of ionized helium, neutral helium, hydrogen, and little else. The reason for their uncluttered appearance is that all the other kinds of atoms they possess in abundance, such as iron, are heavily ionized because of the high temperatures, i.e., their outer electrons are missing. Their inner electrons are so tightly bound to the nucleus that very large amounts of energy are required to move them from their ground state orbits to excited orbits. Consequently, the photon that is emitted when the electron collapses back to the ground state again is highly energetic, with a short wavelength in the deep ultraviolet region of the spectrum. This ultraviolet radiation cannot be seen from the surface of the earth because it is absorbed in the atmosphere. We are learning more about the very hot stars today because we are beginning to perform ultraviolet astronomy with telescopes orbiting the earth in satellites above the atmosphere.

*B Stars.* In stars with temperatures below about 30,000°K, collisions are less violent, and very few, if any, helium atoms will be ionized. Thus, in the spectra of these somewhat cooler stars, you would expect to find the lines of neutral helium but not the lines of ionized helium. If the temperature of a star is still lower, the energies required to excite neutral helium atoms are not available, and the lines of this element do not appear in the spectrum. Astronomers use the presence or absence of the lines of neutral helium as a means of marking off another group of

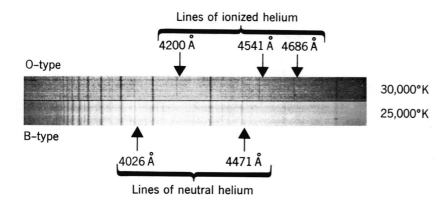

Figure 4.20 *Comparison between O-type and B-type spectra.*

stars. Stars whose spectra contain lines of neutral helium but do not contain lines of ionized helium are referred to as "B" stars or "B-type" stars. The temperatures of B-type stars range from a lower limit of roughly 10,000°K to an upper limit of about 30,000°K.

Figure 4.20 shows the spectrum of an O-type star compared with the spectrum of a B star that is nearly but not quite hot enough to be placed in the O classification. Notice that the spectrum of the B star does not contain the lines of ionized helium that are clearly visible in the spectrum of the O star. However, it does contain the lines of neutral helium.

*A Stars.* In a star with a surface temperature of 10,000°K, the neutral helium lines are just barely visible. Below 10,000°K, they become very faint and finally disappear entirely. Stars that are somewhat cooler than B stars and have no neutral helium lines are called A stars. They fall within a temperature range from 10,000°K down to 8,000°K. Figure 4.21 compares the spectra of a B star and an A star with surface temperatures of 25,000°K and 10,000°K, respectively. The spectrum of the

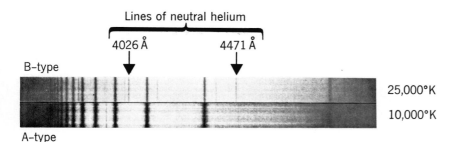

Figure 4.21 *Comparison between A-type and B-type spectra.*

A star is placed below the spectrum of the B star. Notice that the lines of neutral helium show clearly in the spectrum of the B star at wavelengths of 4,026Å and 4,471Å, but these lines are missing from the spectrum of the cooler A star.

Observe that the hydrogen lines are very strong in the A star spectrum. They remain strong down to a temperature of approximately 8,000°K. Below 8,000°K, the strength of the hydrogen lines declines once more, because fewer and fewer collisions have the strength required to move the electron in the hydrogen atom upward to an excited orbit. Of course, if the atmosphere did not block them out, the ultraviolet Lyman lines could be observed in these stars. On the other hand, as we have seen, at temperatures greater than 12,000°K, a large fraction of the hydrogen atoms in the outer layer of a star are ionized, and the strength of the lines decreases for that reason. It happens that the combination of the two opposing factors — degree of ionization and strength of collisions — combines to produce a peak strength of hydrogen lines for temperatures in the 8000°K to 10,000°K range. These intense hydrogen lines characterize the spectra of A-type stars.

*F Stars.* The A-type stars are followed by F or F-type stars, which are defined as stars with surface temperatures ranging from 8000°K down to 6000°K. Spectra of stars in this group possess very strong lines of ionized calcium at the violet end of the spectrum as well as many lines for metals such as iron and titanium. Lines for neutral atoms of metals make their first clear appearance in F stars.

An example of the spectrum of an F-type star is shown in figure 4.22. Polaris, the North Star, is a star of this type. Note that in the spectrum of the F-type stars, for the first time in the descending sequence of temperatures, the simple, uncluttered spectra of the hotter stars give way to a complicated, many-lined spectrum.

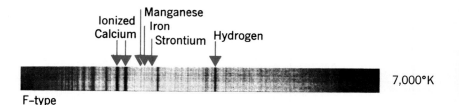

*Figure 4.22   Spectrum of an F-type star.*

Beginning with the F stars, the changes from one type to the next are more subtle, and the divisions more arbitrary, than in the hotter stars.

*G Stars.* Continuing in descending order of temperatures, the stars beyond the F-type are known as G-type stars, and have surface temperatures in the range of 4500°K to 6000°K. The sun, with a surface temperature of approximately 5800°K, is a G-type star. Its spectrum contains thousands of closely-spaced lines (Figure 4.23).

*K Stars.* The next group of stars, with surface temperatures lower than the G-type stars, are known as "K" stars or "K-type" stars. This group contains stars with temperatures that range from 4500°K down to 3500°K. K-type stars have a spectrum that is very densely packed with

*Figure 4.23   Spectrum of a G-type star.*

the absorption lines of metals and other elements.

*M Stars.* After the K stars come the M stars, the coolest stars in the standard classification list. The temperatures of M-type stars range from 3500°K down to 2000°K. In these stars, the temperature is so low that even the relatively fragile bonds holding together the atoms in a molecule can be preserved. Barnard's Star is an example of an M star.

When molecules appear in the outer layers of a star, they produce an absorption spectrum of a very special and easily recognizable kind, quite different from the spectrum produced by an atom. The spectrum of a typical molecule is characterized by many finely-spaced lines crowded together into a band of wavelengths. This type of spectrum is called a *band spectrum.* An example from the emission spectrum of the cyanide molecule is shown in Figure 4.24. Approximately 100 lines are crowded into a wavelength band of 47 Å in this spectrum.

*Molecular Spectra.* The explanation of the bands in a molecular spectrum is as follows: when atoms are bound together in a molecule, the electrons of these atoms can absorb photons and jump upward to higher orbits, just as electrons do on an individual atom. Such a jump or transition should produce one absorption line at a wavelength characteristic of the molecule involved and similar in appearance to the absorption line of an atom. The reason that the spectra of molecules contain dense bands of lines is that a molecule, unlike an atom, has a complicated structure consisting of several "pieces," the pieces being the individual atoms that make up the molecule. This structure can vibrate in and out, and it can also rotate in space. When an electron in a molecule jumps from one orbit to another, the molecule may change its rate of vibration or rate of rotation at the same time that the electron makes its jump. When the molecule's vibration or rotation changes, its energy also changes by a small amount. The difference in energy from one kind of vibration to another, or one type of rotation of the molecule to another, is quite small in comparison to the difference of energy involved in the transition of an electron from one orbit to another. Because of these small differences, the energy of the electron's transition is separated into many closely-spaced energies, and the wavelength of the line accordingly is spread out into a large number of finely spaced lines covering a band of wavelengths. This group of finely spaced lines is the band spectrum of the molecule.

The most intense band spectrum to appear in the spectra of stars is the band spectrum of titanium oxide. The two atoms in this molecule— titanium and oxygen—are very tightly bound to one another, and therefore the molecule appears in the outer layers of stars before any other type of relatively common molecule can do so. Astronomers use the presence of absorption bands for titanium oxide as their criterion for distinguishing M stars from hotter stars. Figure 4.25 shows the spectrum of an M-type star, with several bands for titanium oxide appearing at the right side of the spectrum in the places marked. The spectrum belongs to Barnard's Star, the sun's second nearest neighbor.

Notice that the individual lines in the bands of the titanium oxide

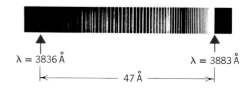

$\lambda = 3836$ Å $\qquad\qquad \lambda = 3883$ Å

$\leftarrow$———— 47 Å ————$\rightarrow$

*Figure 4.24   A molecular band spectrum.*

*Figure 4.25   Spectrum of an M-type star.*

M
(Barnard's
  star)

TiO bands

3,000°K

molecule cannot be seen. The light from most stars is so faint that astronomers cannot spread out any particular part of the star's spectrum to an extent great enough to allow them to photograph the individual lines in a band spectrum. The only star that is bright enough to form an exception to this statement is the sun. Figure 4.26 shows a band spectrum of the titanium oxide molecule in the absorption spectrum of the sun. This spectrum shows three adjacent, overlapping bands of titanium oxide, each band starting at the place marked by the arrow.

*Figure 4.26   Individual lines in TiO band spectra from the sun.*

The appearance of strong absorption bands of titanium oxide in the spectrum of a star is the criterion used by astronomers to separate M stars from hotter stars.

*Cool Stars.* Why does the list of star types stop at stars with temperatures of 2000°K? Are there no stars cooler than this? Cooler stars exist, but their discovery is a relatively recent event in the history of astrophysics; hence, they are not included in the standard classification that was set up in the nineteenth century. These very cool stars were discovered only recently because they emit essentially no energy at all in the visible region of the spectrum, most of their energy being emitted as infrared radiation. Detectors sufficiently sensitive to pick up these infrared stars have only been developed in the last ten or fifteen years.

*Subclasses of Stars.* We have classified each type of star by a range of temperatures. For example, B stars are defined as stars with surface temperatures in the range from 12,000°K to 30,000°K. However, astronomers can measure finer differences in the surface temperatures of stars within such large ranges. They describe these differences by subdividing each type of star into a maximum of ten parts. The B stars, for example, are divided into ten groups ranging from B0 to B9. The surface temperatures of the stars in each of these subdivisions are listed in the table at left.

*The Origin of the Classification.* When the classes of stars are arranged in order of decreasing temperature, with the hottest stars coming first and the coolest last, the letters that designate each group form the sequence: O  B  A  F  G  K  M  The sequence is easy to remember if you associate it with the first letters of the sentence: *"Oh, Be A Fine Girl, Kiss Me."* This sentence has been burned into the memories of many generations of astronomy students.

| Type | Temperature (°K) |
|------|------------------|
| B0 | 30,000 |
| B1 | 23,000 |
| B2 | 21,000 |
| B3 | 18,000 |
| B4 | 17,000 |
| B5 | 15,500 |
| B6 | 13,500 |
| B7 | 14,000 |
| B8 | 12,000 |
| B9 | 10,500 |

Why is this peculiar sequence of letters used? Originally, the stars were classified according to the strength of the hydrogen lines in their spectra. They were divided into 16 groups, with the stars possessing the strongest hydrogen lines listed first and the stars with the faintest hydrogen lines listed last. The groups were lettered in alphabetical order from A through Q (with J missing for some reason). Later, astronomers realized that a classification by surface temperature made more sense, and the stars were rearranged, but the force of tradition prevailed, and the letter designations of the original listing were retained, although they no longer followed alphabetical order.

The stars with the most intense hydrogen lines originally were designated by the letter A, and they are still designated by this letter although, because they are not the hottest stars, they are no longer at the begin-

Figure 4.27  Temperature versus spectral type.

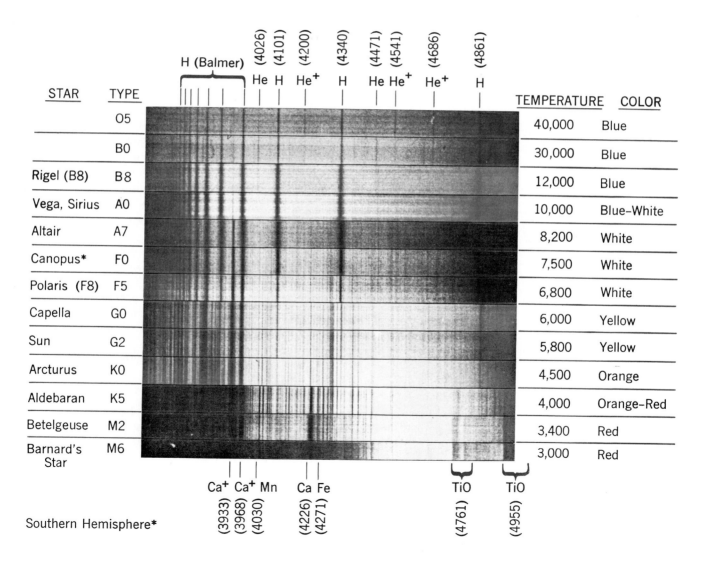

ning of the sequence. Stars with quite faint hydrogen lines, placed near the end of the original sequence and designated with the letter O, are the hottest class of stars, and they are therefore at the head of the modern sequence.

*An Album of Stars.* The photographs of the spectra of all the principal types of stars are collected in Figure 4.27 (page 103). Two subdivisions are included for each type, in order to display the smoothness of the transition in stellar spectra from the hottest to the coldest stars. The most significant lines in the spectra are indicated at the top and bottom of the figure. At the left of each spectrum appears the name of a familiar star of this type, if a familiar example exists. At the right is shown the temperature corresponding to the spectral type and the color of the star.

## DETERMINATION OF THE ABUNDANCES OF THE ELEMENTS FROM STELLAR SPECTRA

The discussion of atomic and stellar spectra earlier in the chapter emphasized that every element leaves its fingerprints on the spectrum of a star in the form of a unique set of spectral lines. In principle, it should be possible to determine the abundances of all the chemical elements in a star[2] directly from these spectral lines. When we looked into the subject in more detail and tried to find out how to measure the abundances of an element such as hydrogen from the intensity of its spectrum, we found that the temperature of the star was a complicating factor. A careful study of the situation revealed that temperature is just as important in determining the strength of the spectral line as the abundance of the element itself.

This line of thought led us gradually away from our original concern with the relation between stellar spectra and the chemical ingredients of a star, and into the area of stellar temperatures.

Now that the relation between the spectrum of a star and its temperature is clear, let us return to the original question: How can we measure the relative abundances of the elements in a star accurately from the analysis of the star's spectrum?

As we saw in the case of hydrogen, this question does not have a

[2] Actually, the spectral lines yield the abundances in the star's *atmosphere,* where the absorption takes place, and not in the *interior* of the star. However, in most stars, those with surface temperatures of 8000 °K or less (type F), the hot gases in the interior of the star boil upward in a turbulent motion that thoroughly mixes the outer layers of the star. In the hotter stars—types O, B, and A—this is not true. However, it is generally assumed that all stars are formed out of a well-mixed, homogeneous cloud of materials and, therefore, that the composition of a stellar atmosphere is representative of the star as a whole. The core of the star is, of course, never homogeneous, since new elements are always accumulating there as a result of nuclear fusion reactions.

simple answer. A faint set of absorption lines for a particular element does not necessarily mean that the abundance of that element is low. In the photograph of the sun's spectrum on page 93, for example, the two intense calcium lines at 3934Å and 3968Å are much more intense than the Balmer line of hydrogen at 6563Å, but the abundance of hydrogen in the sun is several hundred thousand to a million times greater than the abundance of calcium. The explanation is that the intensity of an absorption line of an element depends not only on the number of atoms of the element that are present, but also on the fraction of these atoms that are in the correct initial state to produce that absorption. The 6563Å line in the hydrogen spectrum is produced when an electron moves up from the first excited state of hydrogen to the second excited state. At the temperature of the sun's surface, most hydrogen atoms are in the ground state, and relatively few are in the first excited state. Consequently, the 6563Å line is faint.

If the fraction of the atoms of each element in the ground state, the first excited state, the second excited state, and so on, could be determined, the difficulty would be resolved. The astronomer obtains this information with the aid of a calculation. He uses accurate formulas that predict how many atoms will be in the ground state and in each excited state for a given temperature. In applying these formulas to a star with a given mass and radius, the astronomer first guesses at the value of the temperature on its surface and the abundance of the elements in its atmosphere, then calculates the intensities of the lines in the spectrum with the aid of the formula, and finally compares the theoretical spectrum with the observed spectrum. He adjusts and re-adjusts the values of the temperature and the relative abundances that he had assumed in order to secure the best possible agreement between theory and observation, finally arriving at a value for both the abundances of the elements in the star *and* the temperature at its surface.

The graph of the cosmic abundances of the elements on page 191 is partly based on computations of this kind. The temperatures listed for various spectral types on page 103 are derived from similar computations.

## OTHER INFORMATION YIELDED BY STELLAR SPECTRA

Stellar spectra are a gold mine of facts regarding the structure of stars. In addition to revealing the ingredients of a star and the temperature on its surface, the star's spectrum can reveal whether the star is a small object, like Barnard's star, or a distended red giant like Betelgeuse; it can indicate the speed with which the star is rotating on its axis; and it can determine the strength of the star's magnetic field, if any. The spectrum of a star can even reveal that the star is shedding its outer layers in a massive ejection of material to space.

## The Shape of a Spectral Line

To obtain this information, it is necessary to examine not only the intensity of the line but also its detailed *shape*. If you measure the degree of blackness of an absorption line on a photographic plate, you will find that usually the line is blackest, or most intense, at its center, and shades off gradually to either side of the center, that is, to longer and shorter wavelengths. The shape of the line refers to the precise manner in which the intensity falls off to either side of the center.

The photograph facing page 79 provides a clear example of the gradual decrease in the intensity of the absorption on the short-wavelength and long-wavelength sides of an absorption line. This photograph is an enlargement of a section of Figure 4.27. One segment of the enlargement, containing the hydrogen absorption line at 4101Å, is reproduced in Figure 4.28. If we could accurately measure the degree of blackening of the photograph at various wavelengths in the vicinity of this line, we would obtain a graph similar to Figure 4.29. This figure shows the percent of light transmitted through the star's atmosphere at wavelengths near 4101Å.

Figure 4.29 is the *shape* of this particular line. Because the hydrogen spectrum is relatively intense in an A-type star, its shape is characterized by a flat bottom, indicating nearly complete absorption over a substantial range in wavelength. Weaker lines have the bell-shaped appearance shown in Figure 4.30.

Figure 4.31 shows shapes of several lines of varying intensities ranging from very weak (*a*) to moderately strong (*b*) and very strong (*c*). These line shapes were calculated for the absorption line of singly ionized calcium located in the ultraviolet region of the solar spectrum at 3934Å. This is the first calcium line in the solar spectrum shown in Figure 4.14. The computed line shapes shown in the graph correspond to calcium ion abundances of approximately $3 \times 10^{11}$, $3 \times 10^{14}$, and $3 \times 10^{16}$ ions per cm² in the sun's atmosphere. Line shape (c) agrees approximately with the observed shape of the Ca line in Figure 4.28.

*Measuring Line Shapes.* The shape of a line can be measured accurately by passing a narrow beam of light through the photographic plate on which the star's spectrum is recorded, and determining the amount of light transmitted through the plate with the aid of a photocell mounted below (Figure 4.32).

The narrow beam of light and the photocell are stationary. The photographic plate bearing the spectrum is mounted on a sliding track controlled by an accurately machined gear that moves it slowly through the light beam. Whenever the beam passes through an absorption line in the course of the plate's motion, the photocell records an increase in the intensity of the light passing through the plate. (The intensity increases because the plate is a photographic negative.) The photocell readings, plotted on a wavelength scale, yield a graph similar to Figures 4.29 and 4.30, on which the intensities and shapes of spectral lines are accurately displayed.

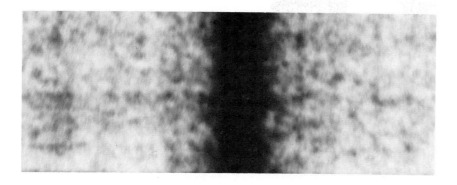

Figure 4.28 Photograph of 4101Å absorption line of hydrogen in the spectrum of an A-type star (Altair).

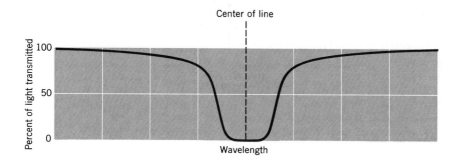

Center of line

Percent of light transmitted

Wavelength

Figure 4.29 Variation of absorption with wavelength for the intense 4101Å line in Figure 4.28 (approximate).

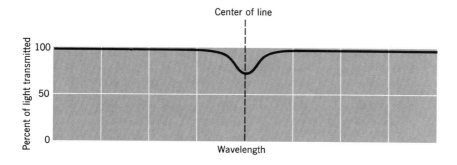

Center of line

Percent of light transmitted

Wavelength

Figure 4.30 Inverted bell-shape characteristic of a relatively weak absorption line.

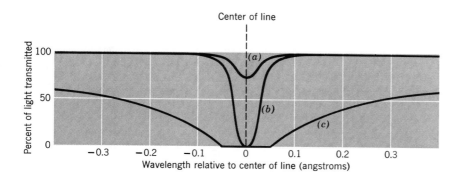

Center of line

Percent of light transmitted

(a)

(b)

(c)

Wavelength relative to center of line (angstroms)

Figure 4.31 Shapes of lines of varying intensity: (a) weak; (b) strong; (c) very strong, calculated for the line of singly-ionized calcium at 3934Å in the solar spectrum.

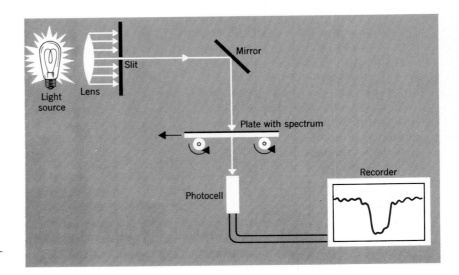

Figure 4.32 Essential elements of a densitometer.

Devices of this kind are basic items of equipment in every astronomical observatory. They are called *densitometers* because they measure the density or degree of blackening of the photographic emulsion. The beam is as narrow as a thousandth of a centimeter in the most precise instruments, which are called microdensitometers.

Figure 4.33 shows the result of a microdensitometer scan of the spectrum of Zeta Ophiuchi, a star at a distance of about 500 light-years. The spectrum was photographed with a one-hour exposure at the Lick Observatory 100-inch telescope. The portion of the spectrum shown in the figure spans the interval from 3960 to 3983Å. The record shows a broad, relatively intense line centered at 3970.07Å with a width of about 20Å. This is a line in the Balmer series of hydrogen produced by absorption in the atmosphere at Zeta Ophiuchi. Broad but somewhat weaker absorption lines appear at 3967 and 3975Å. These lines are probably due to iron.

An exceedingly narrow line, less than an angstrom wide, is also visible at 3968.5Å. This line has been identified as an absorption line produced by atoms of singly ionized calcium, located in the clouds of interstellar gas and dust that occupy a part of the space between our solar system and Zeta Ophiuchi.

Why should the absorption lines produced by hydrogen and iron atoms in the atmosphere of Zeta Ophiuchi be very broad, while the absorption lines produced by atoms of calcium in the space between the stars are extremely narrow? To answer this question, we must try to understand all the major factors that influence the shape of a spectral line. In acquiring this understanding, we will also gain the means of measuring many important properties of stars.

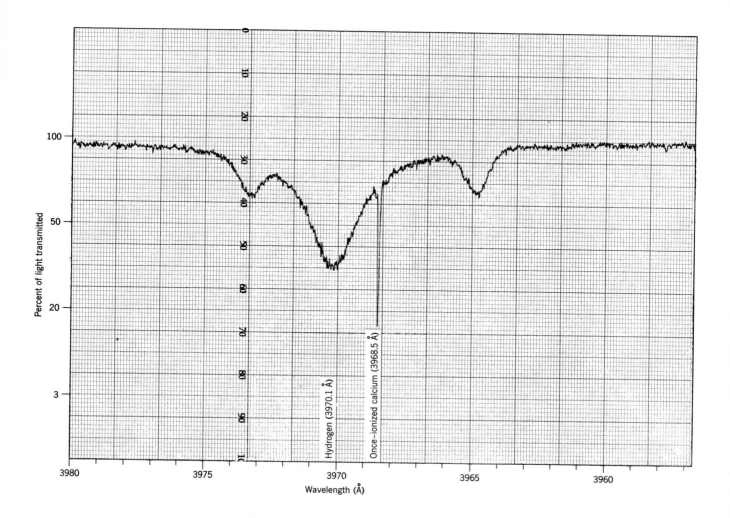

Figure 4.33 The spectrum of Zeta Ophiuchi measured with a microdensitometer, showing broad absorption lines produced in the atmosphere of the star and narrow lines produced by absorption in the low-density interstellar gas.

## Factors Influencing the Shape of a Spectral Line

In the discussion of atomic energy states given on pages 83-84, the energy of an electron in its orbit was described as a precisely defined quantity with no variation in energy from one atom to another. If this were the case, the photon absorbed in a transition from one electron orbit to another would also have a precisely defined energy, corresponding to a unique wavelength, and all spectral lines would be sharp and infinitely narrow.

In fact, however, there is a small degree of blurring of atomic energy states during the transition from one orbit to another, because the atom's structure is disturbed slightly by the absorption of the photon. As a result, the absorption lines produced in the transition are also blurred to a slight degree. That is, the lines are broadened. This blurring of a spectral line, which occurs in the spectrum of every atom, is called the *natural width* of the line. It is very small, being a hundredth of an angstrom in typical cases.

The natural width is the smallest source of broadening of spectral lines. In the sections below, we discuss more important factors that contribute to the broadening of spectral lines and influence their shapes; and we go on to consider the properties of stars that can be deduced from a detailed study of line shapes.

*Collisional Broadening.* When two atoms collide, the electrons on one atom repel the electrons on the other atom, distorting the shapes of their orbits and changing the energies of the atomic states. If one of the atoms happens to absorb a photon during the collision, the energy of the absorbed photon will be affected by this distortion. If the density of the gas is low, collisions between atoms are infrequent, and the absorption of a photon usually takes place when the atom is in an undisturbed state. In this case, collisions will have only a small effect on the width of a spectral line. If the density of the gas is high, an atom will frequently be involved in a collision at the same time that it absorbs a photon, and the energy of the absorbed photons will vary over a considerable range. As a result, the absorption line will be substantially broadened.

This effect on the shape of a spectral line is known as *collisional broadening.* The width of the broadened line is determined by the frequency of collisions between atoms. The collision frequency depends on the gas density and temperature. Thus, if the temperature is known, the width of the spectral lines indicates the density of the gas.

Chapter 5 classifies the population of stars into broad categories, including Main-Sequence stars and red giants. A Main-Sequence star is a "normal" hydrogen-burning star, while a red giant is a star in a later state of its evolution, in which the outer layer of the star has expanded into an enormous, tenuous sphere of low-density gas. A red giant can be distinguished clearly from a Main-Sequence star through the difference in the widths of their spectral lines, even if the surface temperature of the two stars is identical.

For a Main-Sequence star, the width of a collision-broadened line is typically a few tenths of an angstrom. In the outermost layers of a red giant, the density is much less than the density at the surface of a Main-Sequence star, and the absorption lines are correspondingly narrower. Figure 4.34 compares the spectrum of a red giant with a Main-Sequence star. The narrowness of the lines in the red-giant spectrum is apparent. Theoretical investigations of collisional broadening have led to an accurate relationship between spectral line-width and density for a given temperature.

*Doppler Broadening or Thermal Broadening.* The lines in stellar spectra are also broadened through the combination of their thermal motion and the effect of the Doppler shift. Because of the random thermal velocities of the atoms in the star's atmosphere, some atoms move to-

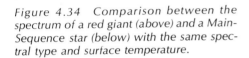

*Figure 4.34 Comparison between the spectrum of a red giant (above) and a Main-Sequence star (below) with the same spectral type and surface temperature.*

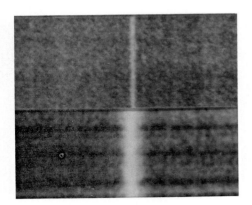

ward us and others move away from us when they emit their photons. These motions produce a Doppler shift in the spectrum. The result is a broadening of the line, called *Doppler broadening*, which adds to the broadening produced by collisions. Since the degree of broadening is determined by the average thermal velocity of the atoms, that is, by their temperature, the effect is also known as *thermal broadening*.[3]

The effects of Doppler broadening or thermal broadening are somewhat smaller than those of collisional broadening in Main-Sequence stars, the width of a thermally broadened line being 0.05 to 0.1Å in typical cases. However, the shapes of the lines produced by the two types of broadening are quite different. In a Doppler-broadened line, the absorption falls off sharply to either side of the center of the line, while in a collisionally broadened line, the absorption diminishes more gradually, forming two broad wings to either side of the center.

Figure 4.35 shows the theoretically computed shape of the absorption line of ionized calcium in the spectrum of the sun at 3934Å. The difference between the two types of broadening is clearly evident in the figure.

*Rotational Broadening.* If a star is rotating, the atoms on its surface move toward or away from the observer during the course of each rotation, unless the axis of rotation is pointed directly at the observer. Let us assume for simplicity that the axis is perpendicular to our line of sight. Assume also that the star is rotating counterclockwise from above (Figure 4.36). Then all the atoms on the left half of the star's disk will be moving toward the earth and all the atoms on the right half of the disk will be receding. These motions produce a Doppler shift in each line of the star's spectrum.

The atoms on the left limb of the star and on its equator will have the largest velocity toward the earth and, therefore the contribution of these

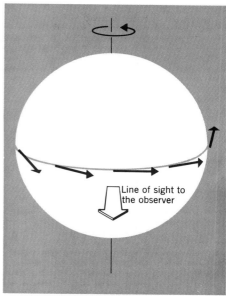

*Figure 4.36 Motions of atoms on the equator of a rotating star giving rise to rotational broadening.*

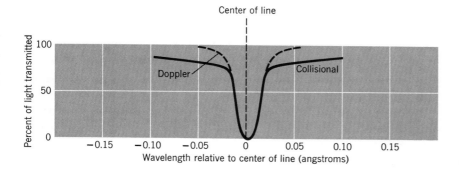

*Figure 4.35 Line shapes for Doppler broadening and collisional broadening.*

[3] Random turbulent motions of the gas in the atmosphere of the star also broaden the lines as a result of the Doppler effect. This *turbulence broadening* produces line shapes similar to those of Doppler-broadened lines.

**111**

atoms to the absorption spectrum of the star will be shifted by the greatest amount toward the blue end of the spectrum. The atoms on the right limb of the star and on the equator will have the greatest velocity away from us and, therefore, their contribution will be shifted by the maximum amount to the red. Atoms at other points on the star will give rise to intermediate values of the Doppler shift. The superposition of these many absorptions will give rise to a broadened line. The extent of the broadening will depend on the rate of rotation and the angle of inclination of the axis of rotation to our line of sight. This effect is known as *rotational broadening.*

How can rotational broadening be distinguished from collisional broadening and thermal broadening? An absorption line looks approximately the same to the eye, regardless of the type of broadening that has produced it. However, theoretical calculations of the shapes of rotationally broadened lines reveal that they differ significantly from other line shapes. The differences are not sharp enough to be evident to the eye when a spectrum is examined, but they can be detected if a densitometer tracing is made of the line shape, as in Figure 4.33. A theoretical formula, representing a combination of the three broadening effects acting simultaneously, can be fitted to the observed line shape as given by the densitometer tracing. By adjusting the relative contributions to the theoretical formula from each of the three broadening effects in order to secure the best possible agreement between the formula and the measured line shape, it may be possible to deduce the separate amounts of collisional, thermal and rotational broadening.

*Determination of Stellar Rotation.* Suppose that we have been able to separate rotational broadening from other types of broadening in the spectrum of a star. Assume for simplicity that the axis of the rotating star is perpendicular to our line of sight. Let the half-width of the rotationally broadened line be $\Delta\lambda$. According to the formula for the Doppler shift (Chapter 2), the velocity, $v$, of the atoms on the limb of the rotating star should be approximately

$$v = c\,\frac{\Delta\lambda}{\lambda}$$

In this formula, $v$ is the speed that the atoms on the limbs of the rotating star have toward or away from us as a result of the rotation, and $c$ is the velocity of light. If the radius, $R$, is known, the period of rotation, $T$, that is, the time in which the star completes one turn on its axis, can be calculated from the relation

$$T = \frac{2\pi R}{v}$$

At the beginning of the discussion we assumed for simplicity that the star's axis of rotation was perpendicular to the observer's line of sight (Figure 4.37a). Usually the axis is tilted at an angle to our line of sight, diminishing the amount of rotational broadening. For example, if a star is rotating very rapidly, but the axis of rotation is pointed directly toward the earth (Figure 4.37b), atoms on the surface of the rotating star will have no velocity toward or away from the observer, and the star's spectral lines will show no rotational broadening.

In general the axis of rotation is tilted at an intermediate angle (Figure 4.37c).

There is no way of determining the inclination of the axis of rotation separately from the speed of rotation. Therefore, the astronomer's only recourse is to assume that the axis of rotation is perpendicular to the line of sight and apply the formulas given above. The speed of rotation calculated in this way will be correct if the axis of rotation happens to be perpendicular to our line of sight. In other cases, the calculated value will be less than the true speed of rotation. Thus, the formulas for rotational broadening on page 112 will always yield a lower limit to the true rate of a star's rotation.

Using this method, it has been found that the hottest stars, Type O and B, rotate most rapidly in general. A-type and F-type stars rotate somewhat less rapidly. A sharp decrease occurs in the rate of rotation in the neighborhood of F-type stars. Nearly all G-type stars, including the sun, rotate relatively slowly.

The difference between the rapidly rotating and slowly rotating stars is very sharp. Many O- and B-type stars rotate on their axes as rapidly as once every 4 hours. This is the case for Zeta Ophiuchi, the star whose spectrum was shown in Figure 4.33. The width of the very broad line at 3970Å in the spectrum of Zeta Ophiuchi is entirely the result of rotational broadening. From the width of the line it can be deduced, by the methods described above, that this massive object is turning on its axis every 4 hours, and possibly even more rapidly, depending on the angle between its axis of rotation and the line of sight to earth. A star rotating on its axis every 4 hours is on the verge of flying apart, like a centrifuge that has been spun at too high a velocity.

The sun rotates on its axis today at the leisurely pace of once every 27 days. Some theories of the origin of the solar system assumed that the sun rotated as rapidly as once every 3 or 4 hours in its youth. According to these theories, the birth pangs of the planets slowed down the rotation of the sun in some way. Carrying this line of thought further, some astronomers believe that if any star is rotating slowly, a possible reason is that a family of planets was formed around it, braking its rapid rotation during the process of planetary birth.[3] If this theory is correct, it provides indirect evidence for a multitude of solar systems in the Cosmos around us, since the majority of stars rotate at the relatively slow rate of the sun.

*The Zeeman Effect and Magnetic Fields of Stars.* When an atom is

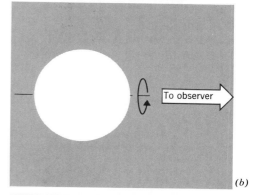

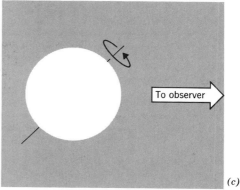

Figure 4.37 Angle between the axis of a rotating star and the line of sight to the observer: (a) perpendicular; (b) parallel; (c) intermediate.

---

[3] The fission of a rapidly rotating star into two stellar masses, forming a close binary, would also slow the rate of rotation.

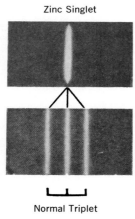

Zinc Singlet

Normal Triplet

Figure 4.38 Magnetic splitting of a line in the laboratory.

placed in a magnetic field, each of its energy levels may be split into several levels by the action of the field on the magnetic properties of the atom. If an energy level is split into three separate levels, for example, the photons emitted in transitions involving these levels can have three distinct energies and wavelengths. Consequently, a single line in the original spectrum will be split into three separate lines when the magnetic field is present.

The splitting of a line by a magnetic field is called the *Zeeman Effect*. It is described in more detail in Chapter 12 on pages 317-318. Figure 4.38 shows the splitting of a line by the Zeeman effect in a laboratory experiment.

A clear splitting of the lines in stellar spectra by the Zeeman effect can usually be observed only in the spectra of sunspots, which contain intense magnetic fields of thousands of gauss. In the spectra of stars more distant than the sun, the light from the entire disk of the star passes through the slit of the spectrograph and many different intensities of magnetic fields, from different areas on the surface of the star, contribute to the observed spectrum. As a result, each line is a blurred average over varying degrees of magnetic splitting. The resultant *magnetic broadening* is an indication of the average strength of the magnetic field on the surface of the star.

The average magnetic field on the surface of a star must be several hundred gauss in order to produce an observable degree of magnetic broadening. Clear indications of magnetic broadening and magnetic field strengths of this magnitude have been found in roughly 150 stars. The largest magnetic field observed in a star thus far is 34,000 gauss. Figure 4.39 shows the spectrum of this star, named HD215441. In this exceptional case the magnetic splitting into distinct lines is discernible because of the very strong field.

Figure 4.39 Magnetic splitting of lines in the spectrum of a star. The absorption lines appear light against a dark background because the spectrum is shown as a photographic negative. The sharp lines above and below the faint stellar spectrum are a known spectrum exposed on the same plate in the observatory in order to determine the wavelengths of the stellar lines accurately.

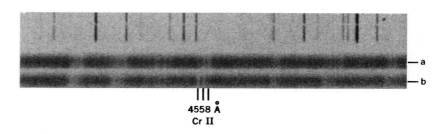

4558 Å
Cr II

## Questions

1. What are the differences between the everyday world and the world of the atom?
2. Describe what takes place in an atom when it is excited by a collision.
3. Explain why the wavelengths of light radiated from the atoms of an element are unique to that element.
4. If an atom of hydrogen is emitting Balmer series lines, what line from

the Lyman series must always accompany them? Why?

5. Describe the three basic types of spectra and the circumstances leading to each type of spectrum in the light radiated from a star.

6. Considering the fact that excited atoms collapse quickly to lower levels, emitting photons with the same wavelength as those absorbed, why do absorption lines appear in the spectra of gases?

7. The sun's atmosphere produces dark absorption lines superimposed on the continuous spectrum of radiation from the underlying photosphere. During a solar eclipse there is a point when the photosphere is covered by the moon, but not the sun's atmosphere. At this point all the dark lines flash into bright lines. Explain.

8. Describe the spectrum you expect to observe if potassium vapor at a high temperature is viewed through an intervening layer of sodium vapor at a low temperature. Explain.

9. Explain why the hydrogen absorption lines become weaker in the spectra of stars whose temperatures are (a) above 12,000°K and (b) below 12,000°K.

10. To what spectral class does a star belong if its radiation is peaked at a wavelength (a) in the far ultraviolet at 1500Å, (b) in the near ultraviolet at 3000Å, (c) in the visible at 5000Å, and (d) in the infrared at 10,000Å.

11. If you looked at the spectrum of what appeared to be a single star and found lines of ionized helium as well as molecular bands, what conclusions could you draw? Why?

12. Collisions between atoms in a gas tend to blur or broaden the individual lines in the spectrum of the gas. The broadening effect is proportional to the number of collisions per second, which is proportional in turn to the temperature of the gas and its density. Suppose you compare the same lines in the absorption spectra of two stars that have the same temperatures and find that one star has broader spectral lines than the other. What can you conclude about the densities in the surfaces of the two stars?

   Barnard's Star and Betelgeuse both have temperatures close to 3000°K, but the lines in the spectrum of Betelgeuse are considerably narrower than those from Barnard's Star. Account for this difference in terms of different structures for these two stars.

13. Why do the absorption spectra of ionized atoms lie mainly in the ultraviolet region, whereas the absorption lines of neutral atoms generally lie in the visible region?

14. Suggest a reason why the absorption spectra of G stars contain many closely spaced lines.

15. Indicate the spectral class and probable temperature of stars whose spectra have the following characteristics:
   (a) Lines of ionized helium weak; exceedingly weak lines of neutral helium and hydrogen.
   (b) Strong lines of neutral helium; weak lines of ionized helium.
   (c) Strong hydrogen lines; no helium lines.

# 5　The Hertzsprung-Russell Diagram

*Classifying People.* A person has two basic properties that stand at the top of any list of his physical characteristics. They are easily estimated by looking at him, and are always used as the basic means of identifying or classifying the individual in a group of people. They are his *height* and his *weight.* Suppose you make a graph of height versus weight and plot a point on it for each member of your class. We have done this for the staff members of the Institute for Space Studies, with the results shown in Figure 5.1.

Unless your class is very unusual, you will find that most of the points lie close to a line. You might call this the "Main Sequence" of physical properties of young adults in the United States. The "Main Sequence" drawn through the middle of these points probably is typical of what would be obtained by plotting the height and weight of any other similar group in the United States. The significance of the "Main Sequence" line is that it gives a connection between height and weight in the population; this connection does not hold true in general, for any *individual* member of the population, it only holds true for the *average* person.

*A Nova outburst in Perseus.*

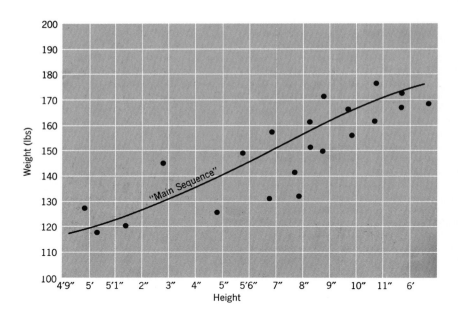

*Figure 5.1   Weight versus height for a sample group of people.*

*Classifying Stars.* Does a star possess observable properties, analogous to height and weight in a person, that may be considered to be its fundamental physical attributes? Does it have two properties that would enable us to classify stars in the sky in the same way that we classify people? A partial answer comes to us immediately: the *temperature* of a star, as evidenced by its spectral type or its color, is one basic property of the star. Is there another? A moment's thought indicates that the intrinsic *brightness* of the star is a second basic property. Is there a third? The answer is no; at least, no other property comes to mind that is as fundamental as the temperature and the intrinsic brightness of a star.

Color and brightness—these are the two properties that every star presents to the naked eye, and they are the only two. Presumably with these thoughts in mind, a Danish astronomer named Hertzsprung and an American astronomer named Russell separately decided to plot data equivalent to color and brightness of all stars for which this information was available. Hertzsprung and Russell had no prior thought of what they might find. Their hope was only that the population of the skies might arrange itself in some way, when plotted on such a chart, so as to seem meaningful to their eyes, and that, thereby, they might acquire a clue to the inner nature of this distant object known as a star.

The two astronomers succeeded beyond their expectations. The diagram they invented, universally known today as the *Hertzsprung-Russell* diagram, has become as essential to the astronomer as the slide rule is to the engineer. As a preliminary to the interpretation of this invaluable graph, we must develop more precise concepts of *color* and *brightness* than the qualitative senses in which these words are normally used in everyday speech.

# COLOR

The lines in the absorption spectrum of a star are determined by the temperature of the star's surface. The surface temperature also determines another property of the star, one which the eye can observe easily without the aid of any special instrument such as the spectroscope. This property is the color of the star.

## Connection Between Color and Spectral Type

We know from everyday experience that objects heated to a very high temperature glow with a white light; objects that are heated to moderately high temperatures are cherry-red in color, and cooler objects are dull red. The same is true of stars. The hottest stars—the ones of the O and B types—radiate most of their energy in the far ultraviolet region. A relatively small amount of their energy is emitted in the visible region, with emphasis on the blue end of the visible spectrum, giving these stars a bluish-white color. A-type stars radiate a more uniform distribution of intensity in the visible region and, therefore, are white in color. As the surface temperature decreases, and the peak intensity moves to longer wavelengths, the color of the star changes from blue-white to white to yellow-white, then to yellow, to orange, and finally to red. F stars are yellow-white, G stars like the sun are yellow, K stars are orange, and M stars are red.

Yet the color of a star does not correspond exactly to its peak of radiation. For example, the sun radiates most intensely in the blue-green part of the visible spectrum, but it appears yellow-white rather than turquoise. Why? The answer has two parts. First, an object radiating at peak intensity in the blue-green region will also radiate appreciably at surrounding wavelengths, in the blue, yellow, and red; the mixture of these wavelengths will produce a sensation of white light with a bluish or blue-green cast. Second, the earth's atmosphere affects the rays of light from the sun in a way that tends to remove the blue-green component. Molecules in the air scatter the sunlight as it traverses the atmosphere. If the atmosphere scattered all wavelengths to the same degree, there would be no effect on the color of the light from the sun, but actually the shorter wavelengths are scattered more strongly than the red. Thus, looking at the sun in the sky, you see light from which a large amount of blue has been removed. This scattering effect eliminates the bluish cast that we might otherwise expect in sunlight, shifting the peak of the spectrum toward the yellow (Figure 5.2).

Incidentally, the scattering effect explains several other phenomena. The sky is blue because the rays of light that enter your eye are the rays from the sun that were scattered downward toward you by molecules of air. Because the blue wavelengths are scattered more strongly than the yellow or the red, the light you see in the sky is primarily blue (Figure 5.3).

Figure 5.2 Scattering of sunlight by the atmosphere.
Figure 5.3 Why the sky is blue.

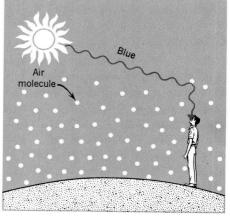

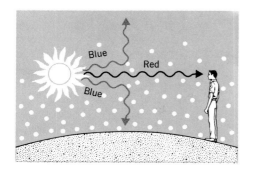

Figure 5.4  Why sunsets are red.

The sun turns orange and then red as it sets because its rays pass through the maximum thickness of air at sunset. The atmosphere scatters all of the blue and green light, leaving only a mixture of yellow, orange, and red in the spectrum of the sun (Figure 5.4).

## Graphs of Radiated Energy versus Wavelength

Scientists have measured in the laboratory the amount of energy radiated at different wavelengths by objects at various temperatures. Theoretical physicists have also calculated the distribution of radiated energy. Both measurement and theory predict a bell curve, with a sharp decrease of energy on the short wavelength side of the peak and a slow falloff on the long wavelength side. The small circles on the graph in Figure 5.5 represent measurements made in the laboratory of the intensity of radiation emitted by an object at a temperature of 1600°K. The solid line in this graph shows a theoretical calculation of the energy radiated by an object at the same temperature. The agreement between the measurements and the theoretical curve is nearly perfect.

We have already mentioned that the surface of the sun has a temperature of approximately 6000°K. The next graph (Figure 5.6) shows the energy radiated at different wavelengths by the sun,[1] or by any other star or object with a surface temperature of 6000°K. The shaded portions of the spectrum lie outside the visible region; the visible part of the spectrum is shown without shading.

This graph illustrates some interesting facts about the sun. Although the

[1] Neglecting the absorption lines.

Figure 5.5  Measured radiation from an object at 1600°K.

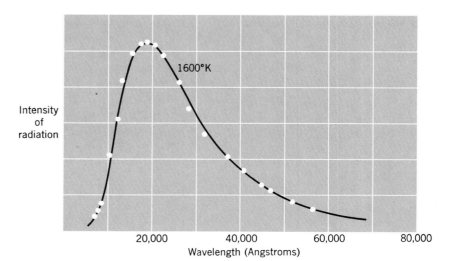

Intensity of radiation

1600°K

20,000          40,000          60,000          80,000

Wavelength (Angstroms)

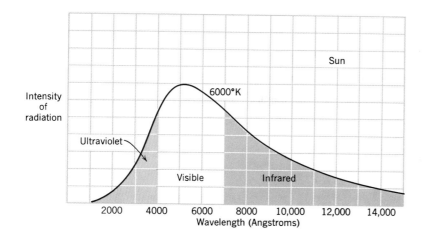

Figure 5.6 Radiation from an object at 6000°K.

peak of the sun's energy emission is in the visible region, the total amount of energy emitted by the sun as visible light is less than the amount emitted outside the visible region. Indeed, the sun emits more energy in the infrared region alone than it does in the visible region. When we are exposed to the sun we can feel the infrared radiation as heat on the skin; the small amount of ultraviolet radiation causes sunburn and suntan, however.

Stars much hotter than the sun emit most of their radiation in the ultraviolet. Vega, with a surface temperature of 10,000°K, is an example of a hot star. The graph in Figure 5.7 shows the relative amounts of energy radiated by Vega at different wavelengths. One-half of the energy radiated lies in the ultraviolet region, about one-third is radiated in the visible region, and the remaining part lies in the infrared region.

The hottest stars in the sky have surface temperatures in the neighborhood of 80,000°K. A star at this temperature emits its peak radiation at a wavelength of 350Å, deep in the ultraviolet region. Such stars shine mostly in the ultraviolet; they emit a relatively small part of their radiation as visible light.

Many stars are cooler than the sun and emit most of their energy in the red and infrared, radiating little energy in the visible region and practically no energy in the ultraviolet. As an example, consider Barnard's Star with a temperature of 3000°K (Figure 5.8). This cool star will emit its peak of radiation at a wavelength of 9600Å, in the infrared region.

Cool, red stars are more numerous than any other kind of star in the sky. They are considerably more numerous than stars resembling the sun. However, very few red stars are visible, partly because red stars tend to be exceedingly dim, and partly because they radiate primarily in the infrared region.

Some red stars are conspicuous exceptions to the rule that red stars tend to be very dim. These stars, known as red giants, are relatively cool, emitting most of their energy in the red and the infrared; nonetheless,

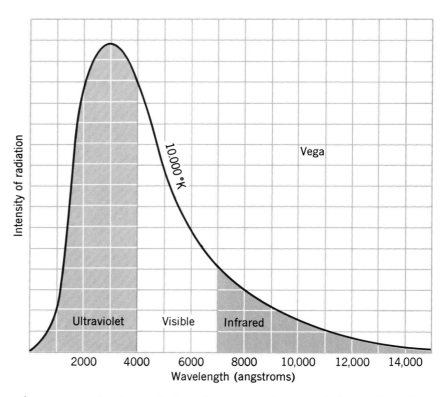

Figure 5.7 Radiation from an object at 10,000°K.

they are very luminous. Antares is an example of a red giant. Although its surface temperature is only 3600°K, it is thirty thousand times more luminous than the sun; that is, it radiates thirty thousand times as much energy into space every second.

## BRIGHTNESS OR LUMINOSITY

The brightness of a star is referred to as its luminosity by astronomers.

Figure 5.8 Radiation from an object at 3000°K.

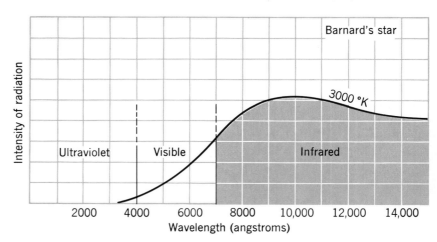

The absolute luminosity of a star is the amount of energy radiated into space each second from the star's surface. This energy is partly in the form of visible light and partly in the form of radiation at other wavelengths, such as the infrared or ultraviolet.

Astronomers and physicists usually measure absolute luminosity in ergs per second. An erg is a unit of energy in the centimeter-gram-second system. It is a very small amount of energy; when a fly collides with a window screen it imparts about one erg to the screen on impact. A 100-watt bulb radiates about 1000 million ergs per second into space.

The sun radiates about $4 \times 10^{33}$ ergs per second into space.[2] Because the absolute luminosity of the sun is a very well-known number to all astronomers, stellar luminosities are often expressed in units of the sun's luminosity, written as $L_\odot$. The circle with a dot in its center is generally used as a symbol for the sun in astronomy (and astrology). Alpha Centauri—the star nearest to our solar system—has an absolute luminosity of $2L_\odot$, meaning that it radiates $8 \times 10^{33}$ ergs of energy into space every second.[3] If the earth were in the same orbit around Alpha Centauri as it is around the sun, its temperature would be close to the boiling point of water.

The brightest stars have luminosities of $10^5 L_\odot$. An example is Rigel, a star in the Orion constellation. If the earth were placed in an orbit around Rigel it would vaporize instantly. Barnard's Star — the star second nearest to the sun — has a luminosity of $10^{-2} L_\odot$. A planet circling Barnard's Star at our distance from the sun would have a temperature of minus 200° Fahrenheit. The dimmest stars observed up to now, called red dwarfs, have luminosities extending down to $10^{-6} L_\odot$. The sun falls in the middle of the range between the brightest and faintest stars.

## Apparent Luminosity and Apparent Magnitude

When you look at the stars in the sky, some seem considerably brighter than others, but this impression does not necessarily correspond to their absolute luminosity. For example, Vega seems brighter than Deneb. Actually it is 1600 times fainter, but Deneb is 54 times farther away. Because the intensity of light falls off as the square of the distance, the apparent luminosity of Deneb is diminished by a factor of 3000, relative to Vega. Consequently, Vega appears brighter than Deneb, although Deneb is intrinsically more luminous than Vega.

Astronomers use a unit called apparent magnitude to measure apparent luminosity. This unit dates back to the second century B.C., when the Greek astronomer Hipparchus drew up a list of stars visible to the naked eye and assigned a number to each one to indicate its relative brightness. He assigned the number "1" to the brightest stars, which came to be

[2] If you were to express this number in words, it would be 4 billion trillion trillion per second.
[3] Alpha Centauri is a triple star. This figure refers to the luminosity of the largest member of the triplet.

known as "first-magnitude stars." In Hipparchus's list, Sirius, Vega, and Deneb were first-magnitude stars. All the stars in the Big Dipper were second-magnitude stars. The faintest stars that the naked eye could see were classified as sixth-magnitude by Hipparchus.

The system devised by Hipparchus is still in use today, but the units of first magnitude, second magnitude, and so on have acquired a more precise meaning. Instead of judging the brightness of a star with the naked eye, astronomers use a photometer to measure the visible radiation from each star more accurately. The photometer readings reveal that, on the average, first-magnitude stars in the Hipparchus scale are 100 times brighter than sixth-magnitude stars. Physiologists find that when the eye observes a linear increase in brightness in a series of light sources, the measured increase in brightness turns out to be geometrical. That is, if the eye records an impression of a uniform *step*-increase, the actual increase is a constant *multiple* from one source to the next. In the case of Hipparchus' scale, the eye observes five uniform steps of increasing brightness from sixth to first magnitude stars. Therefore, the true increase in brightness from the sixth to the first magnitude must be $R \times R \times R \times R \times R = R^5$, where $R$ is the ratio of brightness of a star of one magnitude to a star of the next magnitude. Since the photometric observations show that the ratio of brightness of first- to sixth-magnitude stars is 100, we have

$$R^5 = 100 \qquad \text{and} \qquad R = (100)^{1/5} = 2.51$$

In other words, a first-magnitude star is 2.51 times brighter than a second-magnitude star and $(2.51)^2 = 6.31$ times brighter than a third-magnitude star. The table (left) shows the relative brightness of stars of various magnitudes compared to the brightness of a first-magnitude star.

This table gives the relative brightnesses of the *averages* for each class of stars. *Individual* stars will differ from the average for their class to some degree. For example, Deneb, a first-magnitude star, is a little fainter than the average for a first-magnitude star, although it is definitely brighter than a second-magnitude star. The photometer readings place it between the two, with a relative magnitude of 1.3.

Some first-magnitude stars are brighter than the average for their class. For example, Vega is about 60 percent brighter than the average for a first-magnitude star. Because brighter stars have smaller numbers on the relative magnitude scale, the apparent magnitude of Vega is 0. Sirius—the brightest star—is four times brighter than Vega. The apparent magnitude of Sirius must be *negative*. It turns out to be −1.4. The sun is 100 billion times greater in apparent brightness than the average first-magnitude star. Its apparent magnitude is −26.7.

With the aid of time exposure, the 200-inch telescope on Mount Palomar can photograph stars down to the twenty-third or twenty-fourth magnitude. A twenty-third-magnitude star has the apparent brightness of one candle viewed from a distance of 10,000 miles.

*Absolute Magnitude.* Reference books on astronomy often list the "absolute magnitude" of a star in addition to its apparent magnitude. The

| Apparent Magnitude | Brightness Relative to First-Magnitude Stars |
|---|---|
| 1 | 1 |
| 2 | 1 : 2.51 |
| 3 | 1 : 6.31 |
| 4 | 1 : 15.85 |
| 5 | 1 : 39.82 |
| 6 | 1 : 100 |

absolute magnitude is another way of expressing the absolute luminosity of a star. The absolute magnitude of a star is defined in terms of a unit of distance called the parsec, equal to 3.26 light-years. The explanation of the unit is given in Appendix A on pages 476-478. A star's absolute magnitude is the apparent magnitude that this star would have if moved from its actual position to a new position at a distance of 10 parsecs from the sun. According to this definition, a star with an absolute magnitude of "−1" radiates $1.2 \times 10^{35}$ ergs per second into space. The absolute magnitude of the sun is 4.7. The absolute magnitude of Deneb — a considerably brighter star than the sun is −7.5. The absolute magnitude of Barnard's Star is +10.7.[4]

*Relation Between Absolute Magnitude, Apparent Magnitude, and Distance.* The distance to a star determines the relation between its absolute and apparent magnitudes. If the star is farther away than 10 parsecs or 32.6 light-years, its apparent magnitude is greater than its absolute magnitude. If the star is closer than 10 parsecs, its apparent magnitude is smaller than its absolute magnitude. If the star is at a distance of 10 parsecs, the two quantities are identical. These relationships are contained in the following formula connecting a star's absolute magnitude $M$, apparent magnitude $m$, and distance $d$ measured in parsecs

$$M = m + 5 - 5 \log d \text{ (parsecs)}$$

The apparent magnitude of a star can always be measured. If the star's absolute magnitude or luminosity can be estimated, its distance can be determined from the formula by solving for $d$:

$$d = 10^{(1/5)(m-M+5)}$$

Appendix A discusses the determination of distances to stars and galaxies by this method.

## THE POPULATION OF STARS

Suppose we plot a point on a graph for each star in the sky, plotting surface temperature along the horizontal axis and absolute luminosity along the vertical axis. This graph would be analogous to a graph of height versus weight for a human population. What do you expect to see on such a graph? Will the points representing the entire population of stars lie scattered all over the graph? Or will they lie close to a line, as they do for the properties of height and weight for a population of men and women?

To find the answer to this question, let us plot the luminosity and the surface temperature of all the stars in the neighborhood of the sun (Figure

---

[4] These numbers are bolometric magnitudes, which are based on the total energy radiated by a star at all wavelengths. The numbers in the previous paragraph are called visual magnitudes and are based only on the energy radiated in the visible part of the spectrum.

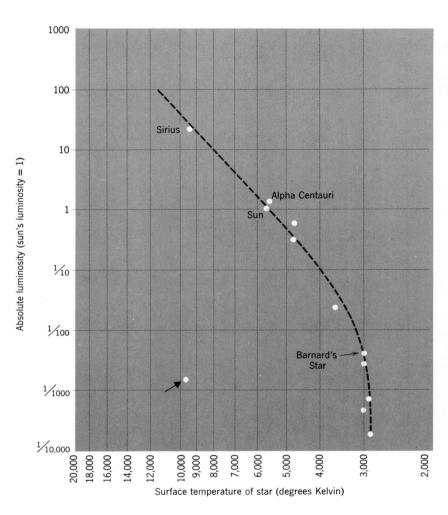

*Figure 5.9    Luminosity versus temperature for stars within 10 light-years of the sun.*

5.9). We limit ourselves to stars near the sun because it is only for these nearby stars that we can measure the absolute distance with precision. We must know the distance of the star to calculate its absolute luminosity from its apparent luminosity. Remember that for the purposes of the graph, the apparent luminosity is of no value, because it is not an intrinsic property of the star but depends on its distance from the sun. It is the absolute luminosity that is the fundamental physical attribute of the star.

Proceeding with the graph, we plot intrinsic luminosity and surface temperature for all stars contained within roughly 10 light years of the sun. Within this volume of space there are 12 stars, including the sun, whose distances are known accurately. The results of the plot are shown above.

There is no particular reason for specifying 10 light years. We could have chosen 30 light years and then would have had a larger number of stars for the plot; or we could have chosen a smaller distance. However, if we choose a distance much smaller than 10 light years, we will have very few stars in our sample, and it will be difficult to tell much from the graph.

Before we try to interpret the graph, attention should be drawn to its two peculiarities. One is that temperatures along the lower axis increase from right to left, rather than from left to right as you would expect. This method of plotting is simply a matter of tradition. The other is that both temperature and luminosity are plotted on *logarithmic* scales, i.e., scales proportional to powers of ten. This kind of graph, known as a "log-log" plot, is necessary because of the extreme variations, both in brightness and temperature, of stars.

Inspection of the graph shows that for stars, just as for people, nearly all the points fall on or near a line running diagonally across the graph; they are not scattered randomly all over the area of the graph. We can immediately draw the following conclusion: there is a connection, on the average, between the intrinsic luminosity and the surface temperature of a star. Furthermore, we can see that the connection is such that the hottest stars are the brightest ones and the coolest stars are the faintest. This result agrees with our everyday experience regarding glowing objects: The hotter a glowing object, the more energy it radiates.

The dashed line drawn on the graph represents a connection between brightness and temperature. This line can be called the Main Sequence of basic properties of the population of stars or, at least, of the stars that are located near us in our Galaxy.

Just as in the case of a graph representing average height versus weight for a human population, the line is not valid for each individual star; it only applies to the average properties of the entire population of stars. That is, most of the stars in the plot do not lie precisely on the line; rather they fall to either side of it by a small amount. There are two reasons for these deviations. First, stars do vary somewhat in their composition, and this variation, although it is not very great, is enough to have some effect on their properties. To some extent the scatter in the points on the graph represents differences in composition among the stars. Second, the graph is based on measurements that are difficult to carry out and are subject to random errors that tend to scatter the points on the graph. If we could carry out the measurements with infinite accuracy, the points on the graph would probably cluster along the line more closely than they do in practice.

In one respect, the star plot differs from the people plot. Although most of the stars cluster along the main sequence line, one out of the 12, marked by an arrow, lies quite far from the line, near the bottom of the plot. Inspection of the graph shows this star has a somewhat higher temperature than the sun but is intrinsically 300 times fainter. No person could possibly deviate from the main sequence line of height and weight in the human population as much as this star deviates from the Main Sequence of stars in brightness and temperature. Discovering such a star at the bottom of the plot is as much a surprise as it would be to discover a thriving person who was 6 feet tall and weighed only a few

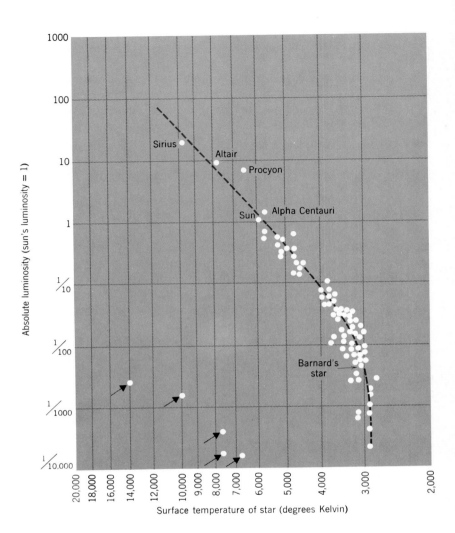

*Figure 5.10 Luminosity versus temperature for stars within 20 light-years of the sun.*

pounds, or one who was 4 feet tall and weighed 1000 pounds.

What can be the explanation for this extremely peculiar star? Perhaps we will find the answer if we plot more stars on the graph. After all, 12 stars are not a large number. If we plot 50 stars, or 100 stars, we may find that several parallel "main sequences" appear on the graph, representing different families of stars. Then the "odd" star would be explainable; it would belong to another "main sequence." That would be a very exciting discovery.

With this thought in mind, we enlarge the volume around the sun to a sphere with a radius of 20 light-years containing 90 stars including the sun. The graph in Figure 5.10 shows the results of a plot of luminosity versus surface temperature for the 90 stars. We see that most of the stars continue to lie very close to the Main Sequence line, just as in the case of

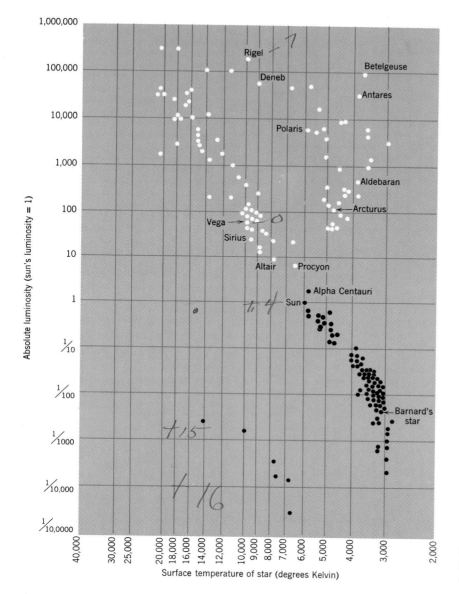

Figure 5.11 Luminosity versus temperature for the 100 brightest stars (light circles), and 90 nearest stars (dark circles).

the smaller sample.[5] However, a substantial number of stars, although still a minority, lie quite far from the Main Sequence. The peculiar star lying at the bottom of the earlier graph is now joined by several others also located near the bottom, well below the Main Sequence. They are

[5] The Main Sequence line shown here is drawn through the middle of the distribution of stars, including stars of all ages. Frequently a theoretical Main Sequence line is drawn for stars of zero age. Since stars move upward and to the right on the Hertzsprung-Russell diagram as they grow older (Chapter 7), the so-called "zero-age" Main Sequence line lies below and to the left of the Main Sequence line in Figures 5.10–12.

indicated by arrows. One of them has a surface temperature of 14,000°K. If this were a normal star following the Main Sequence, it would be 200 times brighter than the sun; yet, the graph shows it to be far fainter!

Figure 5.10 shows that the majority of the stars in our neighborhood are exceedingly faint, only a few being as bright as the sun. We would expect the brightest stars in the sky to be our neighbors, but this turns out not to be the case, for only a dozen of the 90 stars in the list can be seen with the naked eye under normal conditions. The bright stars that make up the familiar constellations of the night sky are for the most part a different group, relatively distant, but so luminous that they outshine the relatively dim nearby stars.

Although stars brighter than the sun are relatively rare, they are exceedingly interesting because they are the giants of the sky in mass and luminosity. Will these luminous stars follow the pattern on the H-R diagram established by the faint stars? The answer appears in Figure 5.11, showing the H-R diagram for the 100 brightest stars in the sky, plotted together with the 90 nearest stars for comparison. Again the majority of the stars are clustered around a line which is an extension of the Main-Sequence line for the faint stars, but, as in the case of the faint stars, one group of stars departs from the general pattern. The unusual stars form a trail leading upward and to the right, into a region of stars that have very low surface temperatures, and therefore are red in color, but nonetheless are enormously bright. Normally a low-temperature, red-colored star is very faint in comparison to the sun. Betelgeuse, however, is a star at the upper right that is deep red in color, with a surface temperature of only 3600°K, and yet it is 100,000 times brighter than the sun!

## Dwarfs and Giants

The stars in the lower-left region of the graph and those in the upper-right region are so remarkable that they have been given special names. The stars at the lower left are called *white dwarfs;* they are named dwarfs because they are very faint, and white because they are white-hot. The stars at the upper right are known as *red giants;* they are called giants because they are extremely bright, and red because they are red in color.

It turns out that the dwarf stars actually are very small, as their name implies. A typical white dwarf has a diameter of about 20,000 miles, only twice the diameter of the earth, although its mass is approximately the same as that of the sun, which has a diameter of almost one million miles. White dwarfs are dense, compact stars, so dense that a teaspoon of matter from a white dwarf weighs about ten tons.

Red giants, on the other hand, are very large in size, typically 100 million miles in diameter, or about 100 times the diameter of the sun. These huge stars have, however, approximately the same mass as the sun in most cases. Thus, the average density of the matter within a red giant is very low.

# THE HERTZSPRUNG-RUSSELL DIAGRAM

A great deal of interesting information can be obtained from star plots of the kind that we have discussed above. Hertzsprung first discussed a star plot of variables equivalent to brightness versus temperature in 1911, and Russell discussed a similar plot in 1913. In the years since then, this type of graph has come to occupy a central position in astrophysics because of the valuable results that have been obtained from it.

In honor of the two pioneers, plots of brightness vs. temperature for stars are known today as *Hertzsprung-Russell diagrams*. For brevity, they often are referred to simply as *H-R diagrams*. The line along which most of the points are clustered is called the Main Sequence in the H-R diagram. The luminous stars that lie near the top of the H-R diagram are called giants. The faint stars near the bottom of the diagram are called dwarfs. The bright stars near the upper end of the Main Sequence, blue-white in color, are called blue giants.* The faint stars at the lower end of the Main Sequence, orange or red in color, are called orange dwarfs or red dwarfs. Barnard's Star is an orange dwarf. The sun, yellow-white in color, occasionally is referred to as a yellow dwarf.

*The Numbers of Stars of Different Types.* About 90 percent of the stars in the sky are Main Sequence stars. The remaining 10 percent is divided between white dwarfs, red giants, and a few odd minor varieties. Among the Main-Sequence stars, the red dwarfs are the commonest variety. The blue giants are the rarest.

## The Significance of the Main Sequence

The stars along the Main Sequence are the "normal" stars in the sky. They are the family of stars to which the sun belongs. The red giants and the white dwarfs, on the other hand, appear to be "abnormal" stars, with a completely different connection between brightness and temperature. Let us put aside these puzzling, peculiar stars for the moment; their nature will be explained in full later in the chapter. In this section we wish to concentrate on the Main Sequence stars, and ask ourselves this question: What is the difference between one Main Sequence star and another? That is, can we discover one single property or characteristic of stars, that is responsible for the distribution of stars along the Main Sequence? Is there a property of stars such that, for example, the cool, red, faint stars at one end of the Main Sequence have a small amount of this property, and the hot, blue, luminous stars at the other end of the Main Sequence have a great deal of it?

The answer is affirmative; the stars along the Main Sequence differ from one another only with respect to their *mass*. The red, faint stars at the lower end of the Main Sequence have small masses; that is, each of these stars is made up of a relatively small amount of matter. The blue, luminous

---

* The very brightest stars are classified as supergiants. Rigel and Deneb are blue supergiants, and Betelgeuse and Antares are red supergiants.

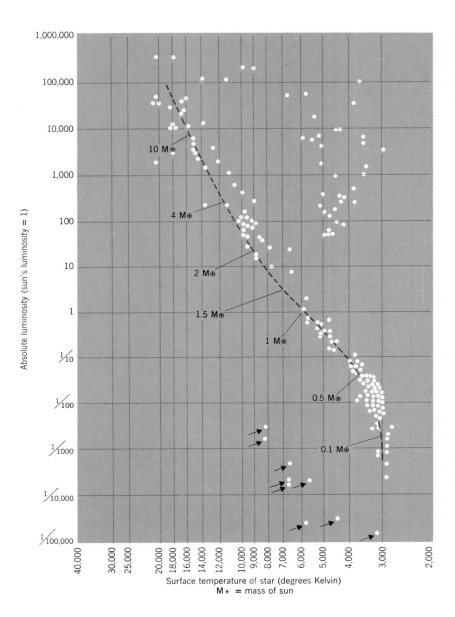

*Figure 5.12 Masses of Main Sequence stars.*

stars at the upper end of the Main Sequence are very massive; that is, each of these stars contains a great deal of material. Of course, even the stars at the lower end of the Main Sequence, which we described as having "small" masses, still are enormously massive. The reddest, faintest stars that appear in the H-R diagram on page 129 contain 10,000 times as much mass as the earth.

The sun lies in the middle of the range between the smallest, least massive stars and the largest, most massive stars. Its mass, which is $2 \times 10^{33}$ grams, is 300,000 times greater than the mass of the earth.

The faintest and reddest star ever observed thus far has a mass about one twenty-fifth that of the sun. The brightest, bluest and most massive stars on the Main Sequence have masses approximately 60 times greater than the mass of the sun.

That is the significance of the Main Sequence: it is a sequence of stars arranged in order of increasing mass, from the smallest stars at the lower end of the sequence to the most massive stars at the upper end. The *mass* of a Main Sequence star, and nothing else, determines its temperature, its brightness, and its place on the H-R diagram.

The graph in Figure 5.12 is the H-R diagram on page 132 with the values of stellar mass marked off along the line. All masses are expressed in units of the sun's mass, which is $2 \times 10^{33}$ grams or $2 \times 10^{27}$ tons. Deneb and Rigel—the brightest stars on the diagram—are 20 times as massive as the sun.

## The Ends of the Main Sequence

Does the Main Sequence extend indefinitely far in both directions? Is it cut off at either end or both ends? It seems likely that there is a lower limit to the masses of stars; that is, if an object is extremely small, it cannot be a star and is more likely to resemble a planet. This guess turns out to be correct; the lower limit for the mass of a star is not known precisely, but is believed to be approximately one one-hundredth the mass of the sun. This is roughly 10 times the mass of the planet Jupiter and 3000 times the mass of the earth.

It is not surprising that there is a *lower* limit to the mass of a star, since a great deal of mass is needed to create the enormous pressure and temperatures that cause a star to burn nuclear fuel at its center. It is a surprise, however, that there also appears to be an *upper* limit to the mass of a star. The most massive stars that have ever been observed are about 60 times as massive as the sun and no starlike object more massive than this has ever been identified.

For a short time after the discovery of quasars, some astronomers theorized that they might be supermassive stars, perhaps as much as one million times as massive as the sun. Other explanations for quasars will be described in Chapter 10.

## The Meaning of the Red Giants and the White Dwarfs

What are the peculiar red giants and white dwarfs? They are the giant stars that lie above the Main Sequence in the upper-right corner of the H-R diagram, and the dwarfs that lie below, in the lower-left corner of the H-R diagram. Are they special stars that were born abnormal? Or are they transient stages through which every star passes in its lifetime? Astronomers were puzzled by these questions for many years, but today we know the answers as a result of laboratory experiments combined with theoretical investigations carried out on high-speed electronic computers. The experiments have revealed how a star derives its energy by burning nuclear fuel, while the computations showed the changes that occur in a star as its fuel is used up.

The combination of laboratory experiments and theoretical studies has provided a complete picture of the life story of a star, which will be discussed in Chapter 7. According to this picture, red giants and white dwarfs are ageing stars that have partly or largely exhausted their fuel and are headed toward extinction. With the aid of the latest theoretical studies, we can trace the path of a star on the H-R diagram as it passes through all the stages of its existence. We are now certain that the giant and the dwarf are stages in the lifetime of a normal star. When a star has burned up a large fraction of its fuel it begins to feel its age and moves off the Main Sequence and upward to the region of the giants. It shuttles back and forth in this region for some time, until finally its end is near. Then, in a relatively short time, it crosses back over the Main Sequence line toward the left and dives downward into the region of the white dwarfs.

The sun will follow this course some billions of years in the future, first increasing in luminosity and turning red, and then moving off the Main Sequence and upward into the red giant region. As a red giant, the sun will radiate enough energy to vaporize the earth. Thereafter, the sun will move down again, this time into the region of the white dwarfs, to become one of the many dying stars that litter the Galaxy.

There is a complication to the story. Stars of modest or average size, such as the sun, usually follow the route described; but a star that is, let us say, 4 times as massive as the sun ends its days in a more spectacular fashion. After its life as a red giant is ended, it does not simply fade away to become a white dwarf; instead, the massive star collapses on itself in a cataclysmic implosion. In some cases it rebounds, blowing its materials out into space in the explosive event known as the super-

*Figure 5.13 (a) and (b).  Light curves for two cepheid variables.*

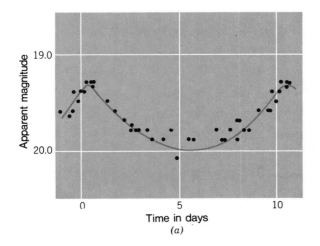

(a)

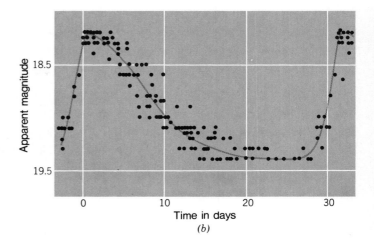

(b)

nova; or, if it is sufficiently massive, it may meet no resistance adequate to halt its collapse, and may contract into one of the so-called black holes in space.

A full account of the fascinating life history of a star requires an acquaintance with the basic facts of nuclear physics, which will form the subject of Chapter 6.

## VARIABLE STARS

One region in the H-R diagram contains a special kind of star that differs in an important respect from all the stars we have discussed thus far. The region containing unusual stars is located approximately half way between the red giant region and the upper end of the Main Sequence (Figure 5.11). The property of these stars that makes them unusual is their variation in light output, which is in marked contrast to the steady, unflickering output of ordinary stars. In some cases these variable stars change in brightness by as much as a factor of two over a period of a few days. The variations follow a regular cycle, with the cycle repeated exactly from one time to the next. The length of the cycle is constant in a given star, but varies from one star to another.

### Cepheid Variables

Cepheid variables are one important class of star displaying a regular cycle of variation in brightness.[6] Examples of the changes observed in the light output for two cepheid variables are shown in Figure 5.13. The first is a cepheid variable (a) with a period of approximately ten days, and the second is a cepheid with a period of one month (b).

The most familiar cepheid variable in the sky is the North Star, Polaris. Its brightness varies by a factor of 9 percent every four days.

The location of the cepheid variables in the H-R diagram is shown in Figure 5.14. This diagram also shows the positions of the red giants and of other important members of the family of stars.

As will be seen in Chapter 7, the stars found in the region between the upper Main Sequence and the red giants are in a relatively advanced stage of their lives. Theoretical studies show that for a brief period in this stage of its life, a star may begin to oscillate, collapsing on itself and expanding outward repeatedly. Its size can change by as much as 30 percent during each oscillation. These rhythmic changes in size are

[6] The name is derived from the fact that the first star of this type to be observed was called Delta Cephei.
Another type of variable star has a light output that flickers irregularly by small amounts. Generally the change in luminosity for this type of variable star is less than one percent. Such stars, called *irregular variables*, are found in a number of regions scattered over the H-R diagram, instead of being clustered in a single region as is the case for the cepheids.

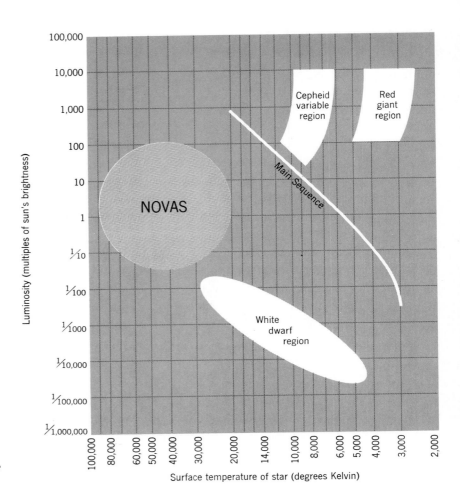

Figure 5.14 The position of variable stars in the H-R diagram.

Figure 5.15 Change in spectral type during the light cycle of a cepheid variable.

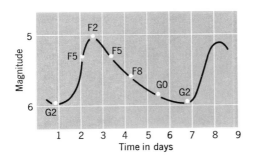

accompanied by large changes in luminosity.

The surface temperature of a cepheid variable also changes during the course of each cycle of variations in brightness. As might be expected, the surface temperature is highest when the brightness is at a peak, and lowest when the brightness is at a minimum. For example, a cepheid variable with a period of six days falls in surface temperature from 7500°K at a peak brightness to 6000°K at minimum brightness. The corresponding change in its spectral type is from F2 to G2. Figure 5.15 left shows the light curve for a cepheid variable with a period of about six days, with the corresponding changes in spectral type indicated along the light curve.

The pulsations of the cepheid continue for a long time, until the star has evolved to a stage where its opacity no longer depends so sensitively on temperature. When that happens, the star's oscillations disappear and it leaves the region of the cepheid variables.

The period during which a star is a cepheid variable may last as long as one million years. Although this seems a very long time, it is a small fraction of the billion-year lifetime of the typical star. That is why only a few of the stars in the sky at a given time are cepheid variables, in spite of the fact that most stars pass through this stage at least once during their lives.

The cepheid variables have an importance in astronomy that is far greater than their small numbers would otherwise indicate. They have become the most important yardstick of the astronomer for measuring distances to other galaxies. Their usefulness in this connection depends on the fact that the period of oscillation of the cepheid variable — that is, the interval of time from the beginning of one cycle of changing light output to the beginning of the next cycle — is uniquely related to the average luminosity of the star. Thus, by measuring the period of the light variations in a cepheid, we can deduce its absolute luminosity. Since we can also measure its apparent luminosity, its distance immediately follows. This method gave the first measure of the distance to nearby galaxies.

## Novas

The family of stars contains another group whose light output varies with time. This group of stars, unlike the cepheid variables, may be found anywhere within a large region on the H-R diagram (Figure 5.14). Also unlike the cepheid variables, which fluctuate in intensity rhythmically according to a fixed pattern, this second group of stars flares up suddenly and unpredictably at intervals of time. On the average, the flare-ups occur every 30 to 50 years for a given star of this type. The flare-ups also differ from the changes of light output displayed by cepheids in the sense that they are very violent, with the total energy output of the star increasing by as much as a factor of a million over a short period of time.

The irregularly flaring stars are called *novas*. They are found on the H-R diagram to the left of the Main Sequence. Novas are believed to be, in fact, stars that have lived out most of their lives and are on the way to becoming white dwarfs. As we mentioned above, the white dwarf is the final stage in the life of all stars of modest size.

Why is the nova different from other ageing stars that are about to become white dwarfs? Do all ageing stars flare up as novas for a time, as they approach this stage? Or, do only a small number do so, as the result of some special condition? From the small number of novas in the sky at any time — no more than one thousand in our Galaxy at this moment — it is clear that the nova must be a very special kind of star, and cannot be a stage through which every star passes during its old age. A clue to the property of novas that makes them special may lie in the fact that many of the novas observed thus far are found to belong to binary stars, that is, they are one of two stars circling closely around one another under the attraction of their mutual gravity. Why the presence of a white dwarf in a binary should lead to a violent flare-up from time to time is not clearly understood, but it is believed by the theoretical astronomers that the outbursts occur when matter is pulled off the partner star and on to the surface of the white dwarf by gravity. This material, deposited on the surface of the white dwarf, produces an instability in the white dwarf that leads to a violent outburst of radiation, or nova outburst. The newly acquired matter is blown out into space by the nova outburst and, as a consequence, the nova subsides into a normal white dwarf state again, until some later time when enough matter has been drawn from its companion star to produce another unstable condition, and trigger another flare-up.

Although novas are a spectacular phenomenon for the astronomer to observe, their occurrence does not contribute any basic new knowledge to our picture of stellar evolution. Novas are more of a curiosity than a central feature in the life story of the stars. The contrary is true for supernovas, which are flare-ups of a star that are even more violent than nova outbursts. Supernova outbursts, as will be seen in Chapter 7, play a critical role in stellar evolution and in the history of the universe. In fact, it can be said that without the supernova outbursts that terminate the lives of the more massive stars in the sky, we would not be here today.

## BINARY STARS

A large fraction of the stars in the sky are *binary stars*, that is, two stars revolving around a common center and bound to each other by their mutual force of gravitational attraction. Binaries are not concentrated in any particular part of the H-R diagram. In fact, the two individual stars making up a binary may appear in widely separated portions on the diagram.

### Mass Determination from Binaries

Binaries are particularly interesting members of the family of stars because they lead to a method of determining stellar masses. Other methods are explained in Appendix A. The method based on observations of binaries is more important than any other because it is applicable to many stars, and covers a large range of masses. The general idea behind the method can be illustrated by considering a special case in which one star in the binary is much more massive than the other. The larger, more massive star controls the movement of the smaller star by its gravitational pull, just as the sun controls the motion of the earth. For an orbit of a given size, the mass of the larger star determines the rate at which the smaller one revolves around it. The more massive the central star is, the greater the pull of its gravity will be, and the faster the small star must revolve around it in order not to be drawn in. Therefore, if the size of the orbit and the period of revolution can be measured for the small star, it should be possible to determine the large star's mass.

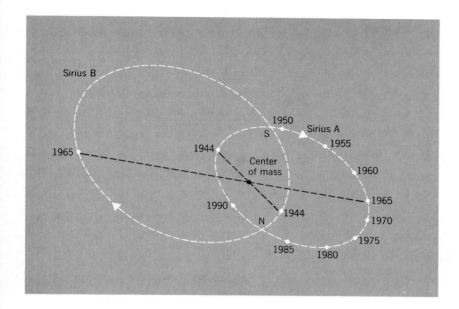

*Figure 5.16   The observed and predicted positions of Sirius A and B between 1944 and 1990. The center of mass of the two stars is the intersection of lines connecting their positions at two different times, e.g., 1944 and 1965.*

The mass of the sun is determined in this way from the size of the earth's orbit and length of the earth's year.

A similar relationship holds if the two stars in the binary have comparable masses. In that case, the theory of motion of objects under gravity shows that both stars revolve in ellipses around their common center of gravity. If the sizes and shapes of the two ellipses and the period of revolution can be measured, the mass of each star can be calculated by an extension of the idea described above. In order to translate this description into an accurate calculation of the mass of the heavier star, it is necessary to know the precise orbit traced by a star under the force of gravity. In general, this orbit is an ellipse. In the special case in which the orbit is circular, the result of the calculation is

$$M = \frac{4\pi^2}{G} \frac{R^3}{P^2}$$

where $M$ is the sum of the masses of the two stars, $G$ is Newton's constant of gravity, which has the value of $6.7 \times 10^{-8}$ in centimeter-gram-second units, $R$ is the radius of the orbit traversed by the lighter star around the heavier one, and $P$ is the period of revolution of the lighter star. Methods for finding $R$ and $P$ are described below.

The visual binary, Sirius A and Sirius B, provides an example of the elliptical orbits traversed by the two members of a binary system. Sirius A is an A-type Main-Sequence star of about two solar masses, and Sirius B is a white dwarf of about one solar mass. The observed orbits of Sirius A and Sirius B are shown in Figure 5.16.

The calculation required to complete the determination of the separate masses of the two stars in a binary depends on whether the binary is *visual* or *spectroscopic*. Visual binaries are close enough to the earth to be resolvable into two individual points of light. Spectroscopic binaries are too distant to be resolved, but are revealed as binaries because their spectra show periodic Doppler shifts as each star alternately recedes from and approaches the earth in the course of its orbit. The magnitude of the Doppler shift gives the velocity of each star along the line of sight to the observer. The repeating pattern of changes in the wavelengths of the lines immediately gives the period of the orbital motion.

The stars resolvable into visual binaries are generally stars of low luminosity and belong to the lower Main Sequence. The spectroscopic binaries are important because they include stars of high luminosity, and permit the determination of masses to be extended to the upper Main Sequence.

*Visual Binaries.* The period of revolution for a visual binary is determined by direct observation of the orbits of the two stars over an extended period of time. The shapes of the orbits and the apparent size or angular size of each orbit are also determined by direct observation. If the distance to the binary is known, the true sizes of the orbits can be calculated from their apparent sizes. The mass of each star is calculated from the period of revolution and the orbital sizes and shapes with the aid of formulas derived from Newton's law of gravity.

The calculation is complicated by the fact that generally the plane of the orbits is tilted at an unknown angle to the observer's line of sight, making it impossible to determine the shape of the orbit accurately. Methods for dealing with this complication are described below.

*Spectroscopic Binaries.* For spectroscopic binaries, the effect of the orbital motion can only be seen indirectly, as a Doppler shift in the spectrum of each star. The magnitude of the Doppler shift gives the velocity of the star along the line of sight to the observer. The spectra of the two stars are superimposed, and since the orbital velocities along the line of sight are different for the two members of the binary, each line is split into two components. (Figure 5.17a). Twice in each orbit, the velocities of both stars are tangential and the Doppler shift for each is zero. At this time, the double lines in the spectrum coalesce into single lines (Figure 5.17b). The period of the cycle of changes in the wavelengths of the lines is the period of the orbital motion.

The magnitudes of the Doppler shifts provide information about the component of the orbital velocity of each star along the line of sight to the observer. With the aid of the relationships provided by Newton's law of gravity, the information regarding the velocities can be converted into information on the sizes of the two orbits. In the special case of a circular orbit, for example, the orbital velocity, $v$, radius, $R$, and period of revolution, $T$, are connected by the formula,

$$v = \frac{2\pi R}{T}$$

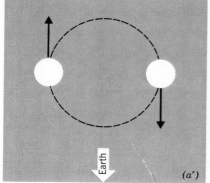 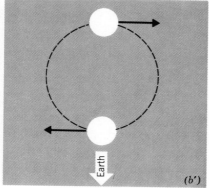

Figure 5.17 Two spectra of the spectro-scopic binary Mizar, one of the stars in the handle of the Big Dipper. The two stars in the binary revolve around a common center every 20.5 days. In the upper spectrum (a) each line is split into two components by the Doppler shift because one member of the binary is moving toward the earth and the other is moving away (a'). In the lower spectrum (b) the lines are not split because at the time this spectrum was photographed the two stars happened to be moving tangentially across our line of sight, with no motion toward or away from the earth (b'). The cycle of splitting and coalescence of the lines is repeated every 20.5 days.

For an elliptical orbit the relation is more complicated. In this case, the observed velocity is plotted as a function of time, and an ellipse is determined that gives the best fit to the observations.

*Tilt of the Orbit Plane.* For both types of binaries, the inclination of the plane of the orbit to the observer's line of sight presents a problem. The problem is readily solved for visual binaries. Suppose for clarity that one star has a considerably smaller mass than the other, and, therefore, describes a large and more accurately measurable orbit. This orbit is an ellipse with the more massive star at one focus. The problem is to find out how the plane of the ellipse is tilted with respect to the observer's line of sight. Figure 5.18 suggests the solution. On the left is the elliptical orbit as it would appear if viewed normal to the plane of the orbit. The focus is indicated by the letter F. The position of the focus is known, being the location of the more massive star. On the right, the same ellipse is shown as viewed by the earth observer, with its plane tilted.

The tilted, foreshortened orbit is still an ellipse, although of greater eccentricity than the original one. However, the focus of the foreshort-ened ellipse no longer coincides with the position of the massive star.

Figure 5.18 Effect of a tilted orbit plane on the shape and position of the focus (F) of an elliptical orbit.

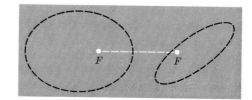

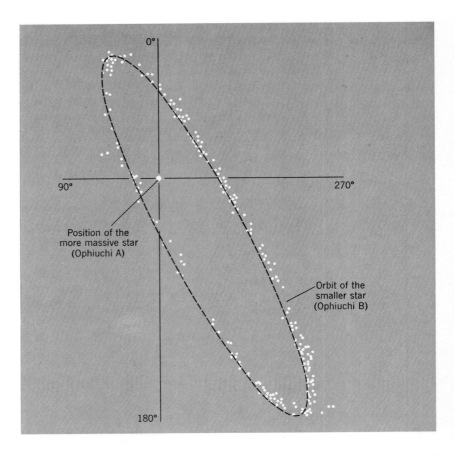

0°

90°    270°

Position of the
more massive star
(Ophiuchi A)

Orbit of the
smaller star
(Ophiuchi B)

180°

*Figure 5.19    Observed orbit of Ophiuchi B.*

In general, the massive star will not even be on the axis of symmetry of the foreshortened ellipse. This asymmetry betrays the tilt. An example is the path of 70 Ophiuchi B around A, shown in Figure 5.19. The curve is the best-fitting ellipse to positions of Ophiuchi B observed between 1850 and 1940. It is clear to the eye that the position of Ophiuchi A is not on the line of symmetry for this ellipse, indicating that the plane of the true ellipse is tilted relative to the observer.

The angle of the tilt can be determined by adjusting the orientation of the orbit until the focus of the ellipse coincides with the position of the massive star.

For spectroscopic binaries, the problem of determining the tilt of the orbit plane is generally insoluble. This is because the Doppler shift only yields partial information about the velocities, namely, their radial components. The best that can be done in this case is to make an assumption regarding the inclination of the orbit plane, and deduce the mass corresponding to this assumption.

If, for example, the orbit is assumed to be precisely edge-on to the observer's line of sight, the analysis can be carried out, and yields a lower limit to the mass of each of the two stars. One can go a little farther by assuming that the orbit planes of binaries are randomly oriented in space. By averaging the mass over this random distribution of orbit planes, we obtain a correction to the lower limit. When this is done separately for binary components of different spectral types, the result is a rough value for the mass of each spectral type.

*Eclipsing Binaries.* For one special type of spectroscopic binary, known as the *eclipsing binary,* the problem of the unknown tilt of the orbit plane disappears. The eclipsing binary is a double star whose orbit plane happens to be nearly edge-on to the observer, so that one star periodically moves behind the other and is repeatedly eclipsed. As a result, the intensity of the light from the binary varies periodically. A periodic variation in the light from a binary usually indicates that it is eclipsing. Thus, the plane of its orbit is known, and the masses of its components can be determined unambiguously.

Eclipsing binaries can yield other interesting information, in addition to stellar masses. The information comes out of a study of the details of the light variation from the binary as one star passes behind the other during the course of an orbit.

Suppose, for example, that the more massive star is a red giant and the less massive star is still on the Main Sequence. This is a plausible situation, since the two stars were formed at the same time, but the more massive star evolves more quickly. Then the less massive star will be smaller in radius and higher in surface temperature. Label the more massive star "A" and the less massive one "B." Star B will be totally eclipsed by Star A once in each orbit. That is, the radiation from B will be completely blocked out for a finite length of time as it passes behind A. In this case the "light curve" for the binary—that is, the plot of measured light intensity as a function of time—has the following characteristic appearance: the light curve is at a maximum when the two stars are separated; once in each orbit, the light curve falls as the smaller star begins to pass behind the larger one; when the smaller star is totally eclipsed, the light curve reaches a minimum, and remains constant at this minimum value until the smaller star starts to reappear on the other side; at that point the light curve begins to rise again.

Later in the same orbit, B passes in front of A and blocks a part of the radiation from A, causing another dip in the light curve. Since B is smaller than A, this eclipse is annular. The dip in the light curve during the annular eclipse has the same shape as during the total eclipse. However, the dip is shallower because, while the same area is blocked off in both cases—that is, the area of the smaller star—the surface radiation *per unit area* is less in the case of the red giant than for its companion.

A light curve for a binary of this type, called a completely eclipsing binary, is shown in Figure 5.20a.

The unmistakable feature of the light curve of this particular type of eclipsing binary is the *flatness* at the bottom of each trough. This flatness indicates that the main eclipse is total; hence one star is considerably smaller than the other.

The radius of the smaller star can be determined from the light curve as follows: the velocities of the stars are known from their Doppler shifts;

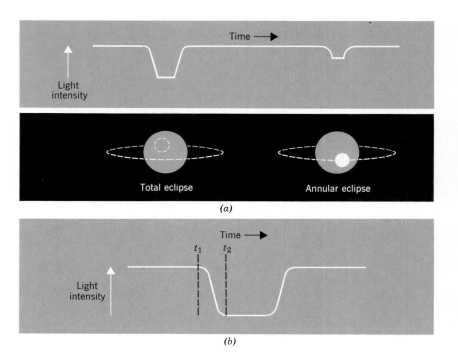

(a)

(b)

Figure 5.20 Light curve for a completely eclipsing binary.

if the sloping portion of the trough in the light curve lasts from time $t_1$ to $t_2$, and the relative velocity of the two stars in this portion of the orbit is $V$, the diameter of the smaller star is $V \times (t_1 - t_2)$. (Figure 5.20b)

If the orbit is tilted slightly to the observer's line of sight, the light curve will display one deep and one shallow trough per orbit, as before, but the troughs will not be flat-bottomed, because the finite tilt of the orbit plane prevents the disk of star B from lying completely within disk of star A at any point during the orbit. Two *partial* eclipses occur in each orbit. Because the eclipses are partial, the troughs are not as deep as in the previous case. Figure 5.21 shows the light curve for a partially eclipsing binary.

Figure 5.21 Light curve for a partially eclipsing binary.

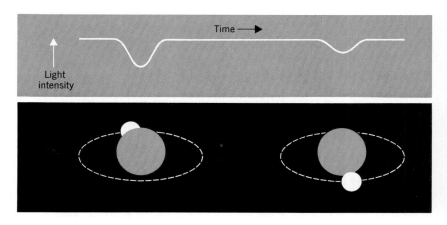

It is clear that the light curve of an eclipsing binary depends on (1) the relative luminosities of the two stars, (2) their radii, (3) the surface brightness of each, that is, the surface temperature, and (4) the precise angle of inclination of the orbit of the binary. The details of the light curve also are influenced by such features as the way in which density and temperature fall off with increasing height in the atmosphere of each star. Information regarding all these properties has been obtained from careful studies of the light curves of eclipsing binaries.

## Questions

1. What determines the color of a star's surface? Sketch a graph of intensity of radiation versus wavelength for each of the major spectral types. Use Figures 5.5 to 5.8 as models.

2. A 100-watt bulb radiates $10^9$ ergs/sec. How many such bulbs are needed to match the luminosity of the sun? If General Electric makes 100 million 100-watt bulbs per year, how many years are required to produce enough bulbs to match the sun's luminosity?

   What is the range of stellar luminosity? How many times brighter are the brightest stars than the dimmest stars in Figure 5.11?

3. Explain the difference between apparent magnitude and absolute magnitude.

4. Two 100-watt bulbs are respectively 50 and 300 feet away from you. Compare their apparent luminosities. What is the difference in their apparent magnitudes? In their absolute magnitudes?

5. Sketch or trace a blank H-R diagram and plot the following list of stars on it. ($L_\odot$ = sun's luminosity.)

   9500°K, 19 $L_\odot$ (Sirius A); 6400°K, 6 $L_\odot$ (Procyon); 12,200°K, 250 $L_\odot$ (Regulus); 3100°K, 0.004 $L_\odot$(Barnard's Star); 26,500°K, 30,000 $L_\odot$ (Mimosa); 5800°K, 1.0 $L_\odot$ (Sun); 2950°K, 2600 $L_\odot$ (Mira); 4600°K, 0.6 $L_\odot$ (Beta Centauri); 5100°K, 0.7 $L_\odot$ (36 Ophiucus); 9000°K, 0.008 $L_\odot$ (Sirius B.)

   From the H-R plot of the stars in question 5, which stars do you immediately recognize as peculiar? How?

6. Betelgeuse radiates $4 \times 10^{38}$ ergs/sec into space. What is its absolute magnitude? The absolute magnitude of Barnard's Star is 10.7. What is its absolute luminosity? Its luminosity relative to the sun?

7. A star whose apparent magnitude is 12 is located at a distance of 50 light-years. What is its absolute magnitude? What is the ratio of its luminosity to that of the sun? Of Sirius? Of Barnard's Star?

8. From the H-R diagram, determine the approximate distance of (a) a K-type Main Sequence star with apparent magnitude of 5, and (b) a G-type white dwarf with apparent magnitude of 15.

9. The relation between absolute magnitude and period is shown in the graph below for the common type of cepheid variable. Using this graph, find the distance to a cepheid with a period of 10 days and an apparent magnitude of 7; to a cepheid with a period of 50 days and an apparent magnitude of 19. Are these stars in the Milky Way Galaxy? If not, in which galaxy?

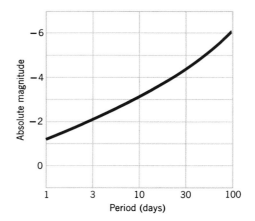

10. Use the formula on page 139 to determine the mass of the sun from the following properties of Jupiter's orbit: average orbital radius (R) = 770,000,000 km, period of revolution (P) = 11.9 years. Compare your result with the accepted value of the sun's mass. Remember that the formula is in centimeter-gram—second units.

11. Why is the observed number of visual binaries fewer than the observed number of spectroscopic binaries?

12. Referring to Figure 5.20, how would you use the light curve to determine the diameter of the larger star in the binary?

# 6  Nuclear Energy Sources

The analysis of stellar spectra and the classification of stars in the Hertz-sprung-Russell diagram provide two valuable clues to the mystery of the stars, but by themselves they are not enough. A third clue is needed. When it is supplied, the puzzle of the Main-Sequence stars, the red giants, and the white dwarfs is resolved; all become a part of a pattern of growth and development in the Universe, in which stars are born, evolve, and die like living organisms.

The third clue concerns the source of the energy that makes stars shine. When that source is identified, we will immediately see that within each star this energy cannot last forever, but must be used up, and that the different kinds of stars we see in the sky are nothing more or less than stars that have used up various amounts of their inner resources of energy—stars that are young, middle-aged, and old.

The amount of energy that comes off the surface of a star staggers the imagination. By using Einstein's law stating the equivalence of mass and energy, $E = mc^2$, and converting the energy radiated by the sun from units of energy to units of mass with the aid of this formula, we find that almost 5 million tons of mass per second are radiated by the sun in the form of electromagnetic energy.

Is the sun wasting away to nothing as a result of this loss of energy? The fossil record indicates that life has existed on the earth for billions of years; hence, the earth and the sun also must have existed for as long a time. In a billion years the sun radiates into space $1 \times 10^{29}$ grams, which is less than one part in $10^4$ or one one-hundredth of 1 percent of its mass. Thus, the sun has lost only an insignificant fraction of its total substance by radiation of energy from its surface.

But $1 \times 10^{29}$ grams is still an enormous amount of matter. It is equal to 17 earth masses. What force can convert all the matter in 50 earths into energy? Asking this question is the same as asking why the sun and stars shine so brightly. The question is the most fundamental one in astronomy. Until it had been answered, modern astronomy could not be said to exist.

Surprisingly, the answer came, not from stargazing, but from studies in the laboratory. Laboratory research in nuclear physics revealed the existence of the only force strong enough to convert matter into energy at the extraordinary rate at which stars pour their energy into space. This force is the nuclear force—the strongest force known to man.

## NUCLEAR FISSION

How do nuclear forces release energy? One way is by fission, that is, by the splitting of a nucleus into two or more pieces. This is the method used in the atomic bomb. The nucleus used in such a bomb usually is uranium or plutonium. The nucleus of a uranium atom or a plutonium atom is like a stick of dynamite; it is unstable, and can explode or disintegrate into several pieces, simultaneously releasing large amounts of energy. The energy released when one uranium or plutonium atom breaks up is roughly 10 million times greater than the energy released *per atom* in a TNT explosion. These nuclei will spontaneously disintegrate if they are left by themselves for a sufficient length of time; for example, $U^{235}$, the isotope of uranium used in the first atomic bomb test at Alamogordo, New Mexico, disintegrates spontaneously in the time of 880 million years. Or, these nuclei will explode immediately if disturbed by an incoming, slow-moving neutron. The explosion breaks the $U^{235}$ nucleus into two fragments, which turn out to be the nuclei of medium weight substances such as barium or krypton. In addition to the two main fragments of the original uranium nucleus, a few free neutrons usually emerge at the same time. These neutrons can trigger the explosions of other uranium nuclei, releasing more neutrons, and so on, setting in motion a chain reaction that can yield enormous amounts of energy, equivalent to the detonation of hundreds of thousands of tons of high explosives in a very short time (Figure 6.1). This is the principle of the atomic bomb.

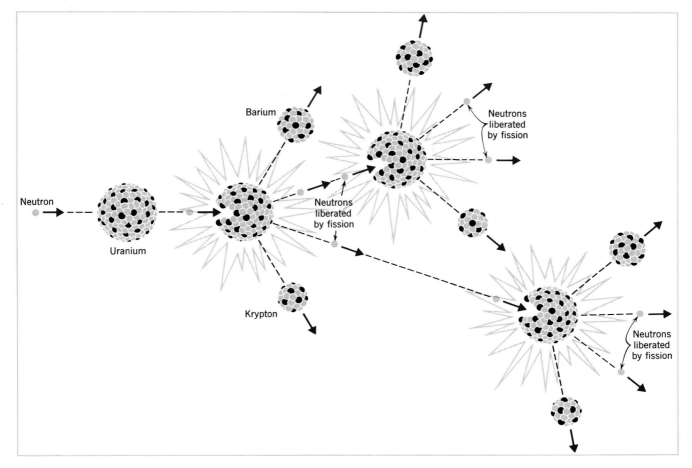

Figure 6.1 *Release of energy by a chain reaction in uranium.*

However, uranium and other unstable nuclei that yield energy by *fission* cannot be sources of the energy within stars, for these heavy elements are present only in minute traces in a star. In the sun, for example, the amount of uranium is only one ten-billionth of one percent.

If uranium is too scarce in stars to be a source of energy, perhaps the energy comes from some other element that is present in greater abundance than uranium. The most abundant element in the stars is hydrogen, but clearly it is impossible for fission to occur in hydrogen, because the hydrogen nucleus consists only of a single proton, and one proton cannot divide into fragments. At least, none has ever been observed to do so. Thus, fission cannot be the source of stellar energy.

## NUCLEAR FUSION

There is a second kind of reaction involving nuclei in which energy is released. It is called nuclear fusion. This reaction consists in fusing two nuclei together, instead of trying to break a single nucleus into two pieces. Two protons can be joined in this way to form a single larger nucleus, with a large amount of energy released in the process. Since protons—the nuclei of hydrogen—are very abundant in stars, it is assumed by all astronomers that the fusion of protons is the main source of energy in stars.

You might ask how two protons can be joined together, since a nucleus consisting of two protons does not exist. (Such a nucleus, which would be chemically like helium because it would have two positive charges, but would have the weight of deuterium, has never been detected.) The answer is that at the very moment of the collision between the two protons, one of them sheds its positive charge of electricity in the form of a positive electron, also called a positron. The removal of the positive charge from the proton leaves behind a neutron which is locked to the other proton to form a deuteron (the nucleus of deuterium or heavy hydrogen.

The positron, accompanied by a massless, electrically neutral ghostlike particle known as the neutrino, carries off most of the nuclear energy released in the fusion (Figure 6.2).

*Figure 6.2    Fusion of two protons.*

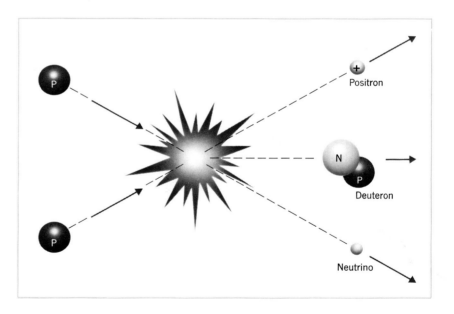

Nuclear physicists have developed a useful shorthand for nuclear reactions like the one shown above. Suppose we let the symbols $d$, $p$, $n$, $e^+$, and $\nu$ represent, respectively, the deuteron, proton, neutron, positive electron or positron, and neutrino. Then the reaction in which two protons fuse to form a deuteron can be written:

$$p + p \longrightarrow d + e^+ + \nu$$

Figure 6.3 presents a glossary of some basic particles appearing in these reactions.

Figure 6.3 Glossary of some particles occurring frequently in nuclear reactions.

| | |
|---|---|
| **p (proton)** | A positively charged particle, relatively massive; it is the nucleus of the hydrogen atom and one of the two basic building blocks of heavier nuclei. |
| **n (neutron)** | Electrically neutral particle about the same mass as the proton; it is the other basic component of atomic nuclei. |
| **d (deuteron)** | A particle composed of a proton and a neutron bound together, containing the same charge as the proton but double its mass; it is the nucleus of heavy hydrogen or deuterium. |
| **e or e⁻ (electron)** | A negatively charged, relatively light particle $1/1840$ times the mass of the proton. |
| **e⁺ (positron)** | Similar to the electron, identical mass, but positively charged. |
| **$\nu$ (neutrino)** | A massless, chargeless particle; it is produced in some nuclear reactions, generally with a positron or electron, and carries off some of the energy released in the reaction; because it has no electric charge and no mass, the neutrino can pass through large amounts of matter, such as the entire body of a star. |
| **$\gamma$ (gamma ray)** | A photon or packet of electromagnetic radiation similar to an ordinary photon, but having extremely high energy and correspondingly short wavelength (less than $10^{-8}$ cm). |

How does the fusion of the two protons in the above reaction release energy? The mechanism for the energy release must be quite different from that in fission, because the deuteron formed by the fusion is not unstable like the nucleus of uranium; a deuteron does not behave like a stick of dynamite, blowing up if disturbed, and releasing large amounts of nuclear energy.

In order to see the reason for the release of energy in nuclear fusion, let us consider what will happen when two protons approach one another on a collision course and come within range of the nuclear force of attraction. Immediately they are seized by this very strong attractive force, and rush violently toward one another; when they collide, owing to the extreme strength of the nuclear attraction, the two particles fuse together, forming a single, heavier nucleus in place of the two separate protons that existed before. At the same time, the energy of their violent collision is released to the surroundings in the form of heat and light. It turns out that the amount of energy released in this way, by the

**153**

fusion of two protons, is nearly as great, *per pound of protons,* as the energy released per pound of uranium in the fission of uranium.

## FUSION REACTIONS IN STARS

Once a large number of protons has fused together within a star to form deuterons (p. 152), a new set of reactions begins. A deuteron produced by fusion can collide and fuse again with another proton, forming the three-particle nucleus, He³, the light isotope of helium (Figure 6.4).

*Figure 6.4 Fusion of a proton and a deuteron.*

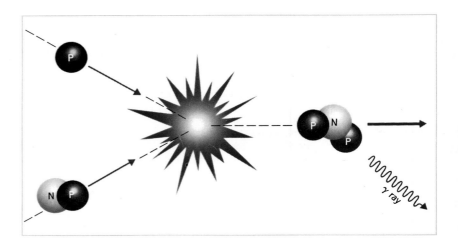

At the same time, the energy released by the fusion escapes in the form of an energetic photon, called a gamma ray.

Since protons are so abundant in stars, this reaction is quite likely to occur once an appreciable number of deuterons has been formed.

The shorthand form for the proton-deuteron reaction is:

$$d + p \longrightarrow \text{He}^3 + \gamma$$

The Greek letter $\gamma$ (gamma) on the right-hand side of the equation is the symbol usually employed to represent the gamma ray. The gamma ray in the above reaction carries off most of the energy released in this particular fusion process, whereas the positron and neutrino carried off the energy in the proton-proton reaction. The lightest particles emerging from a reaction always carry off most of the energy.

Many experiments have been performed in nuclear laboratories in which a target containing deuterium, such as heavy water, is bombarded with fast-moving protons. The experiments show that a He³ nucleus is frequently formed as a result of the bombardment, verifying that the fusion of deuterons and protons does actually occur, and is not just a theoretical possibility.

What other reactions are possible as the result of a collision between a deuteron and a proton? You might think of:

$$d + p \longrightarrow H^3 + e^+ + \nu$$

In theory this reaction also is possible and, in fact, it can occur, but both experiments and calculations show that it is far less probable than the formation of $He^3$.

Nuclear physicists have discovered, as a result of many experiments performed with cyclotrons and other nuclear accelerators, that nuclear fusion reactions take place in *all* the light and moderately light elements, including not only hydrogen and deuterium, but also carbon, nitrogen, and the other elements at the beginning of the periodic table. They have found also that the incoming particle in a fusion process does not have to be a proton. For example, two of the $He^3$ nuclei produced in the reaction on page 154 may fuse together to form a $He^4$ nucleus with two protons emitted in the process:

$$He^3 + He^3 \longrightarrow He^4 + 2p$$

It turns out that an exceptionally large amount of energy is released by this reaction. Most of the released energy is carried off by the two protons.

Suppose that two $He^4$ nuclei collide; will they fuse together to form a nucleus, and what nucleus will that be? Since the resultant nucleus has four positive charges, it is the nucleus of the element beryllium. Thus, the result of the fusion of two $He^4$ nuclei is a $Be^8$ nucleus (four protons and four neutrons):

$$He^4 + He^4 \longrightarrow Be^8$$

But the $Be^8$ nucleus is not stable; although it can exist for a short time, the particles out of which it is made do not attract one another strongly enough, and in less than one-trillionth of a second, it breaks up into two separate helium nuclei again.

However, if other $He^4$ nuclei are present, one of them may collide with the beryllium nucleus before it breaks apart, and fuse with it to form a nucleus with six neutrons and six protons, that is, a carbon-12 nucleus:

$$He^4 + Be^8 \longrightarrow C^{12} + \gamma$$

The carbon nucleus is tightly bound and very stable, and lasts forever once it is formed (unless, of course, it is disturbed by another nuclear collision).

If many $He^4$ nuclei are present, it also is possible for the $C^{12}$ nucleus to be formed in one step, by the fusion of three helium nuclei:

$$3He^4 \longrightarrow C^{12} + \gamma$$

However, a simultaneous triple collision is a very unlikely event. Actually, it is much more likely for the $Be^8$ nucleus to be formed first, and the $C^{12}$ nucleus to be formed immediately afterward by a collision between beryllium and helium nuclei, in spite of the fact that the beryllium nucleus stays together for such a short time.

The fusion of three helium nuclei to form carbon, like the fusion reactions that form helium itself, releases very large amounts of energy. The energy is carried off by the gamma rays emitted in the fusion reaction. If the nuclei of both carbon and helium are present, a collision between the two nuclei may occur and also lead to a fusion. This fusion would create an oxygen nucleus plus a substantial amount of nuclear energy which is carried off by a gamma ray:

$$He^4 + C^{12} \longrightarrow O^{16} + \gamma$$

Similar reactions occur further along in the periodic table. Experiments show that protons fuse with chlorine, for example, to produce argon, simultaneously emitting a gamma ray in the process:

$$p + Cl^{35} \longrightarrow A^{36} + \gamma$$

Or a helium nucleus may fuse with a silicon nucleus to produce sulphur:

$$He^4 + Si^{28} \longrightarrow S^{32} + \gamma$$

A chain of reactions can take place by which heavier nuclei are built up from the original nucleus by the successive addition, one after the other, of "small" particles, such as protons, neutrons, deuterons, and helium nuclei. For example, a neutron striking a fluorine nucleus can be captured by it to produce a heavier isotope of fluorine:

$$n + F^{19} \longrightarrow F^{20}$$

But the heavier fluorine isotope, $F^{20}$, is unstable; that is, it is like an atom in an excited state, and it collapses to its ground state by emitting an electron plus a neutrino:

$$F^{20} \longrightarrow Ne^{20} + e^- + \nu$$

just as an excited atom collapses to the ground state by emitting a photon. The emission of the electron in the above reaction has the same effect as changing one of the neutrons into a proton, thereby converting the fluorine nucleus (nuclear charge = 9) into a neon nucleus (nuclear charge = 10).

A helium nucleus can now fuse with or be captured by the neon nucleus to form magnesium:

$$He^4 + Ne^{20} \longrightarrow Mg^{24} + \gamma$$

All the elements of the periodic table can be built up from the proton in this way, by successively adding small increments of mass and electric charge in the form of protons, neutrons, or small combinations of the two. In fact, the increments of mass and charge need not be small. Two nuclei of fairly substantial size, such as carbon or oxygen, will fuse together if they collide with sufficient violence to form a single and much more massive nucleus. A collision between two carbon nuclei, for example, produces magnesium:

$$C^{12} + C^{12} \longrightarrow Mg^{24} + \gamma$$

and a collision between two oxygen nuclei yields sulphur:

$$O^{16} + O^{16} \longrightarrow S^{32} + \gamma$$

Quite often in these fusion reactions, especially when the reaction involves a collision between two good-sized nuclei, nuclear fragments, such as one or two protons, neutrons, or alpha particles, fly off. In a collision between two carbon nuclei, a proton or a helium nucleus is apt to fly off, leaving behind sodium or neon, respectively:

$$C^{12} + C^{12} \longrightarrow Na^{23} + p$$

$$C^{12} + C^{12} \longrightarrow Ne^{20} + He^4$$

In the reaction involving a collision between two oxygen nuclei, the results shown by the equations below are frequently observed in the laboratory. In addition to the formation of sulphur, phosphorus ($P^{31}$) may be formed.

$$O^{16} + O^{16} \longrightarrow P^{31} + p$$

$$O^{16} + O^{16} \longrightarrow S^{31} + n$$

$$O^{16} + O^{16} \longrightarrow Si^{28} + He^4$$

Reactions between heavy nuclei can also take place if the nuclei are of two different types. For example, carbon and oxygen can collide to produce silicon:

$$C^{12} + O^{16} \longrightarrow Si^{28}$$

## Conversion of Mass into Energy in Nuclear Reactions

The amount of energy released in a nuclear fusion or fission reaction is so great that it causes an appreciable reduction in the masses of the nuclei that are left behind. According to Einstein's law on the equivalence of mass and energy,

$$E = mc^2$$

when an amount of energy $E$ is given off in a nuclear reaction, the mass of the nucleus that is left behind must be smaller by the amount $E/c^2$, where $c$ is the speed of light. For example, consider the reaction in which two protons combine to form a deuteron:

$$p + p \longrightarrow d + e^+ + \nu$$

If we add the masses of the two protons on the left-hand side, and compare their total mass with the masses of the deuteron and the positive electron on the right-hand side (remembering that the neutrino has no mass), we find that the total mass after the reaction is smaller than the mass before the reaction:

$$2 \times \text{mass of proton} = 2 \times 1.67243 \times 10^{-24} \text{ g} = 3.34486 \times 10^{-24} \text{ g}$$

$$\left\{ \begin{matrix} \text{mass of deuteron} \\ + \text{ mass of positron} \end{matrix} \right\} = \left\{ \begin{matrix} 3.34321 \times 10^{-24} \text{ g} \\ + 0.00091 \times 10^{-24} \text{ g} \end{matrix} \right\} = 3.34412 \times 10^{-24} \text{ g}$$

$$\text{difference in mass} = 0.00074 \times 10^{-24} = 7.4 \times 10^{-28} \text{ g}$$

The difference, $7.4 \times 10^{-28}$ g, represents mass that has disappeared in some way during the course of the reaction. This difference seems very small. However, if we compare it with the masses of the fundamental particles, we see that the missing mass is nearly as great as the mass of an electron.

In some way, the equivalent of nearly an entire electron has disappeared during the reaction. This seems impossible. Matter simply cannot disappear into nothing; at least, it never does in ordinary experience. In fact, the conclusion that matter is indestructible became so firmly established in the early years of science that by the beginning of the nineteenth century it had been adopted as one of the basic principles of physics. It was called the Law of the Conservation of Mass.

Yet the law is wrong; and the fact that it is wrong is proved simply by looking at Einstein's formula for the equivalence of mass and energy. Einstein's formula says that mass alone cannot be conserved; if anything, it must be instead the *sum* of mass and energy that is conserved.

Many experiments have proved that this is the correct law. *The sum of mass and energy is indestructible.* In any event occurring anywhere in the world, the total of the masses and the energies of the particles taking part in that event remains unchanged, regardless of the way in which the particles themselves are changed.

In this way, an object can literally disappear from the world, or another object can appear where there was none before; the laws of physics do not place any limitations on such conjuring acts; but they *do* require that whenever an object, large or small, vanishes, an amount of energy equal to $mc^2$ (where $m$ is the object's mass) must appear in its place.

Einstein's genius led him to this result and, furthermore, supplied the famous equation that tells precisely how much energy is yielded when mass disappears, and how much energy must be provided to create new mass.

According to Einstein's equation, every reaction that releases energy must involve a corresponding loss of mass. This is as true for a chemical reaction, such as the explosion of nitroglycerine, as it is for the explosion of a nuclear bomb. In both cases, the masses do not balance before and after the reaction, because some mass is always carried off in the form of energy. In the case of chemical reactions, however, the amount of mass carried off as energy is too small to be measured directly. This is why Einstein's formula was not discovered by laboratory experiments prior to the advent of his theory.

For example, the results of a laboratory measurement show that approximately $10^{-10}$ ergs are released for each nitroglycerine molecule. Dividing by $c^2$ ($9 \times 10^{20}$), this energy release is equivalent to roughly $10^{-31}$ grams. The masses of the molecules involved in the reaction total about $10^{-22}$ grams. Thus, the mass carried off in the form of energy is one-billionth, or 0.0000001 percent, of the masses of the separate molecules. The mass of a molecule cannot be measured with an accuracy sufficient to detect changes as small as this.

## CONDITIONS FOR NUCLEAR REACTIONS IN STARS

Why is nuclear fusion not used as an energy source in everyday life? Energy has been released through nuclear fusion for brief moments in the explosion of the hydrogen bomb, but no one has yet succeeded in fusing nuclei in such a way that the energy can be harnessed for constructive purposes. The difficulty is that enormous temperatures, ranging up to tens of millions of degrees, are needed to produce a significant amount of energy by nuclear fusion.

The need for a high temperature is connected with the electrical forces between nuclei. Two protons, for example, repel one another electrically because each proton carries a positive electric charge. But if the protons approach within a very close distance of each other, the electrical repulsion gives way to the even stronger force of nuclear attraction. However, the protons must be closer together than one 10-trillionth of an inch for the nuclear force to be effective. Under ordinary circumstances, the electrical repulsion serves as a barrier to prevent as close an approach as this. In a collision of exceptional violence, however, the protons may pierce the barrier which separates them, and come within the range of their nuclear attraction. Collisions of the required degree of violence begin to occur frequently in a gas when the temperature of the gas reaches 10 million degrees Kelvin.

Once the electrical barrier between two protons is pierced in a collision, they pick up speed as a result of their nuclear attraction and rush toward each other, fusing together in the reaction described on pp. 152–153.

The fusion of two protons into a single nucleus is only the first step in a series of reactions by which nuclear energy is released during the life of the star. In subsequent collisions, two additional protons are joined to the first two to form a nucleus containing four particles. Two of the protons shed their positive charges to become neutrons in the course of the process. The result is a nucleus with two protons and two neutrons. This is the nucleus of the helium atom. Thus, the sequence of reactions transforms protons, or hydrogen nuclei, into helium nuclei.

Helium does not fuse into heavier nuclei at the ordinary stellar temperature of 10 million degrees because the helium nucleus, with two protons, carries a double charge of positive electricity, and, as a consequence, the electrical barrier between two helium nuclei is stronger than the repulsion between two protons. A temperature of 100 million degrees Kelvin is required to produce collisions which will pierce the helium barrier. If the temperature in a star reaches this level, helium nuclei will begin to fuse in groups of three to form carbon nuclei, releasing more nuclear energy in the process.

The fusion of hydrogen to form helium is the first and longest stage in the history of a star, occupying about 99 percent of its lifetime. In the second stage, which takes up most of the remaining 1 percent of the star's life, three nuclei of helium combine to form the nucleus of the carbon atom. Afterwards, the nuclei of oxygen and other still heavier elements are fabricated, at an increasingly rapid pace, until all elements have been built up. In this way the elements of the Universe are manufactured out of hydrogen nuclei at the center of the star during the course of its life.

These facts of nuclear physics complete the essential body of information needed for an understanding of the life of a star. The star's existence begins in the tenuous clouds of matter that fill all of space. If the atoms in such a cloud come together by accident, the force of gravity pulls the atoms still closer together, forming a condensed pocket of gas. The continuing action of gravity compresses the pocket of gas. As a result of the compression the temperature at the center rises, and after about 10 million years, when the temperature has reached the critical value of 10 million degrees, nuclear fusion of hydrogen commences, with the release of vast amounts of energy. The release of nuclear energy halts the further collapse of the star. The energy passes to the surface and is radiated into space.

The continuing succession of nuclear reactions manufactures all the other elements of the Universe out of the basic ingredient, hydrogen. Eventually these nuclear reactions die out, and the star's life comes to an end. Deprived of its resources of nuclear energy, it collapses under its own weight. Sometimes in the aftermath of the collapse an explosion occurs, spraying out to space all the materials that have been created within the star during its lifetime. There they mix with the primordial hydrogen.

In the course of time, new stars, some with planets around them, condense out of the enriched mixture. The sun and the earth were formed in this way, four and a half billion years ago, out of materials manufactured in the bodies of other stars earlier in the life of the Galaxy, and then dispersed to space when those stars exploded.

The essential feature in this sequence of events is the transmutation of hydrogen into heavier elements. This contained transmutation has never been duplicated on the earth, except in the negligible atom-by-atom quantities of the nuclear accelerator. The difficulty is that no furnace has yet been constructed on the earth whose walls can contain a fire at the temperature of the millions of degrees necessary to produce nuclear fusion. The only furnace that can do this is provided by nature in the heart of a star.

## Questions

1. A 100-watt light bulb radiates about $10^9$ ergs/sec. How many grams of mass does it lose in one second? In one billion years?

2. The mass of the Galaxy is believed to be about $2 \times 10^{44}$ grams. Its luminosity is believed to be $1.3 \times 10^{52}$ ergs/year. How much mass has been radiated away in the $10^{10}$ years of existence?

3. Estimate the luminosity of a star with a mass of $2 \times 10^{34}$ grams from Figure 5.12. If the star is composed of pure hydrogen, how long can it live by burning hydrogen to form helium? On the basis of the same assumption, what is the lifetime of the sun? Of Barnard's Star?

4. How many protons are fused into helium every second to produce the energy radiated by the sun into space?

5. Why must the temperature of a star's interior be hotter for two helium nuclei to fuse than for two protons to fuse?

6. What general conditions of density and temperature are required for nuclear reactions to take place?

7. What would happen to the sun if nuclear reactions were suddenly extinguished?

8. Describe a Universe in which there were no electrical forces; no nuclear forces; no gravitational forces.

# 7 Stellar Evolution

A star's life begins in the swirling mists of hydrogen that surge and eddy in the space between the stars. The photograph on the facing page shows one of these clouds, located about 4000 light years from the solar system in the direction of the constellation Monoceros. In the random motions of such a cloud, atoms sometimes come together by accident to form small, condensed pockets of gas. Stars are born in these accidents.

## BIRTH

Let us concentrate our attention on three atoms of hydrogen that happen to be neighbors of one another in an interstellar cloud. These atoms are labeled by the numbers 1, 2, and 3 in Figure 7.1. Let us suppose that at a given moment the three atoms happen to be moving toward one another, as shown by the arrows in Figure 7.1a. A short time later, the three atoms will have come together as a consequence of their motions, as shown in Figure 7.1b. They now form a small pocket of condensed gas

*Stars in formation.*

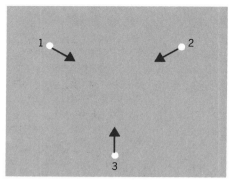

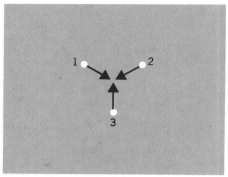

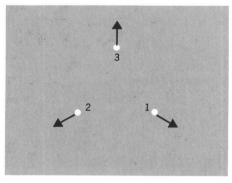

*Figure 7.1 (a), (b), and (c). Random motions of atoms in space.*

in space.

Because the atoms are very small, they rarely collide. Usually they pass by one another, in the course of the same motions that brought them together, and separate again, as shown in Figure 7.1c. As a result, the condensed pocket of gas vanishes.

As the atoms pass one another, however, each atom exerts a small gravitational attraction on its neighbors, which counters the tendency of the atoms to separate. If the gravitational attraction were sufficiently strong, it would hold the three particles together and prevent them from dispersing again into space. The effect of gravity would convert the pocket of gas from a temporary condensation to a permanent one.

Gravity, however, is not a very strong force; it is the weakest natural force known to man. Under normal conditions, the force of gravity that a few atoms exert on each other is not strong enough to slow down their motions and tie them permanently to one another.

But suppose that there are not merely three, four, or five atoms in the pocket of gas, but a very large number. Because the force of gravity extends over great distances, each atom feels the gravitational pull of all the other atoms in the pocket. If the number of atoms is sufficiently large, the combined effect of all these minute pulls of gravity will be powerful enough to prevent any of the atoms in the pocket of gas from leaving the pocket and flying out into space again. The pocket becomes a permanent entity, held together by the mutual attraction of all the atoms within it upon one another.

This is the heart of the theoretical explanation of the birth of stars. According to the theory, a star is conceived when a condensed pocket of gas forms by accident in outer space, and when the number of atoms in the pocket of gas is so great that their own gravity holds them together permanently. The pocket is not yet a star, but it will become one a little later. This tight cluster of atoms formed by accident and held in the grip of its own gravity, is called a protostar.

## Protostars

How large must a cluster of atoms be before its own gravity is strong enough to hold it together permanently? If three or four are not enough, will a million atoms or a trillion atoms suffice? Theoretical astronomers, using pencil and paper and the laws of physics, have calculated the number of atoms that are necessary. Their results show that the answer depends strongly on the temperature of the gas, which controls the speed at which the atoms move about. Clearly, the higher the temperature and the higher the speed of the atoms, the more difficult it is for gravity to hold them together.

Under almost all conditions in space, however, the required number of atoms is much larger than any of the numbers we have mentioned. Under average conditions the theoretical result is that about $10^{57}$ atoms are required. This is a staggeringly large number. We have searched for a comparison that would make the meaning of the number $10^{57}$ clear, such as comparing $10^{57}$ with the number of grains of sand in all the beaches of the world; but even that seemingly uncountable number is only a mere $10^{25}$. In fact, the number of neutrons and protons contained in the nuclei of all the atoms of the entire earth is only $10^{51}$. The trouble is that no number on earth can possibly match this number; it is a number that corresponds to the building blocks of objects the size of stars, not objects the size of planets.

If we translate the $10^{57}$ atoms into grams, assuming that each atom is hydrogen, we find that the cloud weighs approximately $10^{33}$ grams, which is roughly the mass of the sun. Thus, it turns out that the number of atoms theoretically required to hold a pocket of gas together in space has a mass equal to the mass of an average star such as the sun. This agreement provides evidence that the theory is correct — that the process of gravitational condensation we have described is, in fact, the way in which stars are born in space.

Agreement about one number would not by itself be completely convincing evidence. However, photographs of large clouds of gas in certain regions of the sky show small, dark globules of gas that look just like

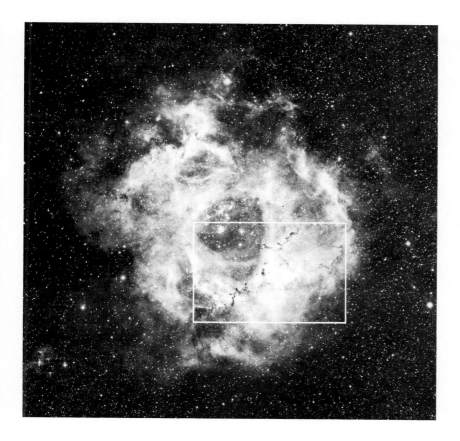

Figure 7.2 The Rosette Nebula, NGC 2237.

stars, or small clusters of stars, in the process of formation. An example is the large cloud of gas and dust in the direction of the constellation Monoceros (The Unicorn), known to astronomers as Nebula NGC 2237. Figure 7.2 shows the entire nebula, which is roughly 100 light-years in diameter. The luminous clouds of the nebula are collections of hydrogen atoms that have been heated by the absorption of radiation from hot, young stars embedded in their midst.

The area outlined in white in Figure 7.2 is shown in Figure 7.3 in an enlarged version. In addition to being enlarged, the second photograph has also been exposed for a shorter time; thus the luminosity of the main cloud does not obscure so many of the fine details of the structure of the nebula. Among these fine details are a number of small, dark spots. These are the clusters of atoms we call protostars. The arrows point to two of the protostars.

*Figure 7.3    Detail in NGC 2237.*

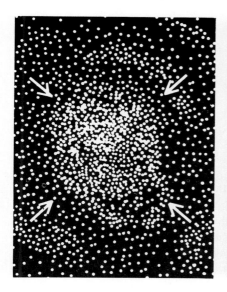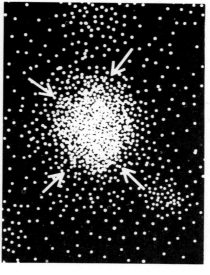

Figure 7.4 Gravity collapses a protostar. (a) The protostar begins to condense. (b) The protostar has collapsed to a small radius and high density.

The famous and beautiful Orion Nebula (Color Plate 4) is still another region that astronomers consider to be a breeding ground for new stars. Embedded in the relatively dense gas of the Orion Nebula are a number of massive and very hot stars that are known to have formed recently. Their ultraviolet radiation strikes the surrounding atoms of hydrogen in the Orion Nebula, heating them and causing them to glow with the beautiful red, blue, and mauve colors of this nebula. The hot stars that can be seen in the Orion Nebula are already formed, but astronomers are certain that many other stars are in the process of formation in this dense concentration of interstellar gas.

## The Rising Temperature of a Protostar

The pockets of gas shown in Figure 7.3 are not yet stars; they are only protostars. How does a protostar—a tenuous collection of hydrogen atoms drawn together out of the cold gas of space—become the dense, flaming sphere of gas we call a star? The answer depends again on the force of gravity, which draws every atom in the protostar toward the center of the cloud (Figure 7.4). The continuing action of gravity, pulling all the atoms toward the center, causes the protostar to shrink in size (Figure 7.4a). As it becomes smaller, its density increases (Figure 7.4b).

This force toward the center is not a mysterious or unfamiliar phenomenon. Every object on the face of the earth is attracted toward the center of our planet by the force of the planet's gravity. In the case of the earth, the solid surface on which we stand prevents us from falling to the center. A protostar, however, is not solid; it is a globe of gaseous matter, and its atoms are unimpeded by the resistance of a solid surface. These atoms literally "fall" toward the center of the protostar. The protostar collapses under its own gravitational attraction.

**167**

As the atoms in the protostar move toward its center they pick up speed, like any falling object. Because the average speed of the atoms in a gas determines the temperature of that gas, the contracting protostar with its accelerating atoms gets hotter.

At the beginning, the temperature of the protostar is the same as the temperature of the interstellar gas out of which it formed. This temperature is roughly 100°K, or 280 degrees below zero Fahrenheit. At 100°K the average speed of an atom of hydrogen is one mile per second. As the gas cloud contracts under the attraction of its own weight, the temperature at the center mounts steadily eventually reaching 50,000° K.

At a temperature of 50,000°K, the hydrogen and helium atoms at the center of the protostar collide with sufficient violence to dislodge all electrons from their orbits around the nuclei. The original gas of atoms, each consisting of an electron circling around a nucleus, becomes a mixture of two gases, one composed of electrons and the other of nuclei.

At this stage the globe of gas has contracted from its original size, which was trillions of miles in diameter, to a diameter of 100 million miles. To understand the extent of the contraction, imagine the Goodyear blimp shrinking to the size of a grain of sand.

When the temperature near the center of the protostar is 50,000° K, the protons in the interior of the globe of gas move at a speed of 20 miles per second. This velocity is still far from adequate to penetrate the electrical barrier and initiate a chain of nuclear fusion reactions. In a sense, the true birth of the star has not yet occurred. Nonetheless, the protostar has already become a luminous object because of the heat generated by its collapse. Some astronomers consider that as soon as the collapsing ball of gas becomes luminous it should be called a star, but we will continue to refer to it as a protostar and reserve the title of star until the time when nuclear reactions first ignite at the center.

After still more time has passed, the protostar has shrunk to 50 million miles, its internal temperature has risen to 150,000°K, and its surface temperature has risen to 3500°K. At this stage, the protostar is a highly luminous object, hundreds of times more luminous than the sun, even though its surface is considerably cooler and redder than the surface of the sun. There is, however, no paradox in the fact that a relatively cool object should be highly luminous. The explanation is that the protostar is 50 times larger in diameter than the sun and has 2500 times more surface area. Each square centimeter of its surface radiates less energy because the surface temperature is lower, but the total amount of energy radiated from the great surface area of the protostar is enormous.

At this point the protostar makes its debut on the Hertzsprung-Russell diagram. The combination of low surface temperature and high luminosity places the protostar in the upper right-hand corner, in approximately the same part of the H-R diagram as the red giants. But it has come to that region by a path very different from that of the red giants, which are not newly-forming stars, but ageing stars.

At a surface temperature of 3500°K, the protostar should emit enough radiation in the visible region of the spectrum for astronomers on earth to see it with a telescope. However, because a protostar stays in this visible stage for such a relatively short time, very few of these objects will be in the heavens at any given time. Therefore, astronomers see them very rarely. When they are seen, they appear as globes of gas glowing with a deep red color.

About 20 years ago there occurred one of the rare instances in which a young protostar may actually have been seen. The evidence is in Figure 7.5. The first photograph (a), taken in 1947, showed several small, luminous globes of gas located in the constellation Orion (The Hunter). They looked like newly forming stars, but there was no definite evidence to prove that this was in fact what they were. However, a photograph of the same region taken in 1954 (b) showed the luminous globes that were present in 1947 and, in addition, another luminous globe that was not visible in the 1947 photograph. Apparently a new cloud had formed and commenced to collapse, heating up sufficiently in the space of seven years to become visibly luminous in the later photographs. Each arrow

*Figure 7.5   Formation of new stars. Left (a), 1947; right (b), 1954.*

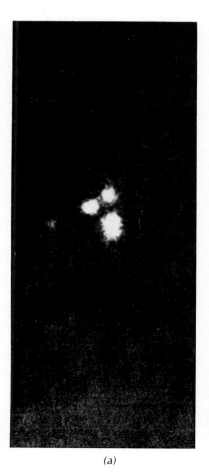

(a)

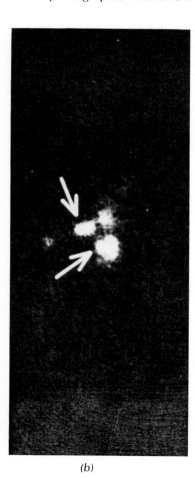

(b)

points to a luminous object—the protostar—that was not visible in the picture taken seven years before.

*Figure 7.6   The track of a new star across the H-R diagram.*

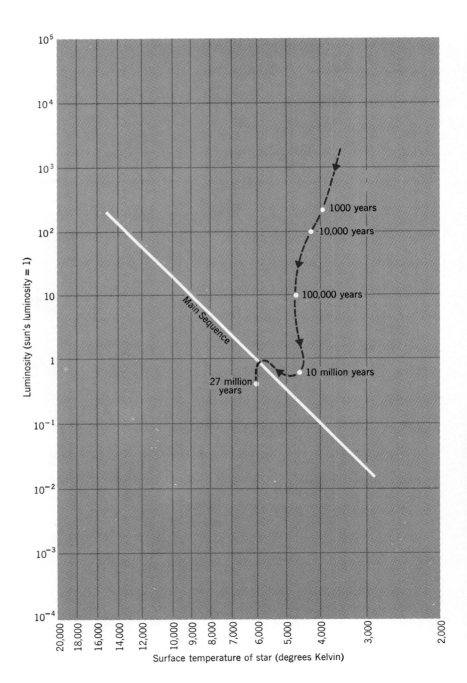

## Protostars on the H-R Diagram

The evolutionary path of a contracting protostar is shown in figure 7.6 as it appears on the Hertzsprung-Russell diagram. This is the path of a star with the same mass as the sun. The age of the protostar is marked at successive points along the diagram, starting three years after the initial formation of the gas cloud and ending when the protostar is ten million years old. You can see that the protostar moves down the H-R diagram very quickly at first, and very slowly later on. The globe of gas reaches the vicinity of the Main Sequence 27 million years after it begins to collapse.

The protostar track plotted on this H-R diagram is not based on direct observations of protostars. As we have noted above, a protostar flashes through the early years of its existence with enormous rapidity and does not linger long enough to be observed, except by a lucky accident. Theoretical astronomers, however, are able to calculate the size, brightness, and temperature distribution of a contracting protostar with the aid of the new high-speed electronic computers. The evolutionary track of the protostar shown in the H-R diagram on page 170 is the result of one of these computations.

Approximately three billion separate additions, subtractions, multiplications, and divisions are needed to calculate an evolutionary track like the one shown here. The latest and fastest computers, with a speed of several million operations per second, can accomplish this task in five minutes at a cost of a few dollars in rental time. If attempted by hand with the aid of desk calculators, the same calculations would take 3000 mathematicians 5 years to finish and would cost more than 10 million dollars. The invention of the high-speed computer is one of the reasons why we have made great progress in our understanding of the birth of stars in the last few years.

## Stars Out of Protostars

Suppose we concentrate on the 10-million-year mark in the evolutionary track. The protostar has been collapsing since its formation, rapidly at first, and then more slowly as its density has built up and the atoms moving toward the center have met with increasing resistance. After 10 million years the protostar has shrunk from its original size of trillions of miles to a diameter of about 1.5 million miles. This diameter is close to, but slightly greater than, the size of the sun. At the same time, the temperature at the center of the protostar has risen to 10 million degrees Kelvin.

Ten million degrees Kelvin marks a critical threshold in the life of the collapsing sphere of gas. Why is this temperature critical? At this temperature, for the first time, the protons at the center of the protostar are finally moving and colliding at speeds great enough to penetrate the electrical barrier and come within the reach of the nuclear force of attraction. At this point nuclear fusion sets in at the center of the sphere. The protostar has become a star.

The surface temperature of the new-born star is 3600°K, and its luminosity is half the luminosity of the sun. With the release of nuclear energy at its center, the star becomes somewhat hotter and more luminous. It also shrinks slightly, to a diameter of one million miles. As these events proceed, the star moves upward and to the left on the Hertzsprung-Russell diagram until, 17 million years after the onset of nuclear fusion and 27 million years after the collapse of the protostar began, it comes to its resting place on the Main Sequence. Here it lives out most of the remainder of its life in a balance between the inward pressure created by the force of gravity and the outward pressure generated by the release of nuclear energy.

## MATURITY

The fusion of hydrogen nuclei produces helium. This reaction dominates the longest single stage in the history of a star, about 99 percent of its lifetime in the case of a star of solar mass. The sun is in the middle of this stage; it came onto the Main Sequence 4.5 billion years ago, and will remain there for another four or five billion years before it moves off to die.

It is fortunate for us that the sun burns its hydrogen so slowly, for the evolution of life on the earth has also been a very slow process. According to the fossil record, simple forms of life, such as bacteria, appeared on the earth sometime during the first billion years of the solar system's existence; advanced forms of life did not emerge until several billion years thereafter. If the sun's lifetime had been, let us say, 100 million years or less, it is doubtful whether intelligent creatures would ever have populated the earth.

Although the sun will live for a total of 9 or 10 billion years, other stars live for as short a time as a million years, and still others may live for as long as a trillion years or more.

Surprisingly, the largest stars live for the shortest time. They have more fuel to burn, but they burn it much more rapidly than the smaller stars. For example, a star ten times as massive as the sun has ten times as much fuel to burn, but it burns it at 1000 times the rate of the sun, so that its lifetime is 100 times shorter; that is, it lives for only 100 million years. Why does a massive star burn its fuel so much more rapidly than a smaller star? The great weight of such a star generates higher temperatures at its center, causing the protons to collide more violently than they do in a lighter star. Under the violence of these collisions, the electrical barrier between protons is more readily penetrated and the nuclear reaction rate goes up. In the more massive stars, the reaction rate climbs so rapidly with increasing temperature that a doubling of the temperature multiplies the reaction rate by a factor of 30,000.

On the other hand, a star with a mass one-tenth that of the sun should live for a trillion years. Barnard's Star is an example. It should still be shining long after the sun has gone out. Stars smaller than Barnard's Star are known to exist and should live for even longer times. These small stars live for such a long time that not a single one has died since the Galaxy itself came into existence, an event that occurred, according to the latest evidence, about 10 billion years ago.

## Ageing Stars

The helium produced by fusion of hydrogen atoms gathers at the center of a star. When an appreciable amount of the hydrogen within the star has been converted into helium, and its center is filled with a core of pure helium, the star begins to show pronounced signs of age.

The first major change involves the conditions in the helium core. Because there is no hydrogen burning at the center, no nuclear energy is released there. The central region of the star, which had been supported against gravitational collapse by the release of nuclear energy, no longer possesses the means for sustaining itself against the inward force of gravity. Under the influence of gravity the helium core shrinks, and its temperature rises, just as the temperature of the entire star rose when it was collapsing at the beginning of its life. The center of the star now consists of a core of helium that is collapsing and steadily heating up. The rest of the star is a shell of hydrogen surrounding the helium core. As the core gets hotter, it heats the hydrogen immediately surrounding it; this hydrogen commences to burn vigorously to form helium.

The structure of the star now differs markedly from its structure during the prime of its life on the main sequence. Whereas formerly hydrogen burned only at the center, now a core of nonburning helium lies at the center, and the only hydrogen that is burning lies in the shell surrounding the helium core.

As the helium core gets hotter under the continued compression caused by gravity, the nuclear reaction rate mounts higher and higher in the shell of burning hydrogen around it. The result is an increase in the amount of nuclear energy created within the star.

It seems surprising that the extinction of the nuclear fire at the center of the star should lead to an increase in the outpouring of nuclear energy. However, you must remember what has happened: when the fire goes out at the center of the star, the helium in the core contracts and heats up. The higher temperature in the core, in turn, raises the temperature of the immediately surrounding hydrogen. The end result is that the hydrogen in the shell around the core blazes more brightly than the hydrogen in the center did before.

Because nuclear energy is now released at a faster rate, we expect that the star will become brighter. The luminosity, however, does not increase immediately. Initially, the increased radiation coming from the center is absorbed by the hydrogen gas in the outer layers of the star before this radiation reaches the surface. The absorption of the radiation heats the gas, which expands outward. An outward expansion means that huge masses of gas are lifted upward against the powerful downward pull of the star's gravity. Most of the increase in nuclear energy is used up in lifting trillions on trillions of tons of hydrogen through distances as great as hundreds of thousands of miles. As a result, the luminosity of the star, which is the amount of radiation actually escaping from its surface every second, remains nearly constant. This is the *constant-luminosity* phase of the star's life.

But the star's appearance changes in another respect: as its size increases and its luminosity remains constant, the amount of energy radiated at each square centimeter of the surface drops, and the surface temperature drops to a value between 3000°K and 4000°K. A star with a surface temperature in this range is distinctly red in color.

As time passes, the helium core continues to collapse and, as a consequence, its temperature continues to rise. Eventually, the rate of hydrogen burning becomes so great, because of the steadily mounting temperature, that the brightness of the star is noticeably affected. The envelope of the star absorbs some of this energy and expands even more rapidly than before, but this outward expansion can no longer absorb the enormous outpouring of radiation that comes from the brightly blazing hydrogen shell. The rate of energy release within the star is now hundreds of times greater than it was when the star was in the prime of its life. Only a small part of this energy can be taken up in the expansion of the star's envelope; most of the energy reaches the surface of the star and escapes as radiation. The luminosity of the star, accordingly, soars upward. The star, still red in surface color, has become brilliantly luminous; it has become a red giant.


### From Young Stars to Red Giants on the H-R Diagram

What happens to the plot of the star on the H-R diagram during these changes? We last left the star on the H-R diagram as it reached the Main Sequence, just after its transition from protostar to star. Let us follow the star's evolution all the way from this point, close to the beginning of its life, up to the time it becomes a red giant.

We will use a star with the mass of the sun as our example. The sequence of changes is similar for stars with other masses, the main differences being in the initial position on the Main Sequence, and the time scale for the changes. If the star is more massive than the sun the Main Sequence position is higher, and if the star is less massive than the sun the Main Sequence position is lower. The time scale is shorter for a star more massive than the sun and longer for one less massive.

The initial radius of a sun-sized star at birth is 750,000 miles. The star lies slightly below the Main Sequence at birth, its luminosity being 70 percent of the present luminosity of the sun. The circled "1" in the H-R diagram in Figure 7.7 indicates the traditional "zero age" point in the star's evolution.

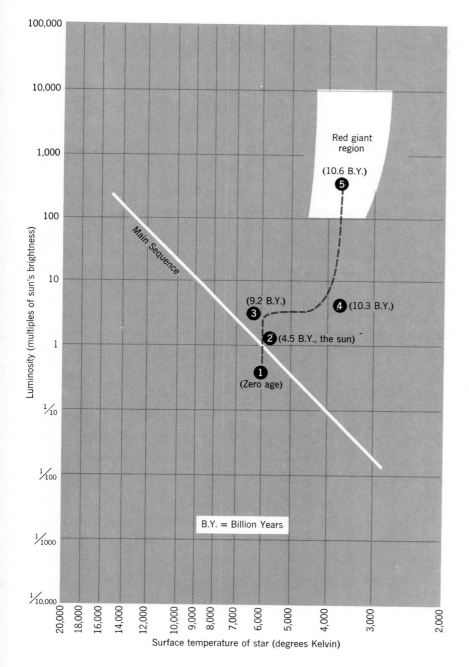

Figure 7.7 The track of a star from the Main Sequence to the red giant region.

During the early part of the star's life, its luminosity slowly increases, and its track moves vertically upward on the H-R diagram. After 4.5 billion years, the luminosity equals the present luminosity of the sun. The star now lies precisely at the present position of the sun, in the middle of the Main-Sequence band of stars. The circled point "2" in Figure 7.7 indicates this point in the star's life.

The diameter of the star also has increased slowly but steadily during this period, and it is now one million miles. The increase in diameter roughly compensates for the increase in luminosity, so that the surface temperature remains approximately constant between point 1 and point 2.

When the star is at point 2 it is still burning hydrogen at its center. Not enough helium has accumulated to form a helium core.

During the next four billion years, the star continues to grow slowly in luminosity and in diameter. When its age is 9.2 billion years, its luminosity is roughly one and one-half times greater than the present luminosity of the sun, or double its luminosity at birth. Its diameter is 1.3 million miles, or 30 percent greater than the sun's present diameter. This stage is shown by the circled "3" in the H-R diagram.

Beyond point 3, the star enters its constant-luminosity phase. The helium core has formed and begun to influence the life of the star. All processes within the star occur very quickly from point 3 onward. Up to point 3 the star grew slowly, taking nine billion years to double in size; but now it doubles in size again in one billion years. That is, the star expands about ten times faster than it did earlier. With this rapid growth, the star takes its first major step away from the Main Sequence. Because its luminosity remains constant and its temperature decreases rapidly at this point, the star shoots horizontally across to the right on the H-R diagram. This movement brings the star to point 4 on the H-R diagram. Its diameter is 2.6 million miles at point 4, and its age is 10.3 billion years.

Figure 7.8 compares the structure of the star at point 4 with its structure

*Figure 7.8(a) and (b). The structure of a hydrogen-burning star.*

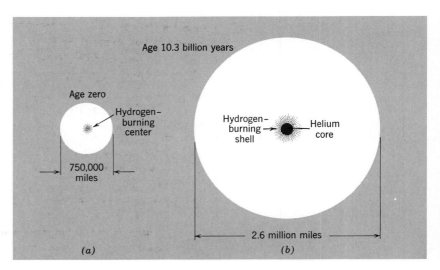

when it started life on the Main Sequence. The size of the helium core is exaggerated to show it more clearly; in reality it is one one-hundredth of the diameter of the star.

Beyond point 4 the luminosity of the star mounts rapidly in response to the greatly increased rate of hydrogen burning. Its diameter also increases very rapidly in this stage, so rapidly that, in spite of the increased luminosity, the surface temperature drops. The star shoots almost vertically upward on the H-R diagram, toward the red giant region. When it reaches point 5, the star is a fully developed red giant, 1000 times brighter than the sun. Its diameter is 100 million miles, 50 times greater than at point 4 and 100 times greater than the present diameter of the sun.

The changes from point 4 to point 5 occur at an extraordinarily rapid rate. Most of the fiftyfold increase in the size of the star between points 4 and 5 takes place during an interval of no more than 100 million years. The mushrooming growth of the star at this stage, and the other changes going on, occur 1000 times faster than the changes that occurred during the first nine billion years of the star's life, when it was on the Main Sequence.

## The Structure of a Red Giant

Red giants are very odd stars. The core of helium at the center of the red giant is enormously compressed, with a density equal to one ton per cubic inch. One-quarter of the mass of the entire star is packed into the core, although its radius is only one one-thousandth of the radius of the star. This core has a diameter of 20,000 miles—about twice the size of the earth—but it weighs nearly 100,000 times as much. Around the core lies a thin shell of burning hydrogen a few thousand miles in thickness. Around the shell of burning hydrogen is an enormously distended and very tenuous envelope of hydrogen gas, 100 million miles across. The average density of the hydrogen gas in the envelope is one ten-millionth of an ounce per cubic inch. This density would be a very good vacuum in a physics laboratory on the earth.

To bring out the peculiar structure of red giants, suppose that we reduce the size of a typical red giant by a factor of about one trillion. Then the star is a sphere the size of a basketball, but the helium core, containing one-quarter of the mass of the star, is a dot at the center no larger than the period at the end of this sentence.

## Helium Flash

What happens after the star reaches point 5? The helium core becomes more and more compressed with the passage of time, and its temperature continues to rise. Therefore, the red giant becomes more and more luminous. Finally, the helium core reaches the critical threshold temperature of 100 million degrees Kelvin. When the core reaches 100 million degrees, a completely new episode begins in the life of the star. At 100 million degrees Kelvin, helium nuclei begin to fuse, producing carbon nuclei. You would expect the fusion of helium to *add* to the luminosity of the star, driving it still further into the red giant region. But instead, a peculiar event occurs that checks the growth of the star's luminosity and sends it *down* the red giant branch.

At this point the helium in the core is packed in with a very high density, equal to many tons per cubic inch. Because the core is also very hot, the helium atoms are entirely stripped of their electrons. Thus the core is made up of separate helium nuclei and electrons. About 40 years ago a great theoretical physicist named Wolfgang Pauli discovered that when large numbers of electrons are packed into a small space, they become virtually incompressible. It is almost as if they were a stiff solid, like steel, rather than a gas of small particles moving freely about. Because of the incompressibility of the electrons in the helium core, the core acts very much as if it were a sphere of solid steel.

Now consider what will happen to the core when the helium nuclei within it start to burn. First of all, the temperature of the core rises. The rise in temperature should cause an expansion of the core. The expansion should cause a drop in the temperature and, therefore, in the nuclear reaction rate—that is, the rate of fusion of helium nuclei into carbon. The drop in the nuclear reaction rate should relieve the pressure on the helium core, stopping its expansion. The star should then live on, burning helium at its center at a rate just sufficient to balance the inward attraction of gravity and the outward pressure produced by the helium burning. This is, after all, the way in which the star lived on the Main Sequence, although then it was the burning of hydrogen, rather than helium, that balanced the inward force of gravity.

But nothing of the sort happens to the core of the red giant of solar mass. It has the properties of a sphere of solid steel and, like all solids, it expands only very slightly when heated. The core expands far less than it would if it were behaving like a normal gas. Because the helium core does not expand appreciably when heated, the star loses the safety-valve feature that used to be built into it. Imagine what must happen in this core without the safety-valve feature of expansion: helium nuclei burn, raising the temperature; the core does not expand; therefore, the temperature stays high; at the higher temperature, helium nuclei burn still faster; the core gets still hotter; it burns still faster; and so on. The reaction runs away with itself. After a time, the core becomes so hot that it literally explodes, just as a steel ball would vaporize and explode if a powerful enough charge of energy were put into it. The center of the star has changed from a controlled nuclear reactor to an uncontrolled nuclear bomb.

It takes a few hours from the onset of helium fusion for the core of the star to reach the explosion point. A few hours may seem like a long time when compared to the time it takes to trigger a hydrogen bomb, but remember that a few hours is the blink of an eye for a star. Because the time is very brief on this scale, the events from the commencement of helium fusion to the explosion of the core are called the helium flash.

## After the Helium Flash

The helium flash does not add to the observed luminosity of the star, for most of the energy of the explosion is taken up in the expansion of the core itself, and practically none of this energy reaches the surface of the star. On the contrary, the explosion changes the internal structure of the star in a way that causes the luminosity to decline. Up to the point at which the explosion occurred, the luminosity of the red giant is fueled by the furious rate of hydrogen burning in the shell surrounding the hot core. As soon as the core explodes, its temperature drops, and the temperature of the surrounding shell also drops. Therefore, the rate of hydrogen burning in the shell decreases. For the first time since the star left the Main Sequence 10 billion years previously, its brilliance diminishes substantially. It moves vertically downward on the H-R diagram, in the general direction of the Main Sequence.[2]

The diameter of the star decreases at the same time, for, with its nuclear energy sources greatly decreased, the red giant lacks the resources needed to keep its envelope distended. Relieved of the enormous outward pressures created by intense hydrogen burning in the shell, the envelope commences to collapse under the attraction of gravity, and the star begins to lose its swollen appearance.

The H-R diagram in Figure 7.9 shows the evolutionary track of the star in the period of its life that we are now describing. At point 5 of the previous H-R diagram the star became a fully developed red giant. At some time shortly after becoming a red giant, the star experiences the helium flash. Point 6 on the H-R diagram marks the helium flash. The helium flash sends the star down toward the Main Sequence again. This slide continues for about 10,000 years, with the luminosity and size of the star continually decreasing.

When the star starts to descend from the red giant region, the helium in the core is not burning, because its temperature has been greatly reduced by the expansion of the core that followed the flash. During the course of the star's descent, however, the helium in the core is again slowly but steadily compressed by gravity, because there are no nuclear energy sources within it to sustain it against this inward force. As always, the

---

[2] Stars that are quite massive, three times as massive as the sun or more, do not experience the helium flash. Instead, the onset of helium burning occurs in a relatively gradual way in these stars. The effect, however, still is to expand the helium core and diminish the overall rate of nuclear energy production within the star, causing it to move back toward the Main Sequence in the same way as the smaller stars.

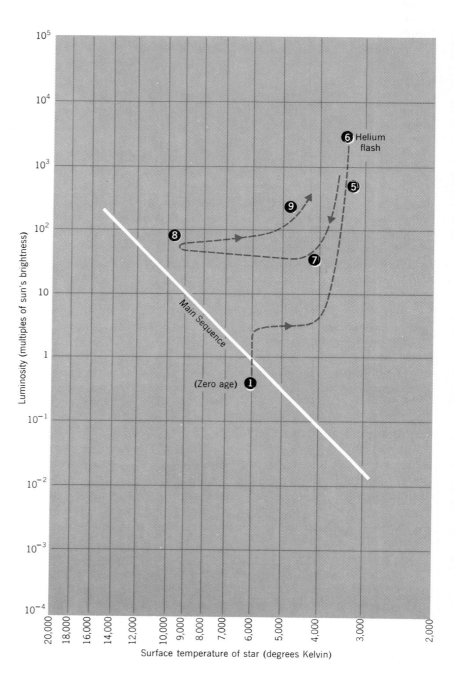

Figure 7.9   The late Evolution of a star of solar mass.

compression heats the core, and its temperature rises. After several thousand years, the temperature gets high enough to start a very small amount of helium burning at the center of the core. By the end of 10,000 years, the temperature of the core has risen enough, and the rate of helium burning has accordingly become great enough for helium burning to constitute a major source of energy for the star. At this point the temperature in the core is roughly 200 million degrees Kelvin. Its structure is shown in Figure 7.10.

The release of substantial amounts of nuclear energy at the center of the star, through the burning of its helium, halts the star's descent toward the Main Sequence. The star now lies at point 7 on the H-R diagram.

This is the second time that the temperature in the core has reached the level required for helium burning. The first time was the helium flash at point 6 on the H-R diagram. But this time no explosion occurs, because the helium core is far less dense than it was before the helium flash occurred; therefore, the electrons in the core are not packed in at such a high density as they were earlier; and consequently they do not give the core that peculiar solidlike incompressibility that it had before. In other words, the helium core now behaves as a proper gas should. If a fluctuation occurs in the rate of burning, and it becomes excessively hot for a short time, the core does not explode. Instead, it expands and cools, dropping the reaction rate and removing the excessive heating.

While the helium burning increases, the hydrogen-burning shell is still suffering the aftereffects of the helium flash. Because the helium-burning core is still not nearly as hot as it was during the helium flash, the burning in the hydrogen shell continues to decrease. From point 7 the increase in helium burning and the decrease in hydrogen burning roughly balance, giving the star a constant luminosity. The star moves horizontally to the left from point 7 to point 8 on the H-R diagram. At point 8 the helium core has an analogous structure to the hydrogen core on the Main Sequence at point 1.

Throughout the movement from 7 to 8, carbon steadily accumulates at the center of the star as the product of the helium burning. Eventually, all the helium at the center of the helium core is burned up and converted to carbon. This carbon now constitutes an inner core within the original helium core of the star. The development of an inner core of carbon halts the helium-burning reactions at the center. The helium burning now occurs in a shell surrounding the carbon core (Figure 7.11). At point 8 the inner structure of the star is analogous to its structure at point 3, when hydrogen was burning in a shell surrounding the newly formed, inert helium core. The carbon core acts as the helium core acted, heating the surrounding shell and increasing the rate of nuclear burning. The heat from the increased burning is absorbed by the cool outer envelope, which keeps expanding. The temperature drops and the luminosity increases.

From point 9 onward, events unfold so rapidly, and the star is so complex inside, that it is difficult to calculate its further course accurately. Theoretical astronomers are not in agreement on the precise sequence of events that follow. But we do know in a general way what happens. At point 9 the star starts up toward the red giant region once again; that is, it swells enormously and becomes very bright, although its surface is relatively cool and red in color. The speed of the evolution is, however, much faster than it was when the star first moved up this path. Everything proceeds about 100 times more rapidly. The entire move toward the red

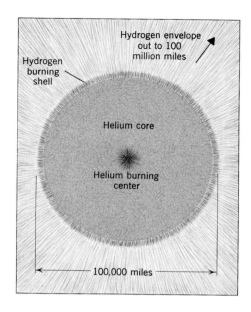

Figure 7.10  The structure of a red giant.

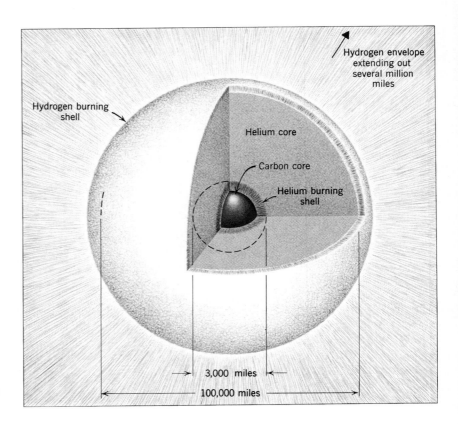

Figure 7.11 The structure of a star in the helium-burning phase.

Labels in figure:
Hydrogen envelope extending out several million miles
Hydrogen burning shell
Helium core
Carbon core
Helium burning shell
3,000 miles
100,000 miles

giant region is completed in a few million years for a star with the mass of the sun, as compared to the several hundred million years required to reach the red giant region the first time.

During its second move toward the red giant region, the star may suffer several "flashes" in the helium-burning shell, somewhat resembling the first flash that occurred in the helium core at point 6. Some theoretical astronomers believe that if these flashes occur, they cause the star to move rapidly to the left toward the Main Sequence and then back to the right again, in two, three, or even more shuttling motions that interrupt the steady climb upward toward the red giant region.

## DEATH

Whether or not the star enters the red giant region again, it is, in any case, close to the end of its life. All stars lead similar lives up to this point. The main variation is in the time scale of the changes they undergo, which is shorter for the large, heavy stars and longer for the small ones. But the subsequent fate of a star depends very much on its mass.

If the star is small, its second approach to the red giant region is its last one. We do not know precisely what happens at this point, but we are sure that shortly thereafter the star moves rapidly to the left on the H-R diagram and then downward, fading out slowly in the lingering death of the white dwarf. This will almost certainly be the fate of the sun.

Large stars, on the other hand, may move into and out of the red giant region several more times; we are not certain; but we are certain of their fate after this relatively brief period of shuttling back and forth on the H-R diagram. When the shuttling stage is over, they do not fade away; instead, they blow up in the gigantic explosion known as the supernova, or disappear from view as a black hole in space.

## Death of Small Stars

The factor that decides between these possibilities is the temperature in the carbon core of the star. In a star with a small mass, the heating produced by the compression under gravity is not as great as it is in a massive star. Consequently, the temperature at the center of a small star never reaches the critical value required to fuse carbon nuclei. According to cyclotron studies, the threshold temperature for carbon burning is 600 million degrees Kelvin. If the temperature at the center of the star fails to reach this value, no significant carbon burning occurs; the carbon core becomes increasingly compressed; and its temperature continues to rise under compression, increasing the reaction rate in the surrounding helium-burning shell. Thus, the star continues to move toward the red giant region, becoming very much larger in diameter and redder in color.

Finally, the outer layers of the star become so red—that is, so cool—that the nuclei in these layers start to capture electrons to become neutral atoms again. For the first time since the birth of the star, a substantial part of its mass is again in the form of neutral atoms, rather than atomic nuclei plus separate electrons.

What happens when an electron is captured by a nucleus to form a neutral atom? The most important consequence is that a photon or quantum of light is emitted. The emitted photon carries away energy with it. The photon travels at the speed of light, but usually it does not get very far from the atom that emitted it before it is absorbed by another atom or particle. Countless photons are created by the formation of neutral atoms and then absorbed shortly thereafter in nearby regions of the star's envelope. Their absorption heats the envelope.

## Planetary Nebulae

The amount of heat produced in the star's envelope by photon absorption is modest compared to the nuclear energy released at its center. Nonetheless, according to one theory of the advanced stages of stellar evolution, this heat triggers profound changes in the star's appearance.

The envelope, heated by the absorption of photons, expands. The outward expansion lowers the temperature of the envelope. At the lower temperature, more neutral atoms form from the separate nuclei and electrons in the envelope, and they in turn release more energy in the form of photons. Most of the photons are, again, absorbed by nearby atoms in the envelope. They heat the envelope, expanding it outward still further. In other words, a runaway process commences in which the capture of electrons by nuclei heats the envelope and causes expansion, which cools the envelope, causing more electron capture, which leads to more expansion. The envelope of the star expands outward faster and faster, until it leaves the star entirely. In effect, the envelope of the star blows off into space and becomes a tenuous, nearly transparent shell of atoms that continues to expand rapidly outward.

The core, which was formerly concealed by the envelope, now stands exposed to view. If someone were observing the star during the course of this entire process, he would see an amazing change in its appearance. At the start the star would seem normal. Then, when the envelope had started to expand but was still dense enough to conceal the core, the observer would see the surface of the expanding, relatively cool envelope, and the star would present the appearance of a large, luminous, red object. When the envelope had expanded out far enough to become more or less transparent, so that the core was exposed, the observer would see

*Figure 7.12 A Planetary Nebula (NGC 7293).*

a small, white-hot object—the core—surrounded by a softly glowing diffuse shell of gas—the blown-off envelope.

Such objects are called planetary nebulae. The name "planetary nebula" came into use because the astronomers who first photographed these nebulae through small telescopes found the images resembled those of planets. We now know that planetary nebulae have no connection with planets or solar systems, but the name has persisted.

Figure 7.12 shows the structure of a planetary nebula clearly. This photograph was taken with the 100-inch telescope at Mount Wilson. A particularly striking color photograph of another planetary nebula is included in the color inset (Color plate 5).

What happens to the core of the star after the envelope blows off? The core is more or less unaffected by the departure of the envelope, and it continues to burn helium in the helium-burning shell at the same rate as before. Therefore, the luminosity of the star, which is controlled entirely by the burning of the helium in the shell, remains constant.

However, the plot of the star's position in the H-R diagram changes dramatically when the envelope blows off, because initially we are plotting the position of the cool envelope of the star (around 3500°K), but after the envelope blows off, we plot the position of the hot core (around 10,000°K) that remains. Thus, on the temperature axis there is a shift from 3500°K to 10,000°K. Because there is no change in luminosity during this increase in surface temperature, the star's evolutionary track shoots horizontally across the H-R diagram to the left. This change is depicted in the H-R diagram in Figure 7.13 (see following page). The star's envelope starts to expand at point 10. At point 11 the hot core of the star is fully exposed. At this point the star, if photographed, would look like the Ring Nebula in the constellation Lyra.

## The White Dwarf

At point 11 on the H-R diagram, the star begins its transition from planetary nebula to white dwarf. The star is now composed of carbon surrounded by a helium-burning shell (Figure 7.14). At this point the temperature of the carbon at the center is still not high enough for the carbon nuclei to fuse, and there is therefore no source of nuclear energy at the center of the star to offset the attraction of gravity and keep the star from collapsing.[3] The star's carbon core continues to contract slowly.

If it were not for the electrons in the star, the collapse would continue and the center of the core would get hotter and hotter until finally, at 600 million degrees Centigrade, the carbon nuclei would begin to burn. Before this can happen, however, the electrons bring the collapse to a halt.

[3] Nuclear energy is produced in the helium-burning shell, but this energy is mostly radiated outward into space without affecting the interior of the star.

Figure 7.13  *The track of a star in the final stages of evolution.*

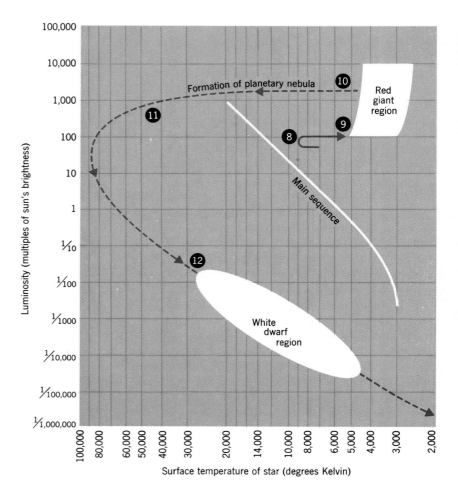

Figure 7.14  *Structure of a helium-burning star in an advanced stage of evolution.*

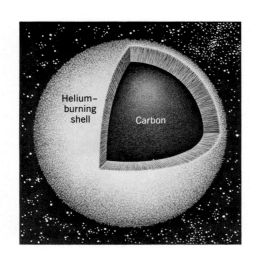

The peculiar "solid-steel" behavior of the electrons comes into play, just as it did at an earlier stage in the star's life when its core collapsed for the first time. As before, the solid-steel incompressibility of the electrons halts the contraction. This event occurs when the star has a radius of about 10,000 miles and a density of about 10 tons per cubic inch. The contraction halts at point 12 on the H-R diagram.

No one knows what happens between points 11 and 12 on the H-R diagram. The theoretical calculations for that stage indicate that any one of a number of different possibilities could take place. Observations of stars do not provide a clue to what actually occurs, because very few stars have been found between 11 and 12.

Once the star reaches point 12, however, its course becomes clear. In the region of point 12 the star is exceedingly dim in comparison to its luminosity in the earlier stages of its life; a star with the mass of the sun, for example, would be 100 times fainter at point 12 than the sun is at the present time. The diameter of the star also is very much smaller at this stage than when the star was in its prime. A star originally the size of the sun would be about 20,000 miles in diameter, which is not much more than twice the size of the earth. The shrunken star is exceedingly dense. Into its relatively small volume, no more than that of a good-sized planet, is packed an enormous mass, hundreds of thousands of times greater than the mass of the earth.

Although the star is now very faint, its surface is quite hot, with a temperature ranging up to 20,000 degrees. The explanation for this seeming paradox is simple. In a normal star the temperature at the surface is considerably lower than the temperature at the center; in the sun, for example, the surface temperature is 5800 degrees and the central temperature is 13 million degrees. This vast drop in temperature is caused largely by the atoms in the outer layers of the star, which absorb the radiation from the star's center before it can escape into space. The radiation is reemitted by the atoms and eventually escapes anyway, but at a much lower temperature than it otherwise would have had.

If there were no atoms in the star, but only separate electrons and nuclei, radiation would not be absorbed and the star's surface would be nearly as hot as its center. This is the case with the shrunken, burned-out stars we are describing now. They are made up almost entirely of electrons and nuclei and contain hardly any atoms. Thus, the shrunken star's radiation reaches its surface nearly unimpeded and without a large drop in temperature. Consequently, the star's central temperature can be relatively low—perhaps 100,000 degrees or so, which is far cooler than the central temperature of the sun—and yet its surface can be at a white heat, and considerably hotter than the sun's surface.

Such stars—small, dense, and exceedingly faint, but white-hot at the surface—are called white dwarfs. The force of gravity on the surface of a white dwarf can be as much as one million times greater than gravity on the earth. Even if we should ever come across a white dwarf whose surface temperature has declined to a comfortable level, we would never be able to land men or even remote-controlled spacecraft on this strange world. A man attempting to land on a white dwarf would weigh 150 million pounds, and he and his spacecraft would literally be flattened by the enormous forces of the white dwarf's gravity.

From point 12 onward, the star—now a white dwarf—shrinks very little in radius. Slowly the white dwarf radiates the last of its heat into space, moving downward in luminosity and temperature as it does so and following the path leading to the dead stars at the bottom of the H-R diagram. Progressively the dwarf turns in color from white to yellow and then to red, until it fades out to a cold, black lump of matter and enters the graveyard of the stars.

## Death of Massive Stars

A very different fate awaits a more massive star at the end of its life. Such a star forms an inert carbon core surrounded by a helium-burning shell, just as a smaller star does. And the carbon core begins to collapse, just as it does in a small star, because no nuclear reactions are going into the core to generate the pressure that could resist the collapse. In a small star the collapse continues until the star is a white dwarf, because the star's temperature never gets high enough to start nuclear reactions in its carbon core. But when a massive star collapses, more heat is generated than in the collapse of a small star. Well before the massive star gets down to the size of a white dwarf, the temperature in its core reaches the critical level of 600 million degrees, and carbon begins to burn, forming magnesium and other elements. The pressure generated by the burning immediately halts the collapse.

After all the carbon has burned up, the center of the star collapses again. The collapse heats it to still greater temperatures at which other nuclear reactions set in, once more halting the collapse, and at the same time creating new elements. In this way, through the alternation of collapse, heating, and renewed nuclear burning, many elements are produced inside a massive star.

The temperature mounts, and nuclear reactions continue to produce new elements, until finally the element iron is reached. At this point the process stops, for iron is a very special element. Any reaction that takes place involving an iron nucleus will use up energy. Thus, when a large amount of iron accumulates at the center of a star, it has the same effect as a fire extinguisher; the iron, instead of providing more fuel to burn, puts the fire out. The center of the star commences to collapse again, but this time, because of the presence of the iron nuclei, the fire cannot be rekindled; it has gone out for the last time, and the entire star commences its final collapse.

The collapse is a catastrophic event. The materials of the collapsing star pile up at the center, creating temperatures of a trillion degrees, pressures of a trillion trillion tons per square inch, and enormous densities. When the density at the center of the star is so great that the nuclei touch one another, the star can be compressed no further. The collapse comes to a halt. The collapsed star, compressed like a giant spring, rebounds instantly in a great explosion. The explosion is as violent as the collapse that preceded it. It hurls out into space the elements that the star has been manufacturing during its lifetime. A large fraction of the material within the star disperses into space in this explosion. At the original location of the star there now remains only the squeezed and burned-out remnant of the core, containing perhaps one-half of the star's original mass. The entire episode lasts a few minutes, from the onset of the collapse to the final explosion. This is a short interval for the demise of an object that may have lived for a billion years or more.

The exploding star seems to resemble a planetary nebula at this stage, with a compact core surrounded by a rapidly expanding shell of gas. The resemblance, however, is superficial. The causes of the events are different and, even more important, they differ enormously in scale. The formation of the planetary nebula is a gentle process in which the luminosity of the star scarcely changes, whereas the explosion of a massive star is a titanic event in which the star blazes up to a brilliance billions of times greater than its normal luminosity. For a short time the exploding star can be as bright as an entire galaxy.

## Supernovas

The exploding star is called a supernova. If a supernova happens to occur nearby in our Galaxy, it appears suddenly as a new star in the sky, brighter than any other and easily visible with the naked eye in the day-time. The last supernova that exploded in our Galaxy was seen in Europe in 1604 and caused a sensation. One of the earliest reported supernovas was a brilliant explosion recorded by Chinese astronomers in A.D. 1054. At the position of this supernova there is today a great cloud of gas known as the Crab Nebula, expanding outward at a speed of 1000 miles per second, which contains the remains of the star that exploded 900 years ago (Color Plate 6).

When a supernova explodes in another Galaxy, it cannot be seen with the naked eye, but it can be photographed through a large telescope. Several hundred supernovas have been photographed since the introduction of the camera to astronomy. Figure 7.15 shows a supernova (Figure 7.15a, arrow) in the Galaxy NGC 7331, photographed in 1959. Months later the supernova had faded to invisibility (Figure 7.15b).

At the very high temperatures generated in the collapse and explosion, some of the nuclei in the star are broken up, and many neutrons and protons are freed. The neutrons and protons are captured by other nuclei, building up the heavier elements, such as silver, gold, and uranium. In this way the remaining elements of the periodic table, extending beyond iron, are manufactured in the final moments of the star's life. Because the time available for making these heavy elements is so brief, they never become as abundant as the elements up to and including iron. In the table of cosmic abundances of the elements (Figure 7.16) there is a rapid drop-off by a factor of 100,000 for the heavy elements beyond iron.

## PULSARS, NEUTRON STARS, AND BLACK HOLES IN SPACE

What happens to the compressed core of the supernova after its outer layers explode into space? The answer to this question was unknown until 1967. In that year, pulsars—the most interesting objects to be found in the sky in many years—were discovered.

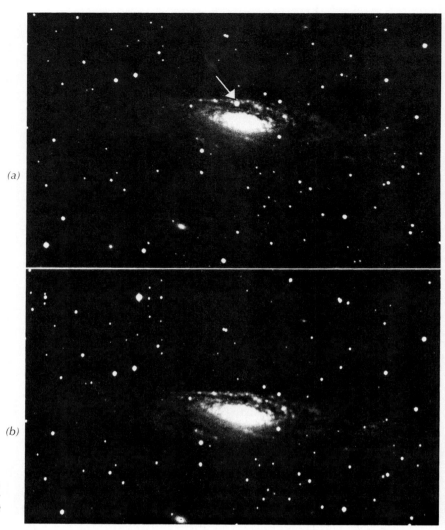

*Figure 7.15 The spiral galaxy NGC 7331, photographed in 1959 (a) during a supernova outburst (arrow) and (b) after the supernova has faded.*

## Pulsars

The discovery came about by pure chance. Jocelyn Bell, an astronomy student at Cambridge University, received the task of investigating fluctuations in the strength of radio waves from distant galaxies. She found unexpectedly that certain places in the heavens were emitting short, rapid bursts of radio waves at regular intervals. Each burst lasted no more than one one-hundredth of a second. The rapid succession of bursts seemed like a speeded-up, celestial Morse code. The interval between successive bursts was about one second and extraordinarily constant. In fact, it did not change by more than one part in ten million. A clock with this precision would gain or lose no more than a second a year.

No star or galaxy had ever before been observed to emit signals as

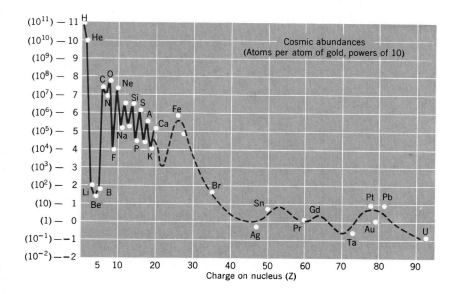

Figure 7.16 Cosmic abundance of the elements. Notice the high abundance of iron (due to its unusually stable nucleus) and the low abundance of elements beyond iron resulting from the fact that they are mostly created in the brief moments of the supernova explosion.

bizarre as these. At first, some astronomers thought that intelligent beings might be beaming a message to the earth, and they referred to the Morse-code stars as LGM's, standing for Little Green Men. But the scientific community soon decided that the radio pulses had a natural and not an artificial origin. One of the main reasons for this conclusion was the fact that the signals were spread over a broad band of frequencies. If an extra-terrestrial society were trying to signal other solar systems, its interstellar transmitters would require enormous power to send signals across the trillions of miles that separate every star from its neighbors. It would be wasteful, purposeless, and unintelligent to diffuse the power of the trans-mitter over a broad band of frequencies. The only feasible way to transmit would be to concentrate all available power at one frequency, as we do on earth when we broadcast radio and television programs.

This cold reasoning dashed the hopes of romantics who believed for a short time that man might have received his first message from outer space. "LGM" disappeared from scientific conversation, "pulsar" took its place, and scientists settled down to search for a natural explanation of the peculiar signals.

The first clue to the answer was the sharpness of the pulses. When an object in space emits a burst of radio waves, the waves from different parts of the object arrive at the earth at different times, blurring the sharpness of the original pulse. The smaller the object, the less blurred the pulse and the shorter its duration. From the fact that each pulse lasted for one one-hundredth of a second or less, astronomers calculated that pulsars were no more than 10 miles in radius.

This is a startling conclusion. Until then scientists thought that the white dwarf—about 10,000 miles in radius—was the smallest, densest star in the universe. How could a star be a thousand times smaller than a white dwarf? The matter in such a star would be one billion times denser than the matter in a white dwarf. If the entire earth were compressed to the same degree as a pulsar, it would fit into the Pentagon. If the Pentagon were compressed as much, it would be the size of a pinhead.

## Neutron Stars

The answer goes back to a prediction made several decades ago. At that time, several theoretical astronomers pointed out that just before a large, dying star explodes as a supernova, it has collapsed to the point where nuclei at the center of the star are touching. The pressures in this condensed ball of matter are so great that individual electrons and protons combine to form neutrons. A pure ball of neutrons forms at the center of the star, only 10 miles in radius, but with most of the star's original mass packed into it. Scientists dubbed the hypothetical ball of neutrons a neutron star.

Starting in 1965, astronomers searched for neutron stars assiduously, investigating with particular care the region at the center of the Crab Nebula, where the squeezed-down core of the supernova explosion of A.D. 1054 should have been located.* But no neutron stars were discovered and interest in them faded.

In 1968 a wave of excitement spread through the astronomical community when a pulsar was discovered at the center of the Crab Nebula at precisely the place where astronomers had previously searched for a neutron star. Suddenly, many items of evidence fitted together like the pieces of a jigsaw puzzle. A neutron star was predicted to exist at the center of the Crab Nebula; a pulsar was found there; and the neutron star and the pulsar are the only objects known that have the mass of a star packed into a sphere with a radius of ten miles. Clearly, neutron star and pulsar are two names for the same thing — a fantastically compressed, super-dense ball of matter, created when a massive star collapses at the end of its life.

One mystery remains to be explained. What produces the sharp, regularly repeated bursts of radiation from which pulsars derive their name? Scientists believe that a pulsar, like the sun and most other stars, is subject to violent surface storms that may last for years, spraying particles and radiation out into space. Each storm occurs in a localized area on the surface of the pulsar and sprays its radiation into space in a narrowly defined direction. When the earth lies in the path of one of these streams of radiation, our radio telescopes pick up the signals that indicate to us the presence of the pulsar.

* A faint star in the Crab had been tentatively identified as the supernova remnant in the 1940's, but no proof could be found that it was a neutron star.

But if the pulsar sprays radiation steadily into space, why do we observe the radiation as a succession of isolated sharp bursts? The reason is probably that pulsars, as most stars do, spin on their axes. In fact, it is entirely possible that pulsars spin as rapidly as several times a second. As the pulsar spun, the stream of radiation from its surface would sweep through space like the light from a revolving lighthouse beacon. If the earth happened to lie in the path of the rotating beam, it would receive a sharp burst of radiation once in every turn of the pulsar.

This theory can be checked, because a spinning object must slow down gradually. Thus, the interval of time between successive bursts of radiation from a pulsar should increase. In 1969 this prediction was confirmed by the discovery that the time between successive pulses from the Crab Nebula pulsar was getting longer, at the tiny but measurable rate of one one-billionth of a second per day.

## Black Holes in Space

With the realization of the connection between neutron stars, pulsars, and supernovas, many astronomers feel that the final pages may have been written in the life story of the stars. But others suspect that at least one surprise is still in store for us, since there is reason to believe that the neutron star or pulsar is not the ultimate state of compression of stellar matter. Under certain conditions, a star may continue to collapse beyond the ten-mile limit of the neutron star, falling inward upon itself faster and faster, until it has contracted to a radius of 2 miles. At this point, the theory of relativity predicts the occurrence of an extraordinary phenomenon.

According to Einstein's theory, a ray of light should possess mass. If Einstein is right, a ray of light emitted from a star will be pulled back by the star's gravity, as a ball thrown up from the surface of the earth is pulled back by the earth's gravity. When the star is normal in size— about one million miles in diameter—the force of gravity on its surface is not strong enough to keep the light rays from escaping, and they leave the star, although with somewhat less energy. But as the star contracts, the force of its surface gravity grows rapidly, and by the time the star's radius has decreased to 2 miles, the force of gravity at its surface is billions of times stronger than the force of gravity on the surface of the sun. The tug of this enormous force prevents the rays of light from leaving the surface of the star; like the ball thrown upward, they are pulled back and cannot escape to space. From this moment on, the star is invisible. It is a black hole in space.

Inside the black hole, the contraction continues, piling up matter at the center in a tiny, incredibly dense lump. According to current knowledge in theoretical physics, no force in nature is strong enough to halt the contraction. The star's volume becomes smaller and smaller; from a

2-mile radius it shrinks to the size of a pinhead, then to the size of a microbe and still shrinking, passes into the realm of distances smaller than any ever probed by man. And at all times the star's mass of a thousand trillion trillion tons remains packed into the shrinking volume.

Intuition tells us that such an object cannot exist. At some point the collapse must halt. Yet, according to the laws of twentieth-century physics, no force, no matter how powerful, can stop the collapse. The implication is that the laws of physics must be modified at extremely short distances in a manner that prevents particles from coming infinitely close together. Here is a hint of the impending discovery of a new law or a new agent in nature, which could produce a profound change in the body of scientific knowledge.

A serious examination of such revolutionary possibilities is premature as long as the black hole remains only a theoretical prediction. It appears that the black hole must retain this doubtful status forever, since, by their nature, it seems impossible to observe one. However, recent evidence obtained from the X-ray satellite Uhuru (p. 65) suggests that black holes actually exist. The data from Uhuru showed a powerful source of X-rays, with unusual properties, in the constellation Cygnus. The source, named Cygnus X-1, was in the vicinity of a blue supergiant, whose spectrum showed that the supergiant was a member of a spectroscopic binary. Apparently, the X-ray source was the other member of the binary. In support of the conjecture, the X-ray source varied periodically in strength, as if it were being eclipsed periodically by the supergiant.

This was the evidence for black holes the theorists had been looking for. When the stars in a binary are relatively close together, the pull of each star's gravity draws matter away from its companion, and streams of gas flow back and forth between the two stars. If one star becomes a black hole, the gas drawn out of the other star will continue to stream toward it, but now, as this gas approaches the boundary of the black hole, it will be accelerated to extremely high velocities by the black hole's gravitational force. The rapidly moving particles, converging on the black hole, will collide with one another and produce an intense stream of X-rays, making the black hole an X-ray source of the kind observed by the Uhuru satellite.

If the binary is of the eclipsing type (p. 143) the X-rays emanating from the vicinity of the black hole will be periodically blocked by its companion as the two stars revolve around one another. The X-ray source discovered by the Uhuru satellite precisely fitted this description, because the intensity of the X-rays coming from it varied regularly every 4.86 seconds.

Is Cygnus X-1 a black hole? According to another theory, the invisible member of the binary could be a neutron star instead, since these stars also emit X-rays. However, the mass of Cygnus X-1 is estimated to be $10 M_\odot$. Calculations on the structure of neutron stars show that the mass of this type of star cannot exceed a few solar masses. The black hole seems to be the most acceptable explanation remaining. There is a general reluctance to accept the existence of black holes because they are such peculiar objects, but the Uhuru observations appear difficult to interpret in any other way.

## EPILOGUE

The life story of the stars has an epilogue. When a supernova explosion occurs and the outer layers of the stars are sprayed out to space, they mingle with fresh hydrogen to form a gaseous mixture containing all 92 elements. Later in the history of the galaxy, other stars are formed out of clouds of hydrogen that have been enriched by the products of these explosions. The sun is one of these stars; it contains the debris of countless supernova explosions dating back to the earliest years of the Galaxy. The planets also contain the debris; and the earth, in particular, is composed almost entirely of it. We owe our corporeal existence to events that took place billions of years ago, in stars that lived and died long before the solar system came into being.

### Questions

1. Why must a large number of particles come together at the same time for a star to form from a gaseous nebula?
2. At what point does a protostar become a star? Why?
3. Why do protostars of large mass contract to stars more rapidly than protostars of small mass?
4. Why does the initiation of nuclear reactions at the center of a newly forming star eventually halt the collapse of the gas?
5. Is there any observational evidence of stars forming? What is it?
6. Where on the H-R diagram does a star spend most of its lifetime? What happens to it when a helium core forms?
7. Why does a star leave the Main Sequence? Describe the sequence of events as the star moves away from the Main Sequence.
8. What is the helium flash? Why does the luminosity of a star decrease following the helium flash? Briefly describe the sequence of events in the life of a star of one solar mass; four solar masses; 0.2 solar masses.
9. Why is a star of one solar mass not able to fuse carbon nuclei in its core?
10. Why should old stars in a galaxy have a lower abundance of heavy elements than young stars?
11. What types of nuclei are formed for the first time in a supernova explosion? What happens to these nuclei?
12. If our solar system had been formed 10 billion years ago, when the Galaxy and the Universe were young, what types of planets would exist in the solar system? Would there be earthlike planets? Could life exist in the Universe 10 billion years ago?
13. If you were observing a star becoming a black hole, what do you think you would see? If you fell into the black hole in space, what do you think would happen to you?

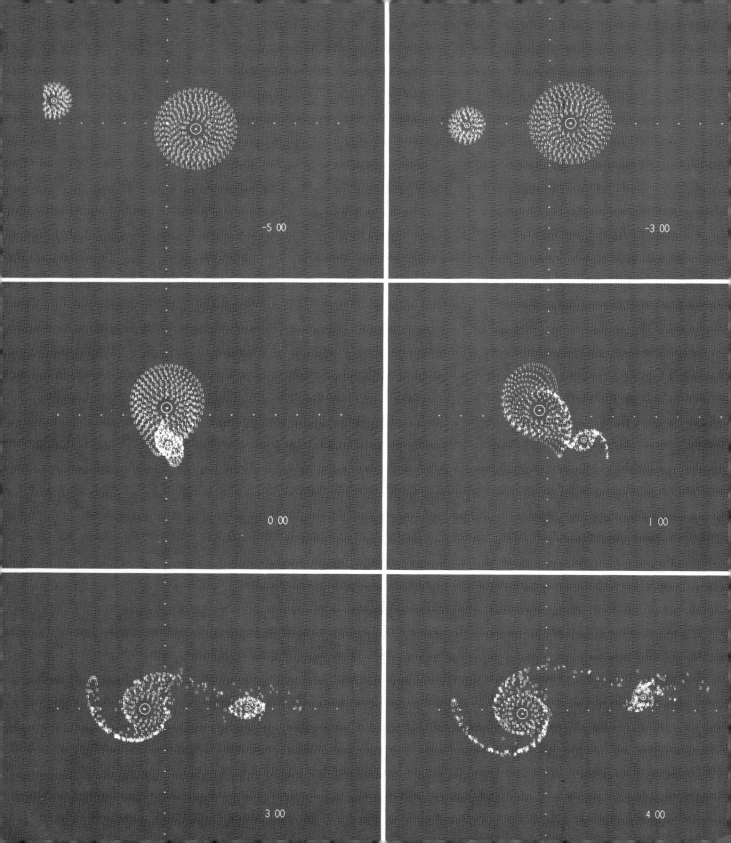

-5.00

-3.00

0.00

1.00

3.00

4.00

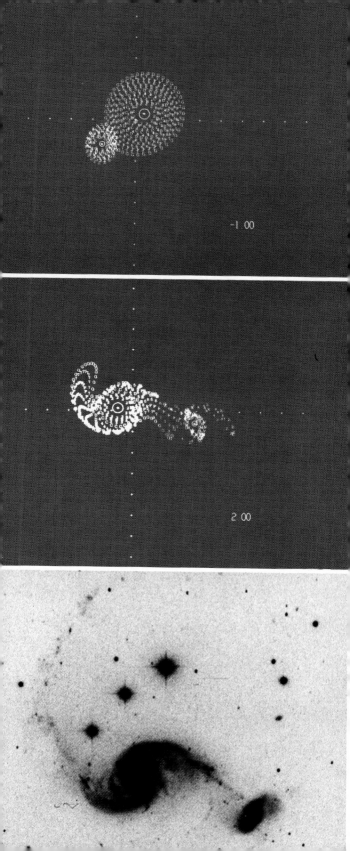

-1.00

2.00

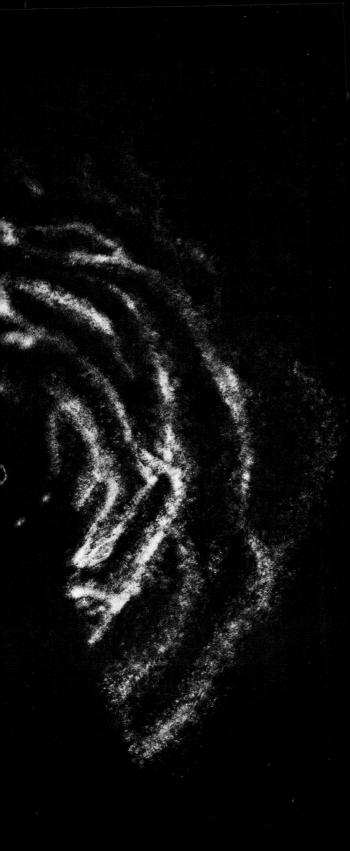

# 8   The Milky Way Galaxy

On a clear night, away from all artificial lights, our Galaxy may be seen in an edge-on view as a luminous band of light stretching from horizon to horizon. This band of light was called the Milky Way by the ancients, and for that reason our Galaxy is usually called the Milky Way Galaxy. Galileo was the first to discover that the Milky Way is composed of myriads of individual stars. He made this discovery in 1610 with the famous one-inch telescope with which he also found the mountains on the moon, the phases of Venus, and the satellites of Jupiter.

In Chapter 1 we described the Milky Way Galaxy as it appears in photographs. It is an aggregate of 100 billion separate stars, held together by their mutual forces of gravitational attraction. This vast number of stars is arranged in the shape of a flattened disk, about 100 thousand light-years in diameter and 5000 light-years in thickness.

Within the disk, the space between the stars is filled with a varying concentration of atoms—mostly hydrogen—as well as molecules and dust particles. Although many of the stars in the disk of the Galaxy

*A radio map of the Milky Way Galaxy (page 204).*

wander through space as individuals, a few are bound together into groups, each containing as many as 1000 members. These groups, called *galactic clusters,* move about in the disk as self-contained units. In addition, more than 100 larger groups of stars move about in the Galaxy above and below the disk. These groups, some containing as many as one million stars, are called globular clusters.

The spherical concentration of stars called the galactic nucleus bulges out of the center of the flat disk; it has a diameter of approximately 10,000 light-years. The galactic nucleus is hidden from our view by intervening clouds of dust in the space between the stars. We are relatively certain of its existence because galaxies that seem to resemble the Milky Way Galaxy in other respects possess such a central concentration of stars.

Some regions of the Galaxy have enhanced concentrations of gas and dust. If these concentrations are very dense they obscure the stars behind them, and become visible as dark clouds (Figure 8.1). Sometimes the atoms in a cloud of gas and dust are excited by the radiation from young, hot stars imbedded in them. The excited atoms reemit their energy in the form of a diffuse glow, forming the so-called emission nebula. Color Plate 7 shows the Trifid Nebula, an emission nebula located at a distance of 2300 light-years in the direction of the constellation Sagittarius. The dark lanes that separate the Trifid Nebula into three lobes are lanes of obscuring dust lying between us and the glowing regions.

Spiral arms radiating out from the center of the Galaxy contain most of the bright stars in the Galaxy along with much of the gas and dust. The sun is located in one of these arms, approximately 30,000 light-years from the center. The entire spiral rotates on an axis through the center perpendicular to the plane of the disk, with the nearer regions rotating at a faster angular rate than the outer ones. The Solar System completes one trip around the center of the Galaxy in 200 million years. It has gone around the Galaxy about 20 times during the 4.6 billion years of its existence.

## THE INTERSTELLAR MEDIUM

The stars of the Galaxy, along with their many accompanying families of planets, move through a tenuous sea of gaseous matter called the interstellar medium. The density of this gaseous material is so low that it constitutes a vacuum one million times better than any that has ever been achieved in any laboratory on the earth.

The interstellar medium is critically important because it provides the raw materials out of which stars and planets are made, as we have seen in Chapter 7.

Most of the interstellar medium—about 90 percent—consists of atoms of hydrogen. If the proponents of the Big Bang cosmology are correct

*Figure 8.1   The Horsehead Nebula, in the direction of the Constellation Orion, is a dense cloud of dust in the Milky Way.*

(see Chapter 11) the Universe had its beginning in an explosive event that occurred approximately 13 billion years ago. Nearly all the hydrogen now in the interstellar medium dates back to that early period. A small fraction of this hydrogen has perhaps been recycled through the sequence of events that take place in the life of a massive star, starting with the birth of the star in a cloud of interstellar matter, and ending with the return of that star's substance to the interstellar medium in the aftermath of its final supernova explosion. Color Plate 6 clearly shows materials from an exploded star spreading outward and mixing with the interstellar medium.

Atoms of helium make up nearly all of the remaining 10 percent of the matter in the interstellar medium. Some of this helium may have been manufactured out of the primordial hydrogen by nuclear reactions going on during the first moments of the Big Bang, when the temperature and density of the Universe were very high. The remainder must have been manufactured out of hydrogen in the bodies of stars and dispersed to space in their final supernova explosions.

It is not known how much of the helium in the world today dates back to the Big Bang, and how much was added subsequently by supernova explosions, but the prevailing belief among astronomers is that both sources were important. One way of answering this question is to measure the helium content of stars that are known to be nearly as old as the Universe. Such stars are probably to be found in the globular clusters. As will be seen below, the stars in these clusters are very old, and they are thought to have been among the first stars to be formed in the Universe. Consequently, the helium in these stars must be entirely primordial. The abundance of helium in the globular clusters is determined by measuring the strength of the helium lines in the spectra from these clusters. The measurements are difficult, and the conclusions are correspondingly tentative. Taken at face value they indicate that much of the helium in the Universe today dates back, like the hydrogen, to the time of the Big Bang.

The remaining 90 elements of the periodic table make up no more than one percent of interstellar matter. Most of this consists of carbon, nitrogen, oxygen, aluminum, and iron. These substances, as well as other, still scarcer elements, ranging up to uranium, have been manufactured in the centers of stars by nuclear reactions, and then sprayed into the interstellar medium after supernova explosions. Presumably these elements, and helium as well, are steadily rising in abundance with time, as more and more hydrogen is drawn into newly formed stars and put through the cycle of nuclear reactions that accompany a star's evolution. By the same token, the amount of hydrogen in the Universe must be steadily diminishing, unless a fresh source of that basic substance exists. At the present time, however, there is no evidence for sources of new hydrogen.

In addition to atoms, the interstellar medium also contains a large variety of molecules. They make up only one millionth of the total amount of interstellar matter, but are more important than their low concentration would suggest. The discovery of this variety of interstellar molecules is one of the more surprising occurrences in modern astrophysics. Until recently, molecular hydrogen, which forms wherever hydrogen atoms exist in relatively dense concentrations, was expected to be the only type of molecule present. Today we know that at least a dozen other molecules exist in the space between the stars, and possibly dozens more are awaiting discovery.

The first of the new molecules to be detected was CN, a molecular ion, or radical as it is called by the chemists. CN is also known as cyanogen or the cyanide radical. The cyanide radical can combine with a hydrogen atom to form the gas hydrogen cyanide, whose chemical formula is HCN. Subsequently, evidence of HCN molecules in space was also detected.

Many molecules were detected in rapid succession after the initial discovery of the CN molecule. Evidence was first obtained for the presence of OH molecules—the so-called hydroxyl radicals—which form water when combined with hydrogen. Shortly thereafter, water itself, and then ammonia gas, were also discovered in space. Most recently, interstellar formaldehyde was detected. Formaldehyde is a particularly valuable molecule for the investigation of the Milky Way

Galaxy, as will be seen below. Searches are also under way for other interstellar organic molecules chemically related to chlorophyll. Other molecular combinations of atoms of hydrogen, carbon, nitrogen, and oxygen have also been found in space.

The discovery of formaldehyde and cyanogen in interstellar space has a bearing on theories regarding the origin of life in the Universe. These molecules have been known for some time to be possible precursors of living matter. From a chemical point of view, they are suitable building blocks for amino acids and other biologically important molecules that provide the foundations of all known terrestrial life.

The discovery of the molecular precursors of life in interstellar space raises a question as to whether the chemical evolution of life may have commenced, not on the earth as many biochemists believe, but in outer space before the solar system existed. Perhaps the evolution of life has been going on in space throughout the history of the Universe, and is still going on today. Unfortunately for these interesting speculations, the interstellar evolution of life is improbable because of the low temperature and density of molecules in space. In these conditions, the collisions between molecules which give rise to the chemistry of life are a rare event, and evolution is almost certainly too slow to produce a living organism, even when measured on a billion-year time scale.

Finally, in addition to atoms and molecules, the interstellar medium also contains numerous small particles of solid matter known as *interstellar dust*. The size of the particles is about one ten-thousandth of a centimeter. This is the same as the size of smoke particles. The composition of the dust is uncertain, but probably consists mainly of particles of iron, carbon, and rocky materials.

Interstellar dust makes up approximately one percent by weight of the matter in the interstellar medium. It is concentrated almost exclusively in the central plane of the galactic disk because that is where the density of all materials is greatest, and is, therefore, the region in which atoms and molecules are most likely to collide frequently and stick together to form solid particles.

## A RADIO MAP OF THE GALAXY

How do we know that the Milky Way Galaxy has spiral arms? It is a simple enough matter to determine the structure of other galaxies than our own by photographing them through a large telescope. But the solar system is immersed in the Milky Way Galaxy, and its view of the multitude of stars around us — except for relatively close neighbors — is blocked by intervening clouds of dust. Charting the shape of the Milky Way Galaxy from the vantage point of our solar system is as difficult as constructing a street map of Manhattan from the middle of Central Park on a foggy night.

The most complete answer has come from observations in radio astronomy. Radio waves, unlike waves of visible light, are not absorbed by the clouds of dust that exist between our solar system and the distant parts of the Milky Way Galaxy. It is this same property that makes radio waves valuable in communications on the earth; if they were not able to pass through clouds of particles, radio broadcasts would be blacked out on every cloudy day. Radio waves in space, reaching the earth with relatively little interference, have provided most of the information that we possess regarding the spiral structure of the Milky Way Galaxy.

Stars cannot be the source of these radio waves. The sun, for example,

*Figure 8.2    A radio map of the Galaxy produced by Leiden Observatory from observations of the 21-centimeter line.*

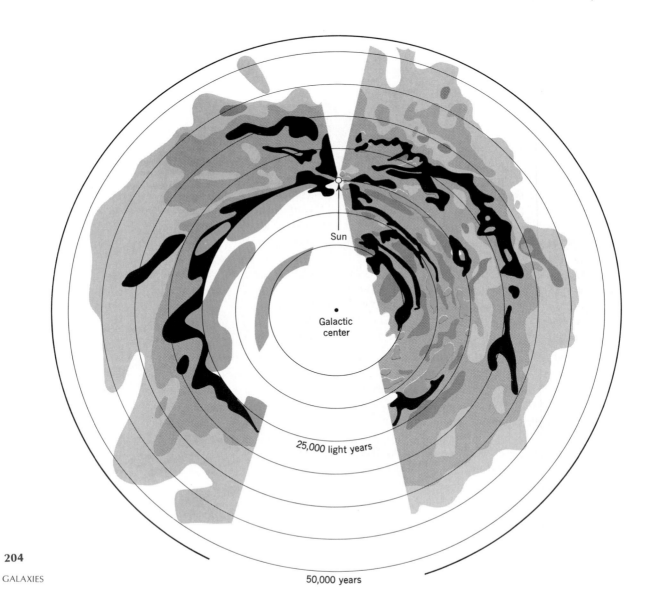

emits only a millionth of its energy in this form. The radio waves in the Galaxy are produced primarily by the interstellar medium. If our eyes were sensitive to radio, all the interstellar hydrogen in the Galaxy would seem to us to be emitting a soft glow. By charting the intensity of this "radio glow" in all directions around us, we could construct a picture of the distribution of hydrogen in the Milky Way Galaxy. Radio astronomers have done this using large antennas. The results indicate that the hydrogen in the Milky Way Galaxy is concentrated in distinct lanes separated by wide spaces containing relatively little hydrogen. The radio picture of the Milky Way Galaxy, shown in Figure 8.2, strongly suggests that the Milky Way Galaxy has a spiral-arm structure, although it is not as clear as photographs of distant spiral galaxies taken in visible light.

## The 21-Centimeter Line

Just as hydrogen atoms radiate energy at characteristic wavelengths in the visible region, they also radiate energy at characteristic wavelengths in the radio region. The most intense hydrogen line in the radio region has a wavelength of 21.1061 centimeters, or about eight inches.[1] Consequently the spiral arms of the Galaxy, where hydrogen is concentrated, emit a strong glow of 21-centimeter radiation.

In most directions in space we receive radio waves from several spiral arms simultaneously. This fact increases the difficulty of mapping the Galaxy by radio. Referring to Figure 8.3, imagine that a radio astronomer has oriented his radio antenna so that it is pointed along the dotted line. As Figure 8.3 shows, he will receive radiation from the hydrogen located in two spiral arms—labeled A and B—which lie in his line of sight when he observes in this direction. How can he separate the two signals? His task is made feasible by the rotation of the Galaxy, which gives the spiral arms a relative motion with respect to our solar system. The relative motion produces a Doppler shift in the wavelength of the 21-centimeter line emitted by each arm. If an arm of the spiral is moving *toward* us, or the sun is moving *toward* that arm, the 21-centimeter line emitted by this arm will be shifted toward shorter wavelengths, and the astronomer will observe a blue shift. If the relative movement is *away* from the sun, the shift will be toward longer wavelengths, that is, the line will display a red shift.[2]

---

[1] Although referred to as a radio wavelength, 21 centimeters is shorter than the wavelengths used in normal radio communications. Twenty-one centimeter radiation lies in the region of the electromagnetic spectrum used by aircraft and ship radars for navigation.
[2] If this radiation were in the visible region, the shifts would be toward the blue end or the red end of the spectrum, respectively. By analogy with the visible region, astronomers refer to these shifts as blue shifts and red shifts, regardless of the region of the spectrum that is actually involved.

Figure 8.3 (a) A drawing of the Galaxy showing the direction of a radio observation of the 21-centimeter signal.

The amount of relative motion, and therefore the amount of the Doppler shift, is usually different for every arm. Thus, the astronomer will receive radiation of two different wavelengths—all in the neighborhood of 21 centimeters—from the two spiral arms A and B that lie in his line of sight.

In the case shown in Figure 8.3, representing our own Galaxy, the rotation of the entire Galaxy is clockwise. Therefore the sun is moving toward the right in this figure. Arms A and B are also moving toward the right, but not as rapidly, because the farther an object is from the center of the Galaxy, the slower is its motion around the center.[3] Thus, the relative motion of the sun is toward arms A and B, and the astronomer should see a blue shift in the 21-centimeter radiation from each of these arms.

The radiation from arm A should be shifted only a small amount toward the blue, because this arm is only slightly farther out from the center of the Galaxy than the arm in which the sun is located, and its velocity is only slightly less than that of the sun. That is, their relative motion is small. The radiation from arm B should have a larger blue shift because this arm is farther out in the Galaxy and its velocity around the center of the Galaxy is considerably smaller. Therefore the sun has a substantial relative mo-

[3] For a body orbiting around a central mass under the attraction of gravity, the orbital speed falls off as one over the square root of the distance from the center (see Chapter XII).

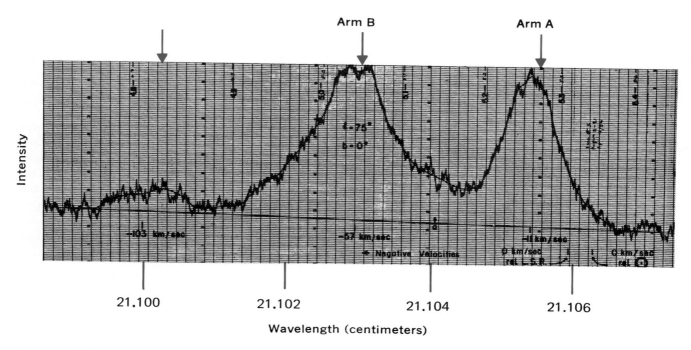

**Arm B**        **Arm A**

Wavelength (centimeters)

tion toward this arm.[4]

Figure 8.4 shows the actual 21-centimeter signal that the astronomers received in a similar case. It has one peak at a wavelength slightly shorter than 21.106 centimeters. This peak is radiation from arm A, shifted a small amount toward the blue as expected. A second peak appears at a still shorter wavelength. This peak comes from arm B. Its shift toward the blue is several times greater than for arm A, indicating that arm B is considerably farther out in the Galaxy than A. A third and exceedingly faint peak is also visible at a still shorter wavelength (arrow). This third peak of radiation must emanate from an arm very far out in the Galaxy. The peak is weak partly because the arm is far away and partly because the density of the matter in this arm is low.

*Figure 8.4    Twenty-one centimeter signal received in a case corresponding to Figure 8.3.*

### Radio Emission from Interstellar Formaldehyde

The 21-centimeter method is a powerful tool for probing the structure of the Galaxy, but it suffers from a major defect. It fails to reveal the regions in the spiral arms where the density of hydrogen gas is greatest, although these regions are the most interesting of all because they are the

---

[4] The blue shift is also influenced by the fact that the various arms cross the line to the sun at different angles, hence different fractions of their velocities contribute to the relative motion.

ones in which new stars, and perhaps new planetary systems, are forming. Observations of the 21-centimeter line cannot detect exceptionally dense clouds of hydrogen because in such clouds the hydrogen atoms combine to form hydrogen molecules. These molecules do not emit 21-centimeter radiation. Thus, the 21-centimeter line becomes weak or undetectable when coming from just the directions where we would expect it to be strongest.

The discovery of a substantial abundance of formaldehyde in space promises to remedy this deficiency. The formaldehyde molecule has a strong absorption line at a wavelength of 6 centimeters. This means that background radiation at this wavelength, coming from distant parts of the Galaxy to our solar system will be absorbed en route by interstellar formaldehyde. The absorption is detectable by radio astronomers as a pronounced dip in the intensity of the otherwise uniform background radiation, at a wavelength of 6 centimeters. As in the case of the hydrogen emission line, the formaldehyde absorption lines generally show two or more distinct peaks at neighboring wavelengths. These peaks are the result of absorption of the 6 centimeter wavelengths as they pass through several spiral arms on their way to us. Unlike the 21-centimeter line, the formaldehyde line is not weakened by concentrations of hydrogen, and is, therefore, the most promising tool known today for the investigation of the denser regions in the Milky Way.

## CLUSTERS OF STARS

Most of the stars in the Galaxy move about freely among their neighbors. However, a few are bound together into groups called clusters, containing as few as five or six stars and as many as one million. Each star in a cluster is bound to the others by the gravitational force of the entire cluster. The cluster as a whole moves about in the Galaxy, with its individual member stars orbiting around the center of the cluster like a swarm of bees. A star cluster is formed when many stars condense simultaneously out of one large cloud of matter. All the stars in the cluster condense out of the cloud in a very short interval of time. They may not all appear at precisely the same moment; there may be a spread of as much as several million years between the first star and the last to form in a cluster; but a million years is still a very short time compared to the lifetime of an average star. Thus, all the stars in a given cluster can be regarded as having been born simultaneously.

Clusters are very valuable because they are groups of stars with a common birthday. If a random sample of the stars in the sky is selected, and their H-R diagrams are plotted, we obtain a very blurred picture of a star's life owing to the circumstance that the sample of the family of stars usually contains stars of all different ages. A sharper picture would result if separate H-R diagrams could be plotted for each stellar age-group. Clusters provide the opportunity to plot such diagrams. As we will learn in the following sections, there is evidence that our Galaxy contains clusters of stars whose birthdates span the full period of years from the formation of the Milky Way Galaxy to the present era. The comparison between the H-R diagrams of Milky Way clusters of various ages and the H-R diagrams calculated from the theory of stellar evolution, described in Chapter 7, provides the most important single observational check on that basic theory, and also provides a means for measuring the age of the Galaxy and the age of some of its stars with a remarkable degree of precision.

## Galactic Clusters

Thousands of star clusters, containing as few as five or six stars in some cases, are scattered throughout the galactic disk. Because they are invariably located within the galactic disk, these star clusters are known as *galactic clusters.* They are also called *open clusters,* but that term will not be used in this book.

An example of a galactic cluster is shown in Figure 8.5, taken through the Mount Palomar 200-inch telescope. This cluster, designated M67, is located in the direction of the constellation Cancer at a distance of 2500 light-years.

*Galactic Clusters and the H-R Diagram.* Significant information can be obtained by plotting the H-R diagram for the stars of a single galactic cluster. The age of the cluster can be measured with surprising accuracy in this way, and at the same time observational confirmation can be ob-

*Figure 8.5   A galactic cluster.*

tained for the entire theory of stellar evolution developed in Chapter 7.

How can a young galactic cluster be distinguished from an old one? The answer can be found in its H-R diagram. You will remember that the stars are arranged along the Main Sequence line of the H-R diagram according to their masses. The most massive stars—the ones that are extremely bright and very blue—are at the upper end of the Main Sequence and the least massive stars—the ones that are dim and red—are at the lower end (see Figure 5.12). The key to determining the age of a galactic cluster lies in the fact that very massive stars in the cluster—the ones that would be plotted at the top of the Main Sequence—have very short lifetimes. For example, a star with 10 solar masses lives for only 20 million years before its fuel is burned up. This is one five-hundredth of the life of the sun. If the H-R diagram for a cluster shows stars at the upper end of a Main Sequence, where stars of 10 solar masses are to be found, it follows immediately that this cluster cannot be more than 20 million years old. It is a very young and newly-formed cluster in comparison to our 10 billion-year-old Galaxy.

Let us carry this reasoning further. Suppose that the H-R diagram for a cluster shows no extremely bright, blue stars but does contain a number of moderately bright stars of bluish-white color, lying fairly far up on the Main Sequence but not at the very top. For example, suppose that the brightest stars in the diagram are approximately 100 times more luminous than the sun, and have temperatures around 10,000°K. Reference to the H-R diagram in Figure 5.12 indicates that these stars would have about 3 solar masses. A star with 3 solar masses lives about 300 million years. If this cluster were considerably older than 300 million years, those stars would not be present on its H-R diagram. On the other hand, if it had lived for substantially less than 300 million years, stars that are still brighter and bluer would be present on its diagram, but they are not. Therefore, the cluster must be approximately 300 million years old.

Of course, the reasoning would be invalid if this cluster had no star greater than 3 solar masses when it was first formed. That is a possible occurrence for a small cluster with, say, only five or six stars. However, if the cluster is relatively large, with, say, 100 stars or more, we can assume that the full range of stellar masses will be represented in it, from the least massive to the most massive stars.

*Three Examples.* As an example of a very young cluster we take the Pleiades, which are a very familiar cluster containing six stars clearly visible to the naked eye (see Color Plate 9).[5] This cluster is 400 light-years away. Examination through a telescope reveals that the cluster contains more than 100 stars in all. The H-R diagram for most of the stars in the

---

[5]In the color insert each of the six stars is surrounded by an intensely blue sphere. This sphere does not belong to the star itself but is dust which reflects the light from the star at its center. The six stars themselves emit most of their energy in the blue and the ultraviolet because they are extremely hot.

Pleiades Cluster is shown in Figure 8.6a. Notice the six extremely luminous and blue stars at the top of this diagram. They are the same six stars that are visible to the naked eye. A comparison with the Figure 5.12 indicates that the brightest of these stars is approximately 6 solar masses. A star of this mass lives for 60 million years. This makes the Pleiades cluster roughly 60 million years old.

Now consider the H-R diagram for the Praesepe cluster, located in the direction of Cancer at a distance of five hundred light years. This cluster also contains approximately 100 fairly bright stars. Its H-R diagram is shown in Figure 8.6b. We see that the extremely luminous and hot stars that marked the youth of the Pleiades cluster are missing from this diagram. The brightest stars it contains are 70 times as luminous as the sun, corresponding to a mass of 3 solar masses. As noted above, a star of this mass lives for 300 million years, which must, therefore, be the approximate age of the Praesepe cluster.

As a final example, consider the H-R diagram for the cluster M67 located in the constellation Cancer at a distance of 2500 light years. This cluster has 80 fairly bright stars. A glance at its H-R diagram, Figure 8.6c, indicates that M67 is far older than the Pleiades or the Praesepe clusters, for nearly all the bright blue stars in the cluster have disappeared,[6] and the brightest stars still present in abundance have clearly peeled off to the right, away from the Main Sequence. When the calculated positions of the stars in the cluster are compared with their observed positions on the H-R diagram, the best fit for the cluster as a whole is obtained for an age of ten billion years, indicating that M67 appeared when the Universe itself was newly formed.

## Globular Clusters

Although the galactic clusters are the most numerous star clusters in the Galaxy, they are not the largest. Clusters containing up to one million stars, known as *globular clusters,* are found outside the plane of the galactic disk in the surrounding region of space known as the galactic halo. Our Galaxy contains more than 100 globular clusters, scattered at random throughout the spherical volume of the halo. Although most are located outside the galactic disk, the globular clusters are still regarded as members of the Galaxy because they are bound to it by the gravitational force of the matter in the galactic disk. Figure 8.7 shows the typical cluster M3.

*The Ages of the Globular Clusters.* All globular clusters are believed to be very old, nearly as old as the Galaxy, because of their location far out-

---

[6] A few hot, blue stars, called blue stragglers, still remain on the Main Sequence. No satisfactory explanation has been given for these stars, whose properties indicate that they long since should have come to the ends of their lives.

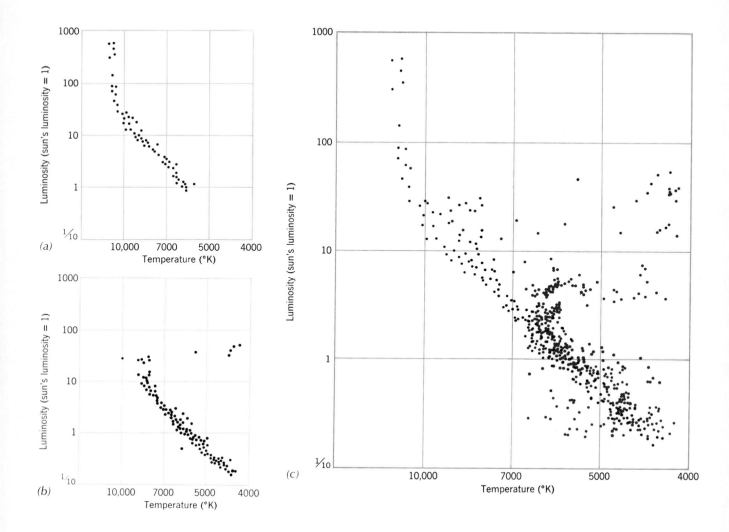

*Figure 8.6 H-R diagrams for galactic clusters:*
*(a) Pleiades*
*(b) Praesepe*
*(c) M67, with the Pleiades and Praesepe clusters added. Luminosity and temperature values are approximate.*

side the galactic disk. This belief is based on our ideas regarding the history of the Galaxy. According to one theory, when the Galaxy was young, it is thought to have had the form of a large spherical cloud of gaseous matter. Clusters of stars could have begun to condense out of the cloud at that time.

As time passed, the Galaxy contracted to its present disk-shaped form. Some groups of stars, which were the first to appear in the Galaxy, were left behind as the cloud contracted. These were the globular clusters. After perhaps $10^9$ years the Galaxy would have completed its contraction and would have the shape it has today, in which nearly all its matter is concentrated in a very flat disk. However, the globular clusters, which had formed during the first $10^9$ years, would have remained outside the disk in the galactic halo, where they were born. From that time to the present, the appearance of the Galaxy has not changed appreciably.

This line of reasoning has led astronomers to the conclusion that globular clusters not only are very old but were, in fact, the first stars to form in the Milky Way Galaxy. Since the Galaxy is believed to be 10 billion years old, that must also be the age of the globular clusters.

*The H-R Diagrams of Globular Clusters.* What will the H-R diagrams of globular clusters look like? If these clusters are really very old, nearly all their stars should have peeled off the Main Sequence line. Inspection of the H-R diagram for M3, a typical globular cluster in our Galaxy, shows this to be the situation (Figure 8.8). All stars in M3 have left the upper part of the Main Sequence and are scattered about on the diagram in various stages of intermediate or advanced evolution. Some stars presumably have disappeared from the diagram in the aftermath of a supernova explosion. The only exceptions are the very small stars at the extreme lower end of the Main Sequence, whose lifetimes range up to a trillion years. These stars will anchor the bottom of the Main Sequence as long as we and our descendants exist in this Galaxy.

M3 is a representative example of globular clusters. We have selected it because it has been more carefully studied than any other globular cluster. All globular clusters in the Milky Way Galaxy have H-R diagrams which resemble that of M3; all have ages, deduced from their H-R diagrams, of approximately 10 billion years. In the case of M3, the precision of the age determination is greater; its age is thought to be 11 billion ± 2 billion years. Presumably this is not only the age of M3, but also the age of the Milky Way Galaxy and perhaps the age of the Universe.

## STELLAR POPULATIONS

When Hertzsprung and Russell plotted the first versions of their now-famous diagram, they used stars that were relatively close to the sun because accurate distances could be most easily obtained for these stars.

Figure 8.7   The Globular cluster M3.

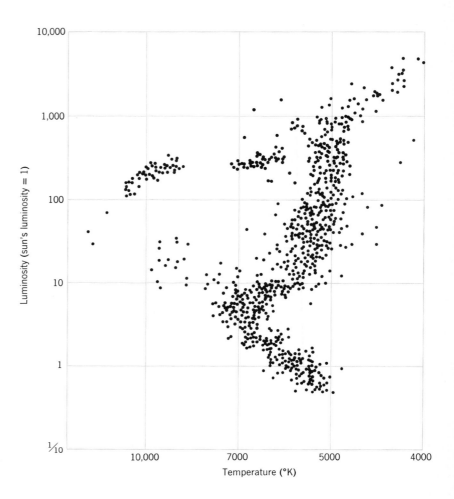

*Figure 8.8   H-R diagram for the M3 globular cluster.*

Their original diagrams looked more or less the same as the one that we have shown on page 128 of the stars within 20 light-years of our solar system, with many stars lying on the Main Sequence and a few stars in the red giant and white dwarf regions. However, the H-R diagram for the globular cluster M3, which we discussed in the previous section, presents a very different appearance.

First, this diagram has hardly any stars in the upper one-half of the Main Sequence above the position of the sun. Second, it has a very densely populated track of stars leading from the middle of the Main Sequence along a curved path upward and to the right into the red giant region. Third, it has a moderately large number of stars running more or less horizontally across the diagram from the red giant region back to the upper region of the Main Sequence. These three major differences indicate or suggest that the stars of the M3 cluster and other globular clusters represent a type of star population different from stars such as the sun and its neighbors. The globular cluster stars are not the only stars that are different, as a group, from stars like the sun and its neighbors. Individual isolated stars that are moving around in the halo have, when their data are collected and put on an H-R diagram, the same kind of diagram as stars in a globular cluster. Finally, many stars in the nucleus of the Galaxy also appear to have the same special property. At least, we conjecture that this is true because we have observed it to be true for the stars in the nucleus of our neighboring galaxy, the Andromeda galaxy, which is believed to be a galaxy very similar in all respects to our own.[7]

These differences are very striking. It is just as though you were to make up a diagram on which you plotted height against weight for the entire population of a country, and were to find that there were two completely distinct distributions showing up in your plot. You would probably conclude that there were two distinct populations or types of people living in that country. Astronomers have come to the same conclusion about the stars. They call the stars whose H-R diagrams resemble that of the sun's neighbors the *population I* (one) stars. The stars whose H-R diagrams resemble those of the globular clusters, galactic halo stars, and many galactic nuclei stars, are called the *population II* (two) stars.

Other differences between the two populations have been observed, in addition to the basic difference in their H-R diagrams. The most important of them is the difference in their chemical compositions or ingredients, as determined from the study of their spectral lines using the methods explained in the chapter on atomic spectra. According to their spectra, population II stars have very little of the metals and heavy elements. In some extreme cases, their spectra show no evidence at all of these heavy elements, indicating that those particular stars are composed entirely of hydrogen and helium. Stars of population I, on the other hand, have substantial amounts of heavy elements, ranging up to as much as several percent in extreme cases. What is the meaning of the difference between the two star populations? What types of stars are contained in each population? A clue to the answer has already been provided in the previous section, where we found that globular clusters are relatively old groups of stars.

Age is the essential difference between the population I and population II stars. All stars in population I are relatively young; all stars in population II are relatively old. By relatively old, we mean dating back to the beginning, or nearly to the beginning, of the Universe. As soon as this difference is clear, all other properties of the two populations fall into place.

If this is so, population II stars were formed out of pristine materials of the original Universe, which were, according to the most widely accepted cosmological theories, a mixture of hydrogen and helium atoms (Chapter 11). Therefore, these stars should contain nothing but hydrogen and helium. As we have seen in Chapter 7, heavy elements, extending from helium, carbon, nitrogen and oxygen up through iron to the heaviest

[7] The dense blanket of dust that lies in the central plain of our Galaxy conceals the nucleus of the Milky Way Galaxy from our direct examination in visible wavelengths.

substances such as gold, lead, and uranium, all are manufactured in the bodies of stars by the nuclear reactions that take place during their lifetimes. These elements are then spewed out to space in final explosions that mark the deaths of these stars, there to mix with the primitive hydrogen and helium gases of the original Universe. Later in the history of the Universe and the history of this Galaxy, other stars are formed out of this mixture, and the contents of these stars will contain some amounts of the heavier elements. The later a star is formed, that is, the closer to the present time, the more of the heavier elements it will have among its ingredients, because the abundance of these elements is rising steadily in the course of time as more and more stars come to the end of their lives, explode, and contribute their contents to the space around them. Another way of looking at the difference between the two populations is to say that population II stars are the first generation of stars to be formed in a galaxy, and population I stars are a mixture of all later generations.

One other point should be mentioned. Population II stars never are associated with gas and dust. For example, the globular clusters are entirely free of gas and dust, in contrast to the stars located within the galactic plane, which are surrounded by, and often embedded in, dense concentrations of gaseous material. Also, population II stars rarely contain blue giants, but population I stars usually include an appreciable number of blue giants. The explanation of this set of properties is also clear when we realize that population II stars are the first to form in a galaxy and are very old. Since they are old, they cannot include blue giants, for blue giants are very hot, massive stars which burn up their resources quickly and disappear in a time that is typically some millions of years or so. By the same token, population II stars should not be embedded in or surrounded by gas and dust, for they are a group of stars that has been around for a long time, and all the gas and dust that was originally present has been used up in forming stars; that is, none remains. Population I stars, on the other hand, exist in regions that are richly abundant in gas and dust and in which star formation is still going on at the present time.

## Questions

1. What is a galaxy?
2. Draw the Galaxy to scale showing the location of the sun. Indicate the shape and size of the nucleus, the disk and the halo, and distribution of the globular clusters.
3. What molecules have been found in the interstellar medium? What is their significance for life?
4. Why are 21-centimeter radio waves more useful than visible radiation in charting the structure of the Galaxy? Why is the 6-centimeter radiation from formaldehyde more useful than 21-centimeter radiation?

**5.** How is the Doppler shift used to determine the structure of the Galaxy?

**6.** What are the major differences between the two types of star clusters found in the Galaxy?

**7.** Suppose globular clusters contained the elements heavier than helium in the same proportion in which they are found in the Cosmos as a whole. What would this imply regarding the importance of stellar evolution in the synthesis of the elements, relative to the importance of the nuclear reactions occurring in the Big Bang? Explain.

**8.** From your study of stellar evolution in Chapter 7, and the discussion of the ageing of clusters in this chapter, plot an H-R diagram showing the successive positions of the stars in a cluster, as the cluster ages from birth to an age of 10 billion years.

**9.** How do the H-R diagrams of the stars of some globular clusters tell us the age of the Galaxy?

**10.** Summarize the differences between the H-R diagram in Figure 5.12 and the H-R diagram for the globular cluster, M3.

**11.** What are the differences between population I and population II stars? Interpret these differences in terms of the history of the Galaxy.

# 9　Galaxies

Stars are the basic units of population in the sky. They are clustered in groups called galaxies, just as individuals are clustered together in nations. Each galaxy is separated from other galaxies by almost completely empty space, containing no stars and no more than a few atoms of hydrogen. We have seen how individual stars are born by condensing out of hydrogen, then evolving to maturity, and finally, expiring. What about galaxies — the nation-states of the heavens? Do they also go through a life cycle of birth, evolution, and death? Do they end their lives in an explosion? Or do they fade away?

As we have seen in Chapter 1, galaxies, like stars, come in many shapes and sizes. In Chapter 7, we found that most of the different varieties of stars are simply stages in the lifetime of a typical "normal" star. Is this also true for galaxies? Are elliptical, spiral, and barred spiral galaxies simply stages in the lifetime of a typical "normal" galaxy? Or are they completely distinct species of galaxies, as unrelated as elephants and giraffes?

In the case of stars, the latest astronomical discoveries have revealed

*The Sombrero galaxy.*

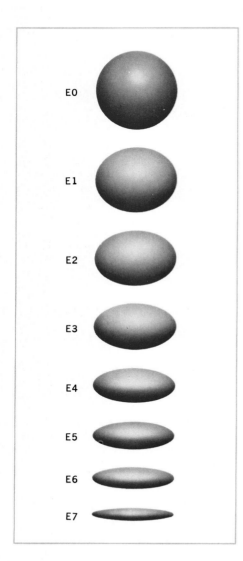

*Figure 9.1    The classification of elliptical galaxies.*

the answer to these questions. But in the case of galaxies, astronomers have made far less progress. They do not even know the answers to the basic questions, such as how a galaxy changes its shape and luminosity during the course of time, or whether it changes at all. Some astronomers think that a young galaxy starts out with a spherical shape, becomes elliptical in its middle age, and ends as a spiral in its old age. Other astronomers think that the pattern runs the other way, starting with young spirals and ending with old spherical galaxies. Still others believe that the various types of galaxies retain the same basic shape throughout their lives. That is, they think that a spiral galaxy, for example, is born with that shape, and remains a spiral galaxy throughout its lifetime. In the study of galaxies, astronomers are 50 years behind their study of the stars. They are at the stage that astronomers were when they drew up the Hertzsprung-Russell diagram, but could not understand its meaning.

In the next few years rapid advances should be made, because we are getting new information about galaxies by studying them in the infrared region of the spectrum, and soon we will be able to study them at other wavelengths by using telescopes in space. We are also learning about galaxies with the aid of fast electronic computers. The computers enable astronomers to calculate the properties a galaxy should have according to the basic laws of physics. With a fast computer, they can bring the galaxy into the laboratory and study it there. In 10 years at most, we will probably know the answers to many of the basic questions about galaxies.

## TYPES OF GALAXIES

Edwin Hubble was an American astronomer who spent years studying galaxies with the 60- and 100-inch telescopes at Mount Wilson. He classified 600 galaxies and found that they could be divided into four major types: elliptical, spiral, barred spiral, and irregular. He published his results in 1926.

## Elliptical

An elliptical galaxy looks like a partly flattened, luminous sphere. Some of these galaxies are nearly perfect spheres; others are moderately flattened spheres; still others are extremely flat objects, like pancakes. Examples of the elliptical galaxies are shown on the following page.

Hubble labeled all elliptical galaxies according to their degree of flattening. He designated the spherical galaxies as ''E0,'' the slightly flattened galaxies as ''E1,'' and so on down to ''E7,'' for the flattest, most elliptical ones. The classification is illustrated in Figure 9.1.

The three elliptical galaxies shown in Figure 9.2 are types E1, E3, and E7, respectively.

Elliptical galaxies were 20 percent of Hubble's sample. Their main property, apart from their shape, is the fact that they contain only old stars. No young stars, such as blue giants, are found in them. Most elliptical galaxies are relatively large and massive, but surprisingly faint in proportion to their mass. Their relative faintness results from the fact that they contain only old stars. Elliptical galaxies were under represented in Hubble's count, and probably constitute the majority of all galaxies.

Elliptical galaxies also appear to have very little gas and dust in the space between the stars. Apparently most of the atoms in an elliptical galaxy already have been gathered together to make stars. This fact, together with the absence of young stars, suggests that elliptical galaxies are relatively old.

## Spiral

Fifty percent of all galaxies were classified by Hubble as spiral. They are the largest single class of galaxies. They have the shape of a flattened disk of matter, with a small, spherical nucleus bulging out of the center. The stars in a spiral galaxy are concentrated in the central nucleus and in the spiral arms that radiate out from the center. These arms give the spiral galaxy its name.

Because of the way the spiral arms are curved, the spiral galaxy gives the impression to the observer that it is rotating, like a pinwheel in a Fourth of July fireworks display. Of course, because of their great size, they rotate too slowly for us to see the wheeling movement during a single night or even a year. In our own Galaxy, a typical example of a spiral galaxy, the pinwheeling motion carries the sun around once every 200 million years. Although we cannot see rotation as slow as this, we can detect the evidence of it by measuring the Doppler shift in the light from stars located near the edge of the galaxy. To a cosmic observer watching the heavens for several hundred million years, the spiral galaxies would look like a fireworks display.

Hubble classified spiral galaxies into three basic types—Sa, Sb and Sc—according to the degree of openness of the spiral arms. Figure 9.3 illustrates the sequence of shapes. The Sa type has a number of spiral arms very close together and almost overlapping, so that the spiral pattern

Figure 9.2 Elliptical galaxies of types E1 (NGC 4278), E3 (NGC 4406) and E7 (NGC 3115).

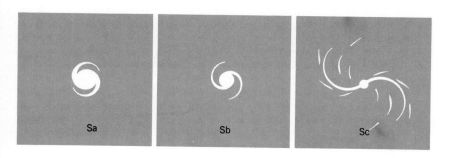

Sa  Sb  Sc

Figure 9.3 The classification of spiral galaxies.

(a)

(b)

(c)

Figure 9.4  Face-on photographs of spiral galaxies of types: (a) Sa (NGC 2811); (b) Sb (NGC 3031); (c) Sc (NGC 628).

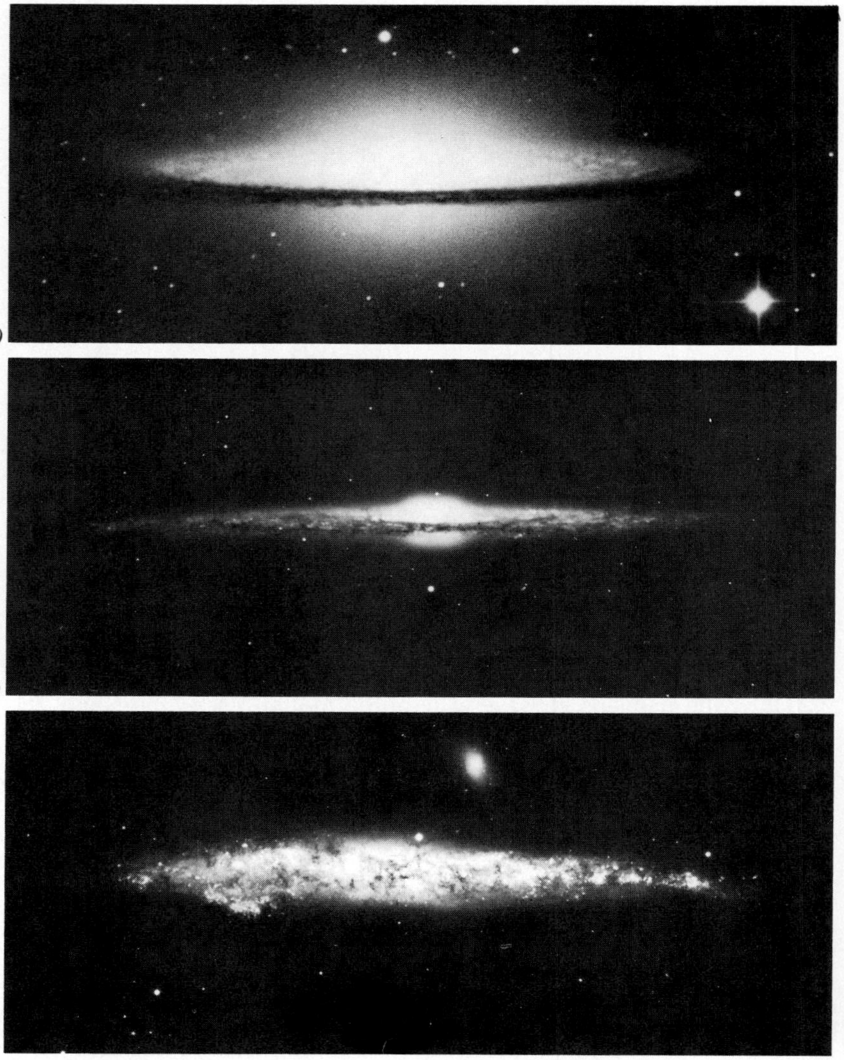

Figures 9.5  Edge-on photographs of spiral galaxies of types: (a) Sa (NGC 4594); (b) Sb (NGC 4565); (c) Sc (NGC 4631).

can hardly be seen. Figure 9.4a shows an Sa galaxy (NGC 2811). In Sb spirals, the arms are well defined, although still relatively close together (Figure 9.4b). In Sc type spirals there are only a few arms, each being clearly separated from the others (Figure 9.4c).

The mass of the galaxy is most highly concentrated in the disk in the case of the Sc type, and least concentrated in the disk in the Sa type. The trend towards concentration of matter in the disk of the galaxy is shown in the sequence of three edge-on galaxies shown in Figure 9.5a,b,c.

(a)

(b)

(c)

All spiral galaxies are large and massive, with about the same mass and number of stars as our Galaxy. The average spiral Galaxy has a mass of $10^{44}$ grams and contains about $10^{11}$ stars, including young stars, middle-aged stars and old stars.

Although most of the stars in a spiral galaxy are located in the nucleus, or in the central disk, a small fraction—about one percent—are found quite far out from the central disk. These stars are concentrated in globular clusters.[1] An average spiral galaxy contains several hundred of these clusters, randomly scattered in the space around the disk, as shown in Figure 9.6. This spherical volume, containing scattered globular clusters plus a few isolated stars, is called the *halo* of the galaxy.

A globular cluster contains about one million stars and is approximately

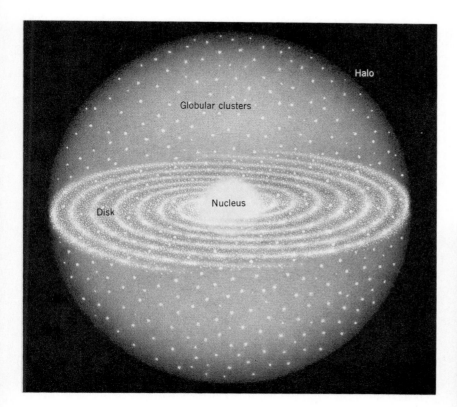

*Figure 9.6 The complete structure of a spiral galaxy.*

one hundred light years in diameter. A photograph of M13, a globular cluster in the halo of our Galaxy about 20,000 light-years from the sun, is shown in Figure 9.7. Many globular clusters may also be seen in the photograph of the Sa/Sb spiral at the top of page 222 (NGC 4594). Most of the points of light immediately surrounding the nucleus of this galaxy are globular clusters.

Globular clusters are believed to be formed at a relatively early stage in the evolution of a galaxy, before the gas of the galaxy has collapsed to a

*Figure 9.7 The globular cluster M13.*

[1] All galaxies, regardless of shape, appear to contain globular clusters, but in spirals only, they do not seem to form part of the general shape of the galaxy.

Figure 9.9 Barred spiral galaxies. From top, SBa (NGC 175), SBb (NGC 1300) and SBc (NGC 1073).

thin disk as the result of rotation. One of the reasons for this belief is the fact that these clusters only contain old stars.

Globular clusters contain no new stars or gas or dust out of which new stars are formed. Apparently all their gas and dust was used up a long time ago in forming stars. As we have seen in Chapter 8, a plot of the stars in the globular clusters of our Galaxy gives a powerful means for determining the age of the Galaxy.

### Barred Spiral

About 30 percent of all spiral galaxies were classified by Hubble as barred spirals. They resemble Sc spirals, with two arms radiating outward. However, instead of coming out of a small central nucleus, the arms project from the ends of a bar-shaped concentration of matter. This bar gives the galaxy its name.

Hubble classified barred spirals into three types—SBa, SBb, and SBc—depending on the degree of openness of the spiral arms, following the same scheme for classifying ordinary spirals. Barred spirals are similar to ordinary spirals in every respect, such as the types of stars they contain, their mass and luminosity, their halos, and their distribution of gas and dust. Figure 9.8 illustrates the sequence of shapes. Figure 9.9 shows barred spirals of the types SBa, SBb and SBc.

### Irregular

A small fraction of all galaxies do not fit into one of these categories. They do not have a clearly defined geometric form. These galaxies are called *irregular*. Most of them are relatively small with one-thousandth of the mass of a typical spiral or elliptical galaxy and only $10^8$ or $10^9$ stars. Irregular galaxies often contain many bright, young stars, and very few old stars. Frequently, an irregular galaxy is attached to a spiral galaxy, held captive by the gravitational force of the massive spiral. The Magellanic Clouds, which are bound to our Galaxy, are examples of captive irregular galaxies. Figure 9.10 shows the large Magellanic Cloud.

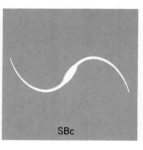

Figure 9.8 The classification of barred spiral galaxies.

*Figure 9.10   An example of an irregular galaxy: the large Magellanic Cloud.*

## THE EVOLUTION OF GALAXIES

According to current ideas in astrophysics, the galaxies were born first in the Universe, and the stars within the galaxies were born afterward. The main reason for believing this to be true is the fact that stars can be seen forming in galaxies at the present time, out of gas and dust. If all the stars were formed first, and then were clustered together later to form the galaxies, there would be no star formation going on today.

If galaxies came into being before any stars existed, then each galaxy must initially have been a great, formless cloud of gas, slowly contracting under the inward force of its gravity. Except for being much larger, it would look very much like a star that had just started to condense. As in the case of the star-cloud, the galaxy-cloud would commence to shrink under the inward force of its own gravity. No astronomer has ever seen a galaxy in the process of condensing, but our knowledge of the properties of gravity leads us to expect that they were formed in this way.

### Protogalaxies

Imagine that we are at an early stage in the life of the Universe, when space was occupied solely by hydrogen atoms, and there were no stars or galaxies yet present. At that stage the entire Universe must have been one supercloud of hydrogen. In the swirling motion of the atoms within this supercloud, now and then a number of atoms would come together by accident to form a momentary condensation. If the condensation were large enough—that is, if it included a sufficient number of atoms—the mutual gravitational attraction of the atoms would hold them together and prevent them from separating. Thus, the condensation, instead of dispersing to space again, would remain bound together as an isolated cloud, distinct from the rest of the supercloud around it. Such a condensed cloud of hydrogen, if it were large enough, would be the beginning of a galaxy.

A galaxy in the process of formation, before any stars have formed in it, is called a *protogalaxy*. In the course of time the protogalaxy contracts and its density rises, as a result of the continuing inward force of its own gravity. Throughout this time, pockets of gas continually form and dissolve in the swirling and eddying motion within the cloud. Whenever the density of one of these pockets of gas is high enough, it turns into a protostar. The protostars contract and heat up until nuclear reactions begin at their centers, at which point they turn into stars. When many stars have been formed, the protogalaxy becomes a galaxy. During the course of time, more and more of the gas of the original protogalaxy is swept up into stars. When much time has elapsed, the galaxy contains many stars and hardly any gas. Throughout this process the more massive stars undergo supernova explosions, returning some of their gaseous matter to space; but supernova debris mixes with the existing gas and is collected once more into the bodies of still more recently formed stars. Eventually, the galaxy consists only of modest sized, slow-burning red stars and nothing else; no massive hot stars, no supernovas, and no gas or dust. When the last faint stars have faded, the galaxy becomes a graveyard of black dwarfs, neutron stars and black holes in space. But a long time is required to reach this state. A small red star like Barnard's Star lasts for 10 trillion years, but the Universe, as we shall see in the next chapter, has probably existed for 10 billion years, which is only a thousandth as long. Thus, there are no completely dead galaxies in the Universe yet; the majority of the galaxies we can see are still in their prime.

**Theories of Galactic Evolution**

These ideas on the birth of a galaxy are very reasonable and most astronomers accept them as correct. However, they give no clue as to why galaxies have the distinctly different shapes—spiral, etc.—that Hubble found.

Hubble believed that each protogalaxy starts out as a more or less spherical cloud. When it turns into a galaxy, it is a spherical or EO type in Hubble's classification. Hubble suggested that in the course of time the spherical galaxies became increasingly elliptical, and then continued to flatten out, eventually becoming spirals or barred spirals. In the very last stage, the spirals broke up into irregular galaxies.

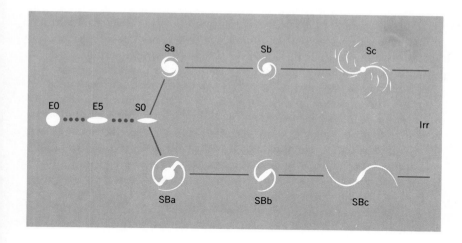

Figure 9.11 The Hubble diagram, illustrating his theory of the relationships among the types of galaxies.

Another theory is similar to Hubble's theory, but operates in the reverse direction. According to this theory, all protogalaxies are irregular in shape initially. Stars begin to form while the protogalaxy is still irregular. In the course of time, the irregulars collapse to a spiral or barred spiral of the Sc or SBc type. From there the arms gradually wind tighter and tighter around the nucleus, until an elliptical galaxy is formed. In time, the elliptical galaxy becomes a sphere. In this theory, all galaxies wind up as sphericals (Figure 9.12).

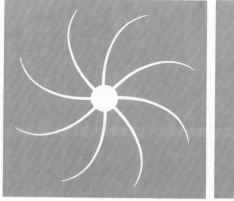

Figure 9.12 A second theory of galactic evolution: the arms wind up with the passage of time.

When Hubble first proposed his theory of galactic evolution—that galaxies proceed from ellipticals to spirals—astronomers knew very little about the types of stars the various galaxies contained. They also knew very little about the birth and death of stars. If Hubble had possessed the knowledge that we have today, he would have known that his theory must be incorrect. The key point is the fact, noted on page 221 that elliptical galaxies contain only relatively old, reddish stars, and no gas or dust in the space between these stars. If the description on page 221 is correct, these galaxies have passed the age of abundant star formation, and are in their declining years. It is not possible for such a galaxy to give rise to a spiral galaxy, containing large quantities of fresh gas and dust in which many recently formed stars can be seen.

Unfortunately, the reverse theory of galactic evolution — proceeding from the irregulars and spirals to the ellipticals and sphericals — is equally unacceptable, although for a different reason. The trouble with this theory is that both irregular and spiral galaxies contain very old stars, as old as the stars in the elliptical galaxies. This fact implies that the irregulars and spirals have been around as long as the ellipticals, and they cannot be newly formed as the reverse theory of galactic evolution suggests.

The only conclusion it seems possible to draw is that the different types of galaxies do not have a parent-child relationship. That is, it is not true that each galaxy evolves from one type or shape to another during its lifetime. Consequently, it is also not true that the galaxies that we see in the sky at the moment are galaxies of many different ages. Thus, we have part of the answer to the basic questions posed at the beginning of the chapter. We know that elliptical and spiral galaxies are not stages in the lifetime of a typical normal galaxy. They are completely different species.

The question now is, what basic property makes them different from one another?

## Explanation of Galactic Shapes

This brings us to the frontiers of research on galaxies. No one is certain of the answer, although there is one explanation that is more widely accepted than any other. It is believed that sphericals are galaxies that were spinning extremely slowly when they were first formed, so that there was no tendency for them to flatten out as they condensed. Ellipticals, on the other hand, were galaxies that were spinning at a moderate rate when they formed, and flattened out to some degree when they condensed.

A spiral, according to this view, would be a galaxy with a still greater amount of spin, which caused some of the matter within it to flatten out into the shape of an actual disk. A disk is, after all, essentially the same as a very flat ellipsoid. The nucleus of the spiral would be, according to this view, composed of the matter that was near the center of the original protogalaxy and, therefore, not much affected by the rapid rotational movement of the outer regions. This part would condense and would shrink down and collapse inward on itself without much flattening, and would be similar in its properties to a small spherical galaxy. In support of this conclusion, it is observed to be a fact that the nucleus of a spiral galaxy like ours has the same types of stars and other properties as a spherical or elliptical galaxy.

Why does the disk of a spiral galaxy still contain gas, dust, hot young stars and stars in the process of formation? If all galaxies are equally old, you would expect the star formation to be as far advanced in the disk as it is in the nucleus, or in spherical and elliptical galaxies. The answer probably is that the density of the matter in the disk is considerably lower than the density in the nucleus or the density in a spherical or elliptical galaxy, because the matter in the disk has been prevented, by its spinning motion, from moving inward to the center. Therefore, it can never condense to the high densities that the matter in the center contains. Thus, stars form relatively slowly in the disk, coming into being only on rare occasions when an exceptionally compressed pocket of gas is created by the buffeting of neighboring clouds.

What is the explanation for the arms in a spiral galaxy? One theory holds that the matter in the disk is concentrated into separate lanes containing a high density of gas and dust and numerous stars. These lanes, radiating out from the center of the galaxy, are curved into a spiral form by its spinning motion.

One of the troubles with this theory is the fact that in a rotating disk of matter, the angular velocity of the spinning disk falls off with the three-halves power of distance from the center. That is, if at a certain distance from the center a part of the disk rotates through 15 degrees in a certain interval of time, twice as far out from the center of the disk the matter will rotate for only 5 degrees in that same interval. As a consequence, as time goes on the outer regions of the arms are left farther and farther behind with respect to the center. After one or two rotations, the arms are tightly wound up around the central nucleus and the spiral form is no longer recognizable.

In other words, a spiral galaxy with an open structure, such as the Sc type, would have a lifetime of only a few hundred million years or so, that being the typical time of rotation of a spiral galaxy. There are far too many Sc galaxies in the sky to be consistent with such a short lifetime as this. This objection is one of the reasons why some astronomers are not satisfied with any theory of the spiral arms which considers them to be individual streamers of gas and dust, as distinct as the arms of an octopus.

The newest theory presents an entirely different picture. In this theory the matter of the central disk is assumed to be spread out in a fairly uniform way throughout the disk. According to theoretical calculations, if the matter is spread out in the disk in this way, density waves or ripples of density will travel through the disk. The theory indicates that the waves will radiate out from the center in a spiral pattern. The difference in density from crest to trough in this spiral pattern might be only 50 percent. However, owing to the very strong tendency of stars to form in regions of higher density, the amount of star formation that takes place in the regions of higher density can be vastly greater than that which takes place in the low density troughs. Where star formation is rapid, and many hot blue stars are being formed all the time, luminosity will be very great. Thus, to the observer recording the visible light from this disk, a very clear pattern of luminous spiral arms will be apparent.

What is the explanation for barred spirals? Some astronomers believe them to be galaxies whose initial rate of spin was exceptionally rapid, even higher than the typical rate of spin of a normal spiral. Theoretical studies of rotating clouds of gas indicate that when the rate of rotation is moderate, the gas flattens out into an ellipsoid or a pancake-shaped object; but when the rate of spin becomes very great, the pancake itself becomes elongated into a thick tube, something like a bar with rounded edges. Suppose that a protogalaxy has achieved this shape, and stars have begun to form out of the gas particles within the bar-shaped concentration of matter. The stars and the atoms of gas will be bound to the neighborhood of the "bar" by the force of the bar's gravity; some will circle around the "bar" in paths such as A in Figure 9.13.

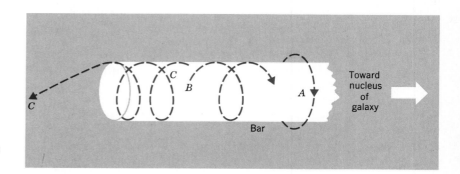

Figure 9.13 The possible paths of stars moving in the "bar" of a barred spiral.

Other stars will spiral toward the center of the "bar" in paths such as B, and still others will spiral, to the ends in paths such as C, depending on their original directions of motion along the "bar".

When a star or an atom reaches the end of the "bar", it suddenly becomes free of the constraint of the bar's gravity, and waltzes off into space, as in the trajectory C in Figure 9.13. If the "bar" is rotating, as is generally the case, the star or atom, moving outward from the edge, is left behind. A stream of stars and atoms will therefore follow a curved path as they leave the ends of the steadily rotating "bar" (Figure 9.14).

## PECULIAR GALAXIES

A small fraction of the galaxies in the sky have highly unusual shapes. These galaxies belong to the group called by Hubble the *peculiar galaxies,* which make up one or two percent of the galactic population of

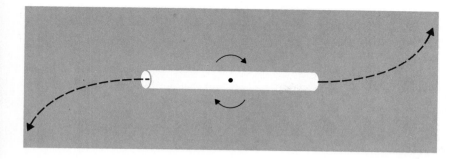

the heavens. Many of the peculiar galaxies look as though they were normal in shape at one time, but were subsequently distorted in appearance by some unusual event. In all the cases that have been examined thus far, the unusual event that altered the appearance of the peculiar galaxy seems to have been either a *collision* with another galaxy, or a gigantic *explosion* within the galaxy that literally blew it apart.

## Collisions Between Galaxies

Several hundred collisions or near collisions between galaxies have been photographed in the past 20 years. Apparently, these collisions between galaxies are much more probable than collisions between stars, for not a single star collision has ever been photographed or observed since the telescope was invented. Stellar collisions are exceedingly rare because the average distance between stars is so great in comparison to the size of the average star. In our Galaxy the distance between stars is about 30 trillion miles, which is millions of times greater than the size of a star.

The average distance between galaxies, on the other hand, is one or two million light-years, which is only 10 times greater than the diameter of a typical galaxy like the Milky Way or Andromeda. Under these circumstances, it is no surprise that when we look into the sky at any given time we see a number of galaxies either colliding with their neighbors, or on collision courses.

What happens when two galaxies collide? Figure 9.15 suggests the answer. These two colliding galaxies look as though they are moving straight through one another without interference. At first sight, this seems puzzling, since each of the two galaxies by itself gives the appearance of a luminous, relatively dense ball of gas. The explanation is, once again, that the stars in each galaxy are so very far apart in comparison to their size. When the two galaxies collide and occupy the same space at one time, the density of stars is temporarily double what it was in each galaxy alone, but even with the density doubled, the distances between stars are still so great that practically no collisions between stars occur throughout the period in which the galaxies as a whole are

Figure 9.15 NGC 2685: this object may be two galaxies in collision.

colliding. However, the transparency of galaxies to one another holds only for the stars they contain. The clouds of gas and dust that accompany the stars in each galaxy do not pass through one another without interference. The atoms in these clouds of gas collide repeatedly as the galaxies come into contact, piling up against one another, and creating pockets of high-density material. It is likely that new stars would form rapidly in these pockets. The birth of new stars is not apparent in this photograph, but it is a reasonable guess that something of that sort is taking place.

### Exploding Galaxies

Some peculiar galaxies look as though they have been disrupted by a violent explosion. In some cases the explosion seems to have hurled a jet of matter out of the galaxy. In other cases the entire galaxy seems to have blown up. The most striking exploding galaxy ever photographed is M82, located roughly 13 million light-years from us in the direction of the constellation Ursa Major. The photo of M82, Figure 9.16, shows long streamers or filaments of hydrogen, up to 14,000 light-years in length, moving outward at speeds as high as 600 miles per second. Tracing the

*Figure 9.16   M82 photographed in the red light of the hydrogen line at 6300Å. Photographs taken at this wavelength reveal the structure of the outward-moving filaments of hydrogen gas with particular clarity. M82 is the most striking example of an exploding galaxy.*

motions of these clouds backward in space and time, we can tell that the "explosion" took place roughly two million years earlier in the life of M82[2] (Color Plate 11).

The peculiar shape of M82 is not its only unusual feature. Recently, astronomers discovered that this galaxy is an exceptionally powerful energy source, sending far more energy out into space than the Milky Way Galaxy. Although astronomers had studied M82 for years, they had not known of its unusually high energy output until recently because they had observed M82 only in the visible band of wavelengths. In 1969 Frank Low measured the *infrared* radiation coming from M82 for the first time, and found that its infrared energy emission is $2 \times 10^{45}$ ergs/sec. The total luminosity of our Galaxy, summed over all wavelengths, is $10^{44}$ ergs/sec. This means that in the infrared alone M82 is emitting twenty times more energy than our Galaxy emits at all wavelengths. Since the mass of M82 is probably less than the mass of our Galaxy, M82 may be putting out at least 20 times more energy, *pound for pound,* than the total energy output of the Milky Way Galaxy.

The energy radiated by the Milky Way Galaxy comes from nuclear reactions going on in the stars of the Galaxy. How can these same nuclear reactions yield hundreds of times more energy—per pound of star-material—in the stars of the galaxy M82 than they do in the stars of our Galaxy? The attempt to explain exploding galaxies stretches astro-

---

[2] Since M82 is 13 million light-years away from us, the explosion must have actually taken place 15 million years ago in our time frame. After the explosion occurred, the light from it took 13 million years to reach us.

nomical theories to their limit.

Several theories have been proposed to explain the energy released at the centers of exploding galaxies like M82. One theory suggests that if stars are packed in tightly at the center of a galaxy, a chain of supernova explosions can occur in which one supernova triggers another, which sets off a third, and so on, releasing a large amount of nuclear energy in a short time. Another theory suggests that the explosion results from the collapse of a superstar, weighing as much as a million solar masses or more. Other theories of a still more unusual character have been suggested. In the next chapter we will see that the radiation from M82 is dwarfed by the energy released from other types of galaxies or galaxylike objects. When we have completed the roster of violent events in the cosmos by our discussion of these objects, we will return to the theories that attempt to explain the source of the energy.

### Questions

1. Briefly describe the geometry of the major types of galaxies. Describe the distributions of gas and dust in each class of galaxy.
2. What population(s) of stars (I or II) are found in each of the major types?
3. How do we know that most galaxies are about the same age?
4. What role may the initial rate of spin (angular momentum) play in determining the shape that a galaxy will eventually take?
5. Describe the three theories of galactic evolution. Give the evidence for or against each theory.
6. Did galaxies form first and stars later, or vice versa? Indicate the reasons for your answer.
7. The average distance between galaxies is much greater than the average distance between the stars within a galaxy. Yet galaxies seem to collide more frequently than stars. Explain.
8. Why should the rate of star formation increase in colliding galaxies?
9. Why do astronomers believe that M82 is exploding?
10. Suggest a possible mechanism for the explosion of an entire galaxy.
11. It is reasonably accurate to assume that in every spiral galaxy the brightest Main Sequence stars that can be seen are the very massive blue giants at the top of the H-R diagram. One such blue giant in our galaxy is Deneb. According to Figure 5.11 Deneb has a luminosity of $2 \times 10^5$ L$_\odot$. Its apparent magnitude is $+1.3$ and its distance is 1400 light-years. In the spiral galaxy NGC 224, the brightest blue giants, similar to Deneb, have an apparent magnitude of 17.3. Using these facts and assumptions, and remembering that apparent brightness falls off as the square of its distance, calculate the distance to the galaxy NGC 224. (*Hint:* remember that an increase of 1 in apparent

234

GALAXIES

magnitude means a decrease of 2.5 in apparent brightness.) This line of reasoning is an accepted technique for estimating distances to all galaxies sufficiently close so that individual stars can be resolved. The technique is effective out to a distance of approximately 100 million light-years.

12. Estimate the angle of inclination of Andromeda Galaxy to our line of sight by measuring the relative dimensions of the disk of the galaxy in the photograph on page 2.

# 10   Radio Galaxies, Seyfert Galaxies, and Quasars

Out of the description of galaxies in the previous chapter emerges a concept of what may be called a normal galaxy. A normal galaxy may be regular or irregular in shape, but it has, in general, the properties that one would expect for a cluster of many millions of ordinary, individual stars. The energy emitted from a normal galaxy is what we would expect to find if we added up the radiation emitted from many separate stars with a range of masses. The distribution of wavelengths in the radiation from galaxies also fits this picture. Furthermore, we occasionally see a supernova flare up in one of the normal galaxies, which suggests that stellar life flows on in its familiar way in these galaxies, each star passing from the Main Sequence to the red giant region and then expiring as a supernova or a white dwarf.

But at the close of the last chapter, we raised the curtain on a different scene, far removed from the pattern of starlike events that seems to produce the energy coming from the Milky Way Galaxy and other galaxies like it. M82 and other exploding galaxies emit far more energy per pound than our Galaxy and, moreover, they emit it at different wavelengths.

*Centaurus A: a Radio Galaxy.*

The impact of these discoveries on astronomy has been heightened by the recent detection of still more anomalous galaxies, or galaxylike objects, which may be emitting a thousand times more energy than a normal galaxy. The energy radiated from these newly discovered objects cannot be accounted for solely in terms of emission from a multitude of separate stars. The explanation of their properties is one of the greatest challenges that has been presented to astronomy in modern times.

## RADIO GALAXIES

The first of the abnormal objects to be discovered were the *radio galaxies*. These are objects which emit intense radio signals, but look like galaxies, although sometimes their appearance is strange in comparison to that of a normal galaxy. The radio emission from the most powerful radio galaxies is equal to, and in some cases greater than, the entire output of energy from our galaxy at all wavelengths.

Grote Reber, one of the two pioneers of radio astronomy, discovered the first radio galaxies in 1940, using a 31-foot antenna that he built with his own hands and at his own expense, in his back yard. Reber observed three separate sources, located in the directions of the constellation Cassiopeia, Sagittarius, and Cygnus. The source of the radio signals from Cassiopeia turned out later to be a supernova remnant — a cloud of agitated gas, blown out into space in the aftermath of a supernova explosion in the Milky Way some time ago. The Sagittarius source turned out to be the center of our galaxy. Subsequently it became known that all normal galaxies, including ours, emit a small amount of radio waves. The radio emission from normal galaxies was not mentioned previously because it is millions of times weaker than the emission in the visible region from a normal galaxy.

The third radio source observed by Reber, located in the direction of the constellation Cygnus, was another galaxy 500 million light-years from us. This source was far more interesting, for when allowance was made for its great distance, the relatively faint radio signals detected by Reber turned out to be equivalent to an outpouring of $5.7 \times 10^{44}$ ergs/sec of energy at radio wavelengths. That emission is three times greater than the total amount of energy radiated into space from the stars in our galaxy at all wavelengths, and three million times greater than the radio emission from our galaxy.

Photographs of the sky taken with the Mount Palomar telescope reveal that at the center of the radio source discovered by Reber is a strange-looking object that resembles two galaxies in collision. A Mount Palomar photograph showing the Cygnus A object is shown in Figure 10.1.

Although thousands of radio sources have been discovered since Reber's time, the radio source he discovered in the direction of the constellation Cygnus remains one of the most powerful known. The

*Figure 10.1 Cygnus A, a strong radio source.*

Cygnus A galaxy is typical of all powerful radio galaxies in its strange appearance. Some look like two galaxies in collision, as in the case of the Cygnus A galaxy, while others are marked by enormous jets of matter that shoot out into space, sometimes in one direction, and, in other cases, in two opposed directions, from the center of the galaxy.

An example is M87, Figure 10.2, in which one jet is clearly visible. Some astronomers believe that a second jet of matter is pointed in the opposite direction. M87 is located in the direction of the constellation Virgo at a distance of 30 million light-years. The main jet, clearly visible in the photograph. is three thousand light-years in length, and is traveling outward at a speed of 15 thousand miles per second.

Another example of a radio galaxy is NGC 4038–9, shown in Figure 10.3. This object may be a pair of galaxies in a close encounter or it could possibly be one single galaxy which has experienced an explosive release of energy.

A third example of a strong radio galaxy is NGC 5128, also called Centaurus A, which is shown opposite the opening page of this chapter. This object presents one of the strangest appearances in the sky. A luminous body, looking like a normal elliptical galaxy, is bisected by a broad lane of dark matter. The dark lane may be a spiral galaxy, viewed edge-on, in the process of collision with the elliptical galaxy behind it. If this is true, NGC 5128 is a pair of colliding galaxies similar to the Cygnus radio

*Figure 10.2 The radio galaxy M87 in a short exposure showing the massive jet of ejected material.*

source. Objections have been raised to this theory, principally because the bisecting dark lane is wider than would be normal for the disk in a spiral galaxy, and also because the lane is more disturbed than the matter in the disk of a spiral galaxy should be. It is possible that NGC 5128 is a single galaxy that has suffered an explosion at its center, with the consequent ejection of fast-moving material out to space. An adequate explanation of the nature of this very unusual object has yet to be given.

### The Radio Image of a Galaxy

The radio image of a galaxy can be formed by plotting the strength of the signals received from each part of the galaxy on a map of the sky. The radio image usually exceeds the diameter of the visible galaxy. In most cases the radio emission comes from two large regions, one on each side of the visible galaxy, and each situated at a considerable distance from the visible object.

The size of each region and the separation of the two regions from each other vary from one radio galaxy to another. On the average, the two radio sources are situated about a million light-years apart, and each source is about 300,000 light-years in diameter. The visible galaxy always lies on or near the line between the centers of the two radio sources.

The Cygnus radio galaxy is an example. A photograph of the visible object associated with this galaxy was shown in Figure 10.1. Figure 10.4 shows the double pattern of radio signals emitted from the Cygnus

*Figure 10.3   NGC 4038-9. A radio galaxy located in the Constellation Corvus.*

galaxy, with the contours shaded in intensity in proportion to the strength of the radio emissions. The visible galaxy, or pair of galaxies, is midway between the centers of the two radio sources.

The radio source associated with NGC 5128 or Centaurus A is also a double source, as the diagram of radio contours in Figure 10.5 demonstrates. A photograph of NGC 5128 has been superimposed on the diagram with the correct location and size to properly relate the optical object to the radio sources on either side of it.

What is the explanation for the twin-lobed appearance presented by the typical radio galaxy? The hypothesis tentatively accepted by astronomers is that a violent event—a collision or an explosion—occurred at the location of the visible galaxy some time ago, ejecting two vast clouds of matter that expanded rapidly outward in opposite directions from the scene of the disturbance. It is thought that these two outward-moving

**241**

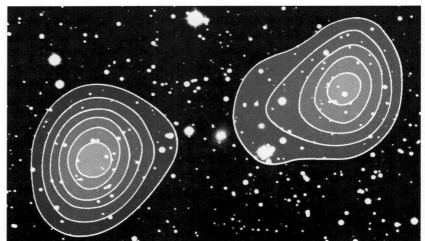

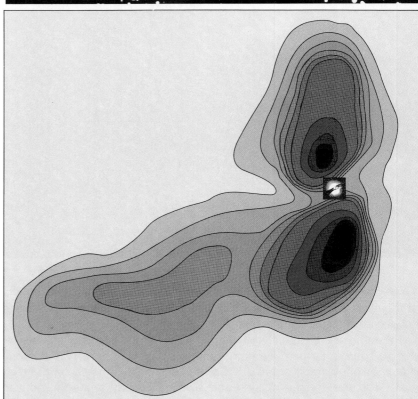

*Figure 10.4   The radio image of Cygnus A superimposed on a photograph of the galaxy.*

*Figure 10.5   The radio image of NGC5128 or Centaurus A.*

clouds of matter, although too tenuous to glow visibly, are betraying their presence by the powerful radio signals that they emit. The twin-lobed radio image of a galaxy such as Cygnus A or NGC 5128 is, according to this theory, the boundary of the double cloud of matter ejected from the visible galaxy.

## Explanation of Radio Emission

Astronomers are not yet certain why the two expanding clouds of matter should generate radio waves. The answer is believed to be that conditions within each cloud are chaotic and disturbed. As a consequence, many atoms are stripped of their electrons, so that the matter within each cloud contains a large number of free electrons. It is also believed that a magnetic field may be present in the cloud, since magnetic fields are commonly present in galaxies.[1] One of the basic laws of electromagnetism states that an electrically charged particle moving in a magnetic field must travel in a circle. If the circular path of the charged particle is viewed edge-on, it appears to be vibrating up and down. In Chapter 2 we saw that a vibrating electrical charge generates a train of electromagnetic waves. This must be the case for the circling electrons in the two radio source regions around a radio galaxy (Figure 10.6).

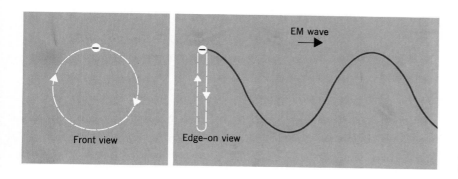

Figure 10.6 *The emission of an electromagnetic wave by a circling electron.*

The stripped atoms carry a positive charge and also generate electromagnetic waves, but because of their greater mass and inertia they vibrate at much lower frequencies that cannot be detected. Nearly all the radio-wave emission is produced by the electrons.

What is the source of the large amount of energy contained in the radio waves? If the radio galaxy consists of two galaxies in collision, this question has a ready answer, since the energy of the collision will be more than adequate to account for the observed intensity of even the strongest radio sources. But in the case of a radio galaxy that shows no signs of having been involved in a collision, the answer is less obvious. In the case of M-87, for example, an internal explosion must have created the jet of matter that shoots out to space from the center of this galaxy. What forces produced the explosion? The answer is not known, no more than it is known for the explosion that seems to have occurred at the center of the galaxy M-82 (Chapter 9).

[1] Although these magnetic fields are relatively weak—no more than one ten-thousandth of a gauss, they play a critical role in the explanation of radio galaxies.

# SEYFERT GALAXIES

The second group of abnormal galaxies are called Seyfert galaxies after their discoverer, Carl Seyfert, who first identified them and pointed out some of their unusual properties in 1944. Seyfert galaxies, like radio galaxies, have properties that cannot be explained readily in terms of the emission of radiation from a multitude of stars. They are distinguished by the fact that their total emission of energy at all wavelengths is extremely large—as much as 100 times the total emission of energy from an ordinary galaxy like ours—and moreover, this energy is emitted from an extremely small and brilliant central nucleus.

It was the property of an exceptionally small and brilliant nucleus that originally attracted the attention of Seyfert to these objects. He noted that, although this group of galaxies have a spiral structure resembling the Milky Way Galaxy, their nucleus appears to be far smaller than the nucleus of our Galaxy and, as noted, to be at the same time of an unusual brilliance, its total output of visible light being greater than the output from our entire Galaxy.

When a Seyfert galaxy is photographed with a short exposure, the brilliant nucleus is the only object to appear on the photograph. It looks like a small, fuzzy, starlike object (Figure 10.7a).

When the exposure is lengthened, the spiral arms in the disk begin to appear (Figure 10.7b).

With a very long exposure, the spiral-arm structure becomes fully visible, and the fuzzy starlike object is revealed as a complete spiral galaxy (Figure 10.7c).

Seyfert galaxies emit up to 100 times more energy than our Galaxy. As in the case of the exploding galaxy M82, most of this energy is emitted from Seyfert galaxies at wavelengths in the infrared part of the spectrum. Although it is undoubtedly significant that the energy of a Seyfert galaxy is emitted predominantly at infrared wavelengths rather than at visible wavelengths, that property of Seyfert galaxies should not distract one's attention from the most important fact about these galaxies, which is that their total energy output is enormous. That fact presents the basic problem: what is the source of the cascade of energy that pours out of a Seyfert galaxy?

You might think at first that Seyfert galaxies emit more energy simply because they are larger and more massive than ordinary galaxies, and contain more stars. But this is not true, for the typical Seyfert galaxy—or rather, the nucleus of the galaxy, which emits most of its energy—is no more than 10 light-years in diameter. This is far less than the diameter of our Galaxy. Furthermore, Seyfert galaxies are not unusually massive, as would be the case if they contained more stars. On the contrary, the available evidence suggests that the mass of a typical Seyfert galaxy could be even less than the mass of our Galaxy. These facts imply that, pound for pound, the Seyfert galaxy may be emitting as much as one hundred times more energy than our Galaxy. Even more so than in the case of M82, this energy release seems difficult to explain in terms of nuclear reactions in stars.

(a)

(b)

(c)

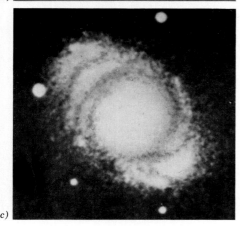

Figure 10.7   A Seyfert galaxy (NGC4151) photographed with (a) short, (b) medium, and (c) long exposure.

## QUASARS

The third goup of abnormal galaxies or galaxylike objects are the quasars. These objects release enormous amounts of energy that dwarf the energy released by Seyfert galaxies. In a sense, quasars combine the properties of radio galaxies and Seyfert galaxies, but they carry the unusual properties of these two objects to an extreme. They are the most difficult objects of all to explain in terms of the properties of stars.

### How Quasars Were Discovered

The discovery of quasars is a fascinating vignette in modern astronomy. The objects that later came to be known as quasars had been showing up on photographic plates for many years as ordinary, relatively faint stars which, everyone assumed, were probably located in our Galaxy. Around 1960 it was first noticed that these otherwise ordinary stars appeared to be the sources of radio waves. While radio waves are frequently observed to come from galaxies and nebulas, no star had been observed to emit strong radio signals. In fact, it seemed impossible to explain how radio waves could be generated in large amounts in any ordinary star.

The strange new objects were dubbed "radio stars." Following the normal procedure in investigating a new phenomenon in the sky, astronomers set about getting the spectra of some of the "radio stars." By 1963, observers at Mount Palomar had succeeded in obtaining the spectra of two of the "stars." Astronomers expected that the spectra would be somewhat unusual, but their expectations were surpassed, for the spectra of the "stars" were different from the spectrum of any star previously observed. The wavelengths of the lines in the spectra did not agree with the wavelengths of the lines for any known element ever observed on the earth or in the heavens.

Did the "radio star" contain an exotic substance, never before seen? Physicists knew enough about atoms and nuclei to be sure that this was almost impossible. But what was the explanation?

### The Quasar Red Shift

Maarten Schmidt, an astronomer at Mount Palomar, found the answer. He looked at the pattern of the lines in the spectrum of 3C273 — the brightest of the "radio stars" — and noticed that the spacing of these lines resembled the spacing of the lines in the Balmer spectrum of the hydrogen atom, but their wavelengths were different. All the lines in the 3C273 spectrum were shifted toward the red end of the spectrum relative to the

Balmer lines. Schmidt wondered whether the mysterious spectrum of 3C273 might be the familiar Balmer spectrum of hydrogen, but with the wavelength of each line displaced toward the red by the same proportionate amount. This would keep the pattern of the lines unchanged, but move each individual line to a new location at a longer wavelength.

Schmidt checked his idea by comparing the 3C273 spectrum with the Balmer spectrum. He found by trial and error that he could multiply the wavelengths of the Balmer lines by 1.158 and obtain the wavelengths of the lines in the 3C273 spectrum. For example, the second line in the Balmer series has a wavelength of 4861Å in the blue region of the spectrum. Schmidt multiplied 4861 by 1.158 and obtained a new wavelength of 5632Å, which would be in the yellow-green part of the spectrum. There is a yellow-green line in the spectrum of 3C273 at just this wavelength. The third line in the Balmer series has a wavelength of 4340Å, placing it in the violet. Multiplying 4340 by 1.158, Schmidt obtained a new wavelength of 5026Å, which would be in the blue-green part of the spectrum. There is a blue-green line in the spectrum of 3C273 at precisely 5026A. In the same way, Schmidt was able to match the fourth Balmer line to the fourth line in the 3C273 spectrum.

The precise agreement of three distinct lines in the two spectra was not likely to be a coincidence. Clearly, the 3C273 spectrum was nothing more than the Balmer spectrum shifted to the red by 15.8 percent.

What was the cause of the shift toward the red? Astronomers had observed red shifts like this in other spectra, and had a ready explanation. Whenever a star or a galaxy is receding from the observer at a high speed, its light is shifted in wavelength towards the red end of the spectrum by an amount proportional to its speed of recession. This shift in wavelength is called the Doppler effect. It was explained in Chapter 2. Applying the theory of the Doppler shift to 3C273, Schmidt concluded that this "star" was moving away from the earth at the enormous speed of 28,400 miles per second. If Schmidt's interpretation was correct, the other "radio stars," which had larger red shifts than 3C273, must be moving still faster. Some of them, in fact, had red shifts that indicated that they must be moving at 90 percent of the speed of light.

If the "radio stars" were moving this fast, they must be relatively distant objects. They could not be in our own Galaxy, since if they were, their rapid motion across our line of sight would cause their positions to change drastically from year to year against the background of the fixed stars. Yet some of these "stars" had been observed for years without displaying any noticeable change in position.

Perhaps the radio stars were outside our Galaxy but located in other galaxies quite close to us, so that they could be picked out as individual stars in these galaxies, but were far enough away so that the apparent motion of these stars would be small. Unfortunately for this idea, none of the "radio stars" seemed to be connected with any of the galaxies in our neighborhood. Thus Schmidt's brilliant explanation of the spectrum of 3C273 eliminated the original mystery, only to replace it with another: Why did these "stars" have such a large red shift?

As we will see in the next chapter, astronomers have discovered a relationship between the red shift or speed of recession of an object and its distance from us. According to this relationship, an object receding from the earth at a speed of 28,400 miles per second must be 2 billion light-years away. If the relationship also holds true for the "radio stars," 3C273 must be two billion light-years from us. Knowing the distance of 3C273, we can calculate its absolute magnitude or luminosity from its apparent magnitude, which is +13. An object of the 13th magnitude, located at a distance of 2 billion light-years, turns out to have a luminosity of $10^{45}$ ergs/sec.

But an object that sustains a luminosity of $10^{45}$ ergs/sec for an extended period cannot be a single star.[2] The only known objects that sustain luminosities comparable to this over long times are galaxies.

All the "radio stars" for which spectra could be obtained turned out to have very large red shifts; hence, they were apparently billions of light-years distant. Therefore, when the apparent magnitudes of these objects were translated into absolute magnitudes, all turned out to be extremely luminous, with luminosities comparable to those of galaxies. It looked as though these "radio stars" were not stars at all, but galaxies located so far away that the details of their galactic structure were not visible in photographs, hence, they looked like stars.

Since the strange objects looked somewhat like stars, but were not stars, they became known as "Quasi-Stellar Sources," or QSS's, soon shortened to *quasars*.

The first quasars that were discovered were identified by their combination of radio and optical properties. After these first discoveries, astronomers found numerous other starlike objects that had the same unusual optical properties as quasars—that is, very large red shifts, and unusual emission spectra—but were not strong radio sources. These objects received the separate name of "Quasi-Stellar Objects", or QSO's.

Some books and articles still refer to quasars as quasi-stellar sources and quasi-stellar objects, or QSS's and QSO's. In this book the two types of objects are grouped together under the common label of quasars.

## Can Quasars Be Nearby Objects?

As we will see in the later sections of this chapter, it is nearly impossible to explain the high luminosity of quasars in terms of our current astronomical knowledge. In an effort to escape from that dilemma, some astronomers have questioned whether quasars are really as far away as their red shifts suggest. If quasars were nearby objects, their calculated

---

[2] Supernovas are one form of star that approaches the brilliance of 3C273; they flare up to $10^{44}$ ergs/sec and stay at that level a few months. But 3C273 had been maintaining its high luminosity for years. It could not be a supernova.

luminosity would not be so great and they would present no problem to the astronomer.

But if quasars are nearby objects, how can we explain the large red shifts in their spectra? The only objects that have large red shifts, similar to those of quasars, are exceedingly distant galaxies. In the case of these distant galaxies, the red shift is known to be a Doppler effect connected with their rapid speed of recession. Is there an explanation other than the Doppler effect for red shifts?

The theory of relativity provides another explanation. In the chapter on pulsars and black holes in space, it was pointed out that all forms of energy, including light, have mass. Since light has mass, and mass is attracted by gravity, the light escaping from a star tends to be held back by the star's gravitational pull. During the collapse of a star the pull of gravity at its surface mounts, until finally it becomes great enough to hold back the rays of light entirely and prevent any radiation from leaving the star. At this point the star becomes a black hole in space. As the collapsing star approaches the radius of the black hole, but before it actually reaches this radius, the backward pull of gravity is sufficient to diminish the energy of the outgoing light rays appreciably, although not great enough to hold the rays back entirely.

The loss of energy suffered by the photons leaving the star is the clue to the relativistic explanation of the red shift. A relationship exists between the energy of a photon and its wavelength such that high energy means a short wavelength, that is, blue light, while low energy means a longer wavelength, that is, red light. A reduction in the energy of the photons escaping from a star, therefore produces an increase in the wavelength of the star's radiation. That is, it produces a shift toward the red end of the spectrum.

Looking at the spectrum, there is no way to distinguish this so-called "gravitational red shift" from the red shift produced by the Doppler effect. Thus, the gravitational effect could be the explanation of the red shift in the quasar spectrum. However, very few astronomers believe in this explanation of the red shift. One important reason for their scepticism is (as discussed later in this chapter) the evidence that quasars are galaxies. Assuming that quasars are galaxies, if a quasar were located very near us it would have a strong gravitational effect on members of the Local Group. Yet this effect is not seen.

Another objection is that quasars have emission lines in their spectra of a kind that can only be produced in a very tenuous envelope of gas. If a quasar were sufficiently compact and dense to produce a strong gravitational red shift, it would be too dense to produce these emission lines.

A third reason is that one quasar—labeled Parkes 2251 + 11—seems to be associated with a cluster of ordinary galaxies in the direction of the constellation Pegasus. The Parkes quasar has the same red shift as the galaxies in the Pegasus cluster, and is situated near to them in the sky. These two facts are not likely to be a coincidence. The implication is that this quasar and the cluster of galaxies are actually close together in space and are receding from us at very closely the same speed. But if this is true, it means that the red shift for the Parkes quasar is a true Doppler shift indicating a high speed of recession and a great distance from us, and not a gravitational red shift. If that is true for this one quasar, it is presumably true for all quasars; their red shifts all indicate a large velocity of recession and a great distance from us.

These arguments have convinced many astronomers that quasars are exceedingly distant and, therefore, exceedingly luminous objects.

## The Visible Image of a Quasar

As we have seen above, although 3C273 and other quasars are intensely luminous objects, emitting far more energy than our Galaxy, they appear as very faint objects in the sky because of their great distance from us. None of the quasars is visible to the naked eye, nor can they be photographed except with the aid of large telescopes. The photographs reveal them to be somewhat fuzzy starlike objects. Figure 10.8 shows quasar 3C196, located approximately 6 billion light-years from us.

An interesting feature appears in photographs of the closest and brightest quasar, 3C273. In the photograph of 3C273 in Figure 10.9, obtained with the 200-inch telescope, a jet of matter may be seen extending to the lower right. This jet, whose presence has not been explained, is 160,000 light-years long.

*Figure 10.8    Quasar 3C196.*

## The Radio Image of a Quasar

When quasars were thought to be nearby "radio stars," they were regarded as relatively weak radio sources. As soon as their great distances were established by the measurement of their red shifts, it was realized that they must be extremely powerful emitters of radio energy.[3]

When the strength of their radio emissions is corrected for their distance from us, some quasars outrank the most powerful radio galaxies known. The question immediately arises: Do quasars, as radio sources, have the striking twin-lobed appearance of a typical radio galaxy? The answer is that many of them do. An example is Quasar 3C47, located 2.5 billion light-years from our Galaxy. A plot of contours of equal radio brightness from the quasar, shown in Figure 10.10, is its radio image. The "X" placed on the radio image marks the position of the optical object. As in the case of radio galaxies, the radio image is a double source, with the optical object located close to the line running between the centers of the two sources. The centers of the two radio sources in the radio image are 700,000 light-years apart.

Although quasars have some unusual properties not possessed by radio galaxies, the similarities in the appearances of their radio images suggest that there may be a kinship between the two classes of objects. Additional evidence linking quasars to galaxies is discussed in a later section.

*Figure 10.9    Quasar 3C273.*

[3] The remarks in this section refer only to the quasars that are strong radio sources. Many quasars do not emit a detectable radio signal.

## Quasar Luminosity

When the apparent luminosities of quasars are corrected for their distances, the output of visible light from these objects ranges up to $10^{45}$ ergs per second. This luminosity is 10 times greater than the total output of energy from large spiral galaxies, such as Andromeda and the Milky Way Galaxy, and equal to the output of energy from the giant ellipticals, which are the most luminous "normal" galaxies known.

The substantial energy radiated by quasars at visible wavelengths turned out to be only a small fraction of their total energy output. Observations by infrared astronomers indicate that most of the energy of quasars is radiated in the far infrared region of the electromagnetic spectrum, at wavelengths ranging between 10,000 and 100,000 angstroms. In the case of 3C273, the energy output in the far infrared is at least $10^{48}$ ergs per second, indicating that this quasar produces 1000 times more energy than the most luminous galaxies known.

Their intense emission of infrared radiation is a characteristic which quasars share with Seyfert galaxies and exploding galaxies such as M82. The ratio of infrared energy to visible light is at least 1000:1 for quasar 3C273, about 300:1 for Seyfert galaxies, and 100:1 for M82. On the other hand, in a normal galaxy such as ours the same ratio is only 1:30. With respect to their high output of infrared energy, quasars seem to lie at the far end of a sequence of successively more unusual objects, extending from ordinary galaxies through exploding galaxies to Seyfert galaxies. Other evidence linking quasars to Seyfert galaxies and possibly to other galaxies will be presented on pp. 252–253.

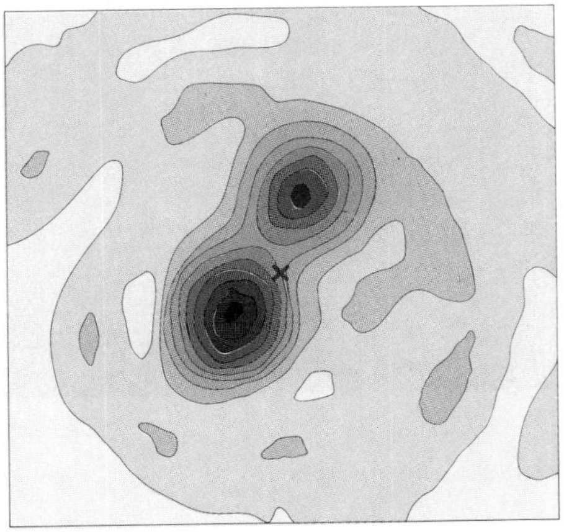

## The Small Size of Quasars

Quasars have another remarkable property, in addition to their great distance and exceptional output of energy at all wavelengths. They are extraordinarily small, far smaller than ordinary galaxies, and even smaller than Seyfert galaxies. Quasar 3C273 is less than a light-year in diameter, and at least one quasar may be as small as one light-week in diameter.

The evidence for the small size of quasars comes from the fact that large changes in brightness of quasars have been observed over short periods of time. Two examples of the variations in the luminosity of quasars are shown in Figure 10.11. Figure 10.11a shows the measured intensity of Quasar 3C273 over the last 80 years,[4] and indicates changes of nearly one magnitude (a factor of 2.5) over periods of one year or less. Figure 10.11b shows even more rapid changes in the luminosity of Quasar

*Figure 10.10   The radio image of Quasar 3C47. As with radio galaxies, the radio image of the quasar consists of two sources lying on either side of the optical object.*

[4] This quasar has been photographed by Harvard College Observatory astronomers since 1855, in the belief that it was an ordinary faint blue star. Only in 1960 did they realize that the "star" had unusual properties.

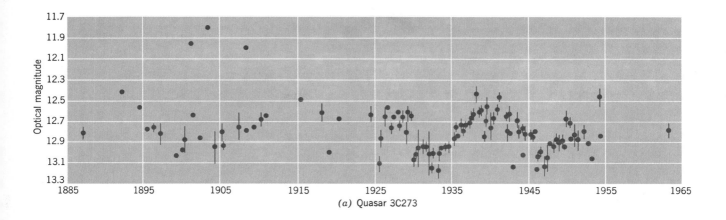

3C454.3. At one point, marked by the arrows, a change of a full magnitude occurs over an interval of less than one week.

If one part of the emitting region from Quasar 3C273 were more than a light-year away from other parts, the streams of radiation from these separate parts of the quasar would arrive at the observer at least a year apart in time, and rapid variations in luminosity would be smoothed out. Even if a very large variation occurred in a time shorter than one year, it would not be observed because of the smoothing effect. The fact that such variations are observed proves that 3C273 cannot be more than one light-year in diameter. Similarly, quasar 3C454.3 cannot be larger than one light-

*Figure 10.11(a) Rapid time variation in the energy output from quasar 3C273.*

*Figure 10.11(b) Rapid time variation in the energy output from quasar 3C454.3.*

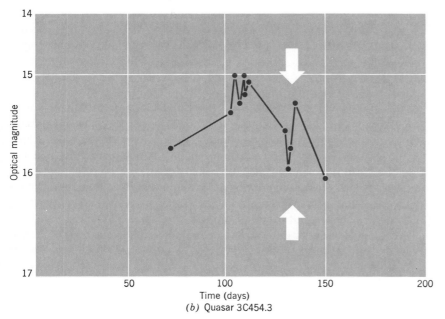

(b) Quasar 3C454.3

week.[5]

Because of this combination of properties—enormous energy output combined with small size—the quasar is more difficult to fit into the body of scientific knowledge than any other object in the sky.

## The Spectrum of a Quasar

Quasars have still another peculiar property in comparison with ordinary galaxies. When the light from a galaxy is broken up into a spectrum, it usually shows the dark absorption lines that are characteristic of the spectra of stars. This is because most of the radiation from a galaxy is made up of light emitted from its stars. The spectrum of a quasar, on the other hand, shows not only absorption lines, but a number of emission lines as well. In addition, the light from quasars has a bluish color with an excess of ultraviolet radiation compared to normal galaxies.

## Are Quasars Galaxies?

At first, it seems obvious that quasars cannot be galaxies, because they are hundreds or thousands of times brighter, far smaller, and have a completely different type of spectrum—many emission and fewer absorption lines in a quasar spectrum versus many absorption and few emission lines in the spectrum of a galaxy.

But in spite of these great differences, many astronomers think that quasars may be galaxies after all. The Seyfert galaxies are one of the principal reasons for this belief. There is no question that Seyfert galaxies are true galaxies, and yet they also resemble quasars. In fact, if ordinary galaxies, Seyfert galaxies, and quasars are arranged in a sequence, the Seyfert galaxies completely bridge the gap between ordinary galaxies and quasars. This is true for any one of the three main properties mentioned above. With respect to *brightness,* the brightest Seyfert galaxies are as bright as some quasars, while the least bright Seyfert galaxies have about the same brightness as ordinary galaxies. With respect to *size,* the nucleus of the typical Seyfert galaxy is intermediate between quasars and the nuclei of ordinary galaxies. With respect to its *spectrum,* the Seyfert galaxy is again intermediate; it has more emission lines than an ordinary galaxy, but not as many as a quasar; and it has less absorption lines than an ordinary galaxy, but more absorption lines than quasars.

Even with respect to the way they look, Seyfert galaxies bridge the gap between quasars and ordinary galaxies. As we noted above, if a Seyfert

[5] That is, the luminous part of the quasar cannot be larger than this. The quasar could have a large, relatively faint region surrounding the small, brilliant nucleus.

galaxy is photographed with a short time exposure, the only part of it that shows up on the photographic plate is its brilliant nucleus, which appears as a fuzzy, starlike object, very much similar to a quasar. But if the exposure time is increased, we see that the Seyfert galaxy resembles a normal spiral galaxy. The sequence in Figure 10.12 shows the smooth transition in appearance from quasars to Seyfert galaxies and then to spirals.

Finally, the similarity between quasars and radio galaxies with respect to their twin-lobed radio images also suggests that quasars belong to the family of galaxies (see page 242).

These facts—a similarity to the Seyfert galaxies in some respects, and a similarity to radio galaxies in others—provide fairly convincing circumstantial evidence for the view that quasars belong to the family of galaxies. Apparently the properties of galaxies vary over a wide range, and quasars seem unusual only because their properties place them at one end of this range of variations.

*(a) Quasar 3C273*

*(b) Seyfert galaxy NGC4151*

*(c) Spiral galaxy NGC488*

## A Supernova Theory of Quasars

If quasars resemble normal galaxies, they must be composed of billions of individual stars. Since stars provide the energy output for normal galaxies, presumably they also are the source for the energy of quasars.

We can see immediately that ordinary stars, using up their nuclear fuel at a normal rate, would not be adequate to explain a quasar's energy production. The stars in our Galaxy produce only $10^{44}$ ergs/sec of energy, which is hundreds of times smaller than a quasar's output. One explanation would be that quasars have more stars than normal galaxies and are very bright for that reason. However, there is no evidence available which indicates that quasars are more massive than a normal galaxy.

Another possibility is suggested by the fact that quasars are so small. Because of the small size of the quasar, the stars it contains must be packed in very tightly and, therefore, must collide frequently. Whereas in normal galaxies (as mentioned in Chapter 13) a collision between two stars occurs only once every 10 billion years, in a quasar stars may collide as often as once a day.

The collision will excite the gaseous envelope of each star and make it flare up into a high luminosity. If the collisions occur with sufficient frequency, the energy output from the quasar can be increased by many orders of magnitude over the energy radiated from a normal galaxy.

In addition, theoretical studies show that when two stars collide, they may combine into a single, more massive star. The massive star then evolves quickly to the supernova stage, as described in Chapter 7. During the supernova explosion, the energy output of the supernova rises until it is 100 billion times greater than the normal energy output from the star, and stays at that high level for several months. If many supernova explosions occurred one after the other in rapid succession among the densely-packed stars of a quasar, they could contribute to its enormous energy

*Figure 10.12 This sequence of photographs illustrates the transition in appearance from the photograph of a quasar (a) to the photograph of Seyfert galaxy (b) and a normal galaxy (c).*

production. The energy released from the supernova explosions could be bolstered by the energy radiated from the pulsars or neutron stars that would form at the centers of some of these supernovas. A pulsar can produce more energy during its lifetime than the supernova implosion that created it. If one pulsar per day were formed in a quasar, the average energy production resulting from the pulsars would be $10^{47}$ ergs/sec, which is sufficient to account for all but the most powerful quasars.

Thus the colliding-star theory may account for the properties of the average quasar. However, it is difficult or impossible to stretch this theory far enough to account for the brightness of the most powerful quasars, such as 3C273.

Another theory suggests that the entire quasar is one giant pulsarlike object, with a mass equal to hundreds of millions of solar masses, supercompressed to the extraordinarily high densities of a pulsar or a neutron star. The radius of this highly compressed, massive body would be less than one billion kilometers, or one light-hour, which is consistent with the evidence that quasars are extremely small in diameter. The source of the energy released by the giant pulsar would be its own gravity rather than a series of nuclear reactions. As the giant pulsar collapsed, its gravity would pull particles toward its center with increasing force, causing collisions among those particles, and liberating large amounts of energy. If the collapse proceeded past the neutron-star radius for this huge mass, and approached the black-hole radius, as much as one-third of the matter in the quasar could be converted to energy in this way. This is 50 times more energy per pound than is released by nuclear reactions in stars during the lifetime of a normal galaxy. Again the release of energy would be adequate to account for the output from many quasars.

But these ideas are not supported by observational evidence. They are no more than desperate efforts by the astronomer to take the most luminous single objects that he has ever discovered and scale these objects upward in size and mass by factors of a millionfold or more, without any valid theoretical reason for doing so, in order to arrive at a hypothetical energy output that could match the output of a quasar.

The most accurate assessment of the quasar problem is that no satisfactory explanation has been found for the existence of these objects, whose puzzling properties place them beyond the limits of current astronomical knowledge. But suppose that we do not limit ourselves to theories of stellar evolution, and to astronomical knowledge in a narrowly defined sense. Can we find anywhere else in science a hint of a mechanism sufficiently powerful to account for the most luminous quasars?

**Anti-Matter Theories**

An idea has emerged from research in nuclear physics that could conceivably explain the energy released by the brightest quasars. This idea involves *antimatter,* an exotic substance created by the nuclear physicist out of pure energy in experiments carried out with nuclear accelerators.

Antimatter resembles matter in every way except that it is, in a sense, the photographic negative of normal matter. An *antiproton,* for example, is identical with a normal proton in mass and magnitude of electric charge, but it has a negative electric charge in place of the proton's positive charge. An *antielectron,* usually called a *positron* — is identical with an ordinary electron in nearly every way, except that it has a positive electric charge. Since the positron and the antiprotons carry opposite electrical charges, they are attracted to one another by an electrical force. Thus the positron can circle around the antiproton in orbit, making up an atom of antihydrogen. The antihydrogen atom will have the same chemical properties as an atom of ordinary hydrogen, provided that it is joined in chemical compounds with other atoms of antimatter. However, if an antiatom collides with an ordinary atom, the two will immediately annihilate one another in a burst of energy. They disappear as atoms, leaving behind only pure energy in the form of radiation.

How much energy is released when a particle of matter annihilates a particle of antimatter? The annihilation of a proton by an antiproton, for example, yields 931 Mev — the rest mass of the proton — from each of the two particles. The release of energy in nuclear reactions within stars, on the other hand, amounts to 7 Mev per proton. Pound for pound, matter-antimatter annihilation releases more than 100 times as much energy as is released in nuclear reactions in stars. The potency of matter-antimatter annihilation makes it seem like a promising candidate for explaining the brilliance of quasars.

But if antimatter is the source of energy released by quasars, where does this antimatter come from? You might guess that the quasar has contained the antimatter from the time of its formation, but that is unlikely because antimatter cannot exist in the company of matter for a long time. The antimatter and matter annihilate one another almost immediately, and the antimatter disappears. Faced with this problem, some astronomers and physicists have proposed that antimatter, and perhaps matter as well, are continuously created *out of nothing* at the center of quasars. Other scientists have suggested that what we see as quasars are simply holes in our Universe, through which matter is pouring from that other universe. These ideas show the extremes to which science has been driven in its effort to explain quasars.

## Questions

1. What is the main characteristic of a normal galaxy?
2. Turn to page 121 and look at the radiation curve for normal stars. Now draw the radiation curve for a galaxy emitting an unusually large amount of radio noise.

3. Describe the radio image of a typical radio galaxy. What seems to be happening in the radio image of the Centaurus A radio source?

4. How many years have elapsed since the jet in M87 was formed, assuming that it has always been moving outward at a speed of 15,000 miles per second?

5. How are the radio waves (probably) produced in intense radio sources?

6. What are the characteristics of a Seyfert galaxy?

7. What are the reasons for believing that quasars are extremely distant objects? Discuss the basic problem that is created for the astronomer by the fact that quasars appear to be at a great distance.

8. Look at the photograph of radio galaxy M87 (Figure 10.2) and quasar 3C273. What do they have in common? What other features do radio galaxies and many quasars have in common?

9. How does the spectrum of a quasar differ from the spectrum of an ordinary galaxy? What are the principal reasons for believing that quasars belong to the family of galaxies?

10. List the following objects with the total rate of energy output from each. Express in units of the energy output from the Milky Way Galaxy (that is, Milky Way Galaxy = 1): M82, Andromeda Galaxy, sun, Deneb, typical supernova, Seyfert galaxy (maximum), quasar (maximum). List in order of increasing energy output.

11. Various theories have been offered to explain the energy output of quasars. Which theory do you prefer? Why?

12. As we look outward in space, we look backward in time. For example, if Quasar 3C273 is 2 billion light-years away we see this quasar as it was 2 billion years ago. How would our Galaxy look if viewed from a distance of 2 billion light-years? 10 billion light-years?

13. A famous Soviet astrophysicist named Ambartsumian has proposed a theory that galaxies were originally dense concentrations of matter and energy that exploded outward and gradually approached their present form. This theory runs directly counter to the widely accepted view that all galaxies began as tenuous clouds of gas which evolved by contracting inward under gravity. Based on your reading of Chapters 8, 9, and 10, give reasons — observational or theoretical — for or against the Ambartsumian theory of galactic evolution.

# 11 Cosmology

In this book we have followed the astronomer as he traces the history of the basic ingredient, hydrogen, through a series of events in which, first, galaxies are formed out of the parent cloud of gaseous hydrogen, then stars are formed out of the hydrogen within each galaxy, and last, the heavier elements are formed out of hydrogen within each star. This history answers many questions about the origin of the world. It explains how stars come into being, how they obtain their energy, and how the elements of the Universe are created. But the very success of the astronomer in reconstructing the life story of the stars increases our desire to know the answers to even more fundamental questions: How did the Universe start? Who or what created the hydrogen in the Universe at the beginning? And how will the Universe end? What will happen when the supply of hydrogen is exhausted, and the old stars go out, one by one?

These matters are the domain of the field of scientific investigation known as *cosmology*. Cosmology is concerned with the nature and origin of the entire Universe — its structure today, its past, and its future. To study cosmology, you must stretch your concepts of space and time even more

*E. P. Hubble observing at the prime focus of the 200-inch telescope.*

than you have done earlier in this book. You must adopt a point of view so broad that the tremendous span of a galaxy seems a mere detail, and the passage of a billion years is like a day.

To the cosmologist, the birth of each star is a minor incident in the life of the Universe. He reflects on the innumerable births and deaths of all the stars that have existed since the Universe began, and asks himself the meaning of this pattern of details. Does the life story of a single star have a significance for the Universe as a whole?

When we consider the fact that our Galaxy contains 100 billion stars, and billions of other similar galaxies exist around us, it seems at first thought that a single star cannot possibly tell us anything significant about the entire Universe. But this conclusion is incorrect. Every star that has lived has unalterably changed one aspect of the entire Universe. This aspect is the amount of hydrogen that exists in the Universe. Hydrogen is the essential cosmic ingredient. It is the primary source of the energy by which stars shine, and it is also the source of all the elements in the Universe. As soon as a star is born, it begins to consume some of the hydrogen in the Universe, and continues to use up hydrogen until its death. Once hydrogen has been burned within that star and converted to heavier elements, it can never be restored to its original state. With the passage of time, and the appearance of successive generations of stars, the supply of hydrogen in the Universe must grow smaller.

Globular clusters provide evidence for the slow change of hydrogen into heavier elements in stars. As we noted in Chapter 8, the H-R diagram for these clusters indicates that they were formed early in the history of the Universe. At that time, only a relatively few supernova explosions could have occurred, and the abundance of heavy elements must have been very low. The spectra of the globular cluster confirm this idea. They show that the stars in these clusters have only one-tenth as many heavy elements as the stars in the spiral arms of the Galaxy, which were formed later.

The implications in these observations are clear: Hydrogen is disappearing. As the old stars go out one by one, fewer and fewer new stars can be formed to replace them. Stars are the source of energy by which all beings live. When the light of the last star is extinguished, life must end throughout the Universe.

The depletion of the total supply of hydrogen is the only feature in the life story of the stars that is of interest to the cosmologist. Minute by minute and year by year, the supply of hydrogen continually decreases as a consequence of nuclear reactions that occur in stars. As a consequence, the Universe is running down and changing irreversibly.

Reflecting further on the situation, the cosmologist turns the clock back in his imagination and asks himself what the Universe must have been like billions of years ago. Clearly, there must have been more hydrogen in the Universe at that time than there is today, and less of the heavier elements. At the present time approximately 75 percent of the mass in the Universe consists of hydrogen. A billion years ago this number would have been slightly different; there would have been more hydrogen and less of the heavier elements, because some of the stars that have contributed to today's abundance of those elements had not yet been born. Four-and-one-half billion years ago, around the time when the sun and earth were formed, there would have been still more hydrogen and still less of the heavier elements. Turning the clock back still further, we would eventually come to a time when the Universe contained nothing but hydrogen—no helium, no carbon, no oxygen, and none of the other elements out of which, for example, the earth and the creatures on it are composed. This point in time must have marked the beginning of the Universe.

In other words, projecting the present conditions in the Universe backward in time, we are forced to conclude that it had a beginning, and projecting the present conditions forward in time, we can see that the Universe must eventually come to an end. That is the cosmological significance to the life story of the stars.

Other evidence suggests that the Universe has been changing in an irreversible way. In the 1920's the American astronomers Humason and Hubble discovered that all distant galaxies in the sky seemed to be moving away from us and from one another at very high speeds. Those most distant from us are receding at the extraordinary speed of 150,000 miles per second, which is close to the velocity of light. The Universe appears to be blowing up before our eyes, as if we were witnessing the aftermath of a gigantic explosion.

When did the explosion occur? Knowing how far apart the galaxies now are, and how rapidly they are moving away from one another, we can calculate backward in time to the moment at which the expansion began. In this way some cosmologists have arrived at the conclusion that the Universe began its existence in a fiery outburst between 10 and 20 billion years ago.

Others have disagreed. Based on the single observation of the receding motion of the galaxies, modern science has produced three cosmologies, each with its own school of scientific supporters. In three completely different ways, these schools of scientific thought proceed to build up a picture of the history of the Universe, and to make predictions regarding its future.


## THE BIG-BANG UNIVERSE

One cosmology—the Big-Bang theory—asserts that the Universe had a beginning and is approaching an end. The Big-Bang school of cosmologists proposes that the motions of the galaxies are, in fact, the consequence of an actual cosmic explosion that took place a long time ago. Father Lemaître, a Belgian astronomer educated as a Jesuit priest, and George Gamow, a Russian-born physicist who emigrated to the United States in 1936, are the scientists most prominently associated with this theory. According to the Big-Bang cosmologists, the Universe began its existence as an extremely hot and dense concentration of matter. Later Gamow named this primordial substance "ylem"—

**261**

the name that Aristotle gave to the basic substance out of which the Greek philosophers believed all matter was derived. Probably the particles of the Universe were packed together as densely as the matter in the atomic nucleus at that time. Gamow based this assumption on the fact that the matter in the nucleus is the densest form in which matter is known to exist.

It is interesting that at the density of ylem, all the matter in the observable Universe would fit into the solar system. The earth itself, if squeezed down to the density of the nucleus, would fit into a sphere 200 feet in diameter.

The temperature of the ylem would also have been extremely high at this point, ranging up to trillions of degrees. As the Universe expanded the temperature must have decreased, dropping first to billions of degrees and then down to millions of degrees. When the temperature had decreased to around 10 million degrees, protons or hydrogen nuclei would have begun to stick together in groups of four to form helium nuclei through nuclear reactions like the ones discussed in Chapters 6 and 7. The amount of helium that could have been formed at that time is not accurately known, but calculations suggest that at least 5 to 10 percent of the hydrogen in the Universe could have been transformed into helium in the early stages of the Big Bang.

It might be expected that after helium had been formed, other heavier elements would be built up by more complicated nuclear reactions, until the whole periodic table of the elements existed. However, the calculations indicated that for two reasons, this does not occur. One reason is that the temperature in the Universe continues to drop, making nuclear reactions less and less probable. The other is that a wide gap in nuclear properties exists between helium and the next stable nucleus, which is the nucleus of lithium. This gap is difficult to cross at temperatures in the neighborhood of 10 million degrees, and virtually impossible to cross at the substantially lower temperatures that existed later.

Thus, the Big-Bang picture explains the presence of hydrogen and helium in the Universe, but it fails to explain the existence of the other 90-odd elements of the periodic table. Only the theory of stellar evolution discussed in Chapter 7 can explain the existence of these elements in today's Universe.

Who or what put the primordial hydrogen into the Universe? Why did it begin with a bang as a hot and highly compressed globule of matter? The Big-Bang cosmologists do not attempt to answer these questions, nor do they comment on the conditions that might have existed in the Universe prior to the start of the explosion. However, assuming the starting conditions, described above, they can predict with confidence what happened thereafter. After the helium was formed, the expansion of the Universe continued, and its temperature continued to drop. The matter of the Universe was in the form of hydrogen nuclei, helium nuclei, electrons, and radiation. Atoms did not yet exist, because whenever an electron was captured into an orbit around a nucleus to form an atom, it was knocked out of the orbit almost immediately under the smashing impact of the violent collisions that occur at such high temperatures.

However, by the time the Universe was, perhaps, 100 thousand years old, the temperature had dropped to 5000 degrees, and from that point onward, neutral atoms began to form in increasing numbers. These atoms consisted almost entirely of hydrogen.

With the further passage of time, the expanding materials cooled and condensed into galaxies and, within the galaxies, into stars. The formation of galaxies may have been limited to the early years of the Universe, or it may have been a process stretched out over a long span of time and, perhaps, still continuing today. In either case, the formation of stars probably began shortly after the formation of the first galaxies, when the Universe was about 100 million years old.

After about 13 billion years of continuing expansion, the Universe reached the state in which it exists today.[1] According to the theory, the expansion will continue indefinitely in the future, with the distances between the galaxies continually increasing, and space growing steadily emptier. At the same time, as the hydrogen within each galaxy is used up, the galaxies themselves will grow dimmer and eventually fade out entirely. In the end, the Universe is devoid of matter, energy and life (Figure 11.1).

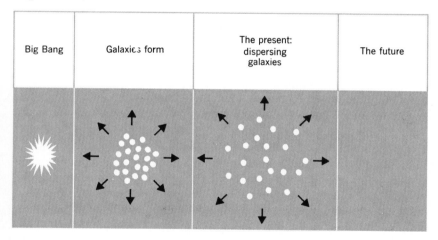

| Big Bang | Galaxies form | The present: dispersing galaxies | The future |
|---|---|---|---|

Figure 11.1   The Big-Bang cosmology.

## THE STEADY-STATE UNIVERSE

Some years ago, Thomas Gold, then a graduate student at Cambridge University, made a proposal that voids these morbid predictions. He suggested that fresh hydrogen is steadily created throughout the Universe

[1] If the Universe had expanded at a constant rate since it was formed, it would require 20 billion years to reach its present state. However, allowance must be made for the continual slowing down of the expansion because of the backward pull of gravity on the outward-moving galaxies. This effect reduces the estimate of the age of the Universe, because it means that the rate of expansion was greater when the Universe was young than it is today. Since the rate of slowing down has not yet been measured accurately (see page 270), we cannot give a precise value for the age. The value of 13 billion years is based on the assumption that the Universe will slow down until all its components are at rest infinitely far apart from each other.

*out of nothing.* The freshly created hydrogen would provide the ingredients for the formation of new stars to replace the old. Also it would fill up the spaces left by the movement of the galaxies away from one another. Thus, the creation of matter out of nothing, as proposed by Gold, could restore the Universe to a state of perpetual balance, without beginning and without end (Figure 11.2). According to this cosmology—the Steady State theory—the Universe has been unchanged and will remain unchanged throughout eternity.

Gold mentioned his idea to Herman Bondi and Fred Hoyle, two English astronomers, who joined him in working out its consequences. They asked themselves, how much hydrogen should be created per year in order to keep the density of matter constant everywhere, as the Universe expands? According to their calculations, the expanding Universe remains in a steady state, with a constant density of matter, if one hydrogen atom is created per year in a volume equal to that of the Empire State Building.

This is a very modest rate of creation, but it violates a cherished concept in science—the principle of the conservation of matter and energy—which states that matter or energy can be neither created nor destroyed. Matter can be produced from energy and vice versa, but the sum of all matter and energy in the Universe must remain the same. It seems difficult to accept a theory that ignores such a firmly established fact of terrestrial experience. Yet, the proposal for the creation of matter out of nothing possesses a strong appeal, since it permits us to contemplate a Universe that extends into the past and the future without limit, a Universe that renews itself *in perpetuum.*

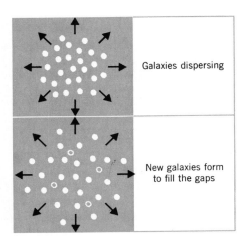

Galaxies dispersing

New galaxies form to fill the gaps

*Figure 11.2   The Steady-State cosmology.*

## THE OSCILLATING UNIVERSE

The Big-Bang theory and the Steady-State theory express different philosophies. One assumes a definite beginning resembling in a loose sense the act of creation that marks the beginning of the world in the Judao-Christian tradition. The other, seemingly akin to the religions of the East, assumes a universe that stretches infinitely far into the past and future with many cycles of birth, death, and rebirth, but no permanent change. It seems impossible to combine these basically different philosophies in a single cosmology. Nonetheless, a third theory has been proposed that reconciles the philosophy of the Big-Bang with the philosophy of the Steady-State in a single, unified world view. This theory suggests that the Universe has forever been, and will forever remain in a state of oscillation, passing from expansion to contraction and back again repeatedly (Figure 11.3). We happen at the moment to be in the expanding phase of the cycle. Going backward in time, the theory predicts that the Universe must have been denser and hotter some billions of years ago than it is today. If we go sufficiently far back in time we will reach a state of density and temperature in the Universe that represents the maximum

| Big Bang | Galaxies formed | Present: galaxies dispersing | Future: galaxies halt | Galaxies fall | Big-Bang cycle repeats |
|---|---|---|---|---|---|

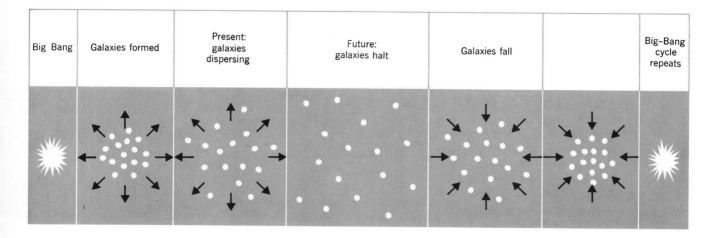

Figure 11.3   The Oscillating cosmology.

compression to which matter can be subjected. This point marks the end of the contraction phase of the previous cycle, and the beginning of the expansion phase of the current cycle.

The conditions at the onset of the expansion phase are identical with the conditions predicted by the Big-Bang theory for the beginning of the Universe. However, the Oscillating theory has the advantage over the Big-Bang theory of being able to answer the question: What preceded the Big-Bang? The answer offered by the Oscillating theory is that prior to the Big-Bang the Universe was in a state of increasing density and temperature. By the time the maximum density and temperature had been reached, all the elements that had been made within stars during the preceding cycle were melted down, so to speak, into the basic hydrogen out of which they had originally been manufactured. Thus, at the moment of maximum compression, the Universe was born anew, composed entirely of hydrogen.

These are the predictions of the oscillating theory for the past. What are its predictions for the future? If the Oscillating theory is correct, the expansion must be slowing down, and will eventually come to a halt and reverse itself. Thereafter the Universe will contract once more, until eventually it reaches the state of maximum compression of matter that will mark the Big-Bang initiating the next stage of expansion.

## THE HUBBLE LAW

Big-Bang, Steady-State, or Oscillating Universe—which cosmology is correct? Was there a beginning? Will there be an end? Or is the Universe destined to experience an infinite number of beginnings and endings. The inquiry into these matters is one of the most interesting areas in human thought.

As in the case of the three cosmological theories themselves, the starting point in the investigation is the observation that distant galaxies are receding from us. Hubble was the first person to notice a special pattern in the velocities of recession. This pattern has since come to play a critical role in all cosmological investigations.

Hubble determined the distance to a number of the galaxies whose velocities had been measured, and found that there was a surprisingly simple correlation between the velocity of recession and the distance of a galaxy: *the more distant a galaxy, the faster it moved*. Hubble's measments indicated that the relationship was a simple proportion. That is, if one galaxy is twice as far away from us as another, it will be moving away twice as fast; if it is three times as far, it will be moving away three times as fast; and so on.

This relation is known as *Hubble's Law*. It can be stated mathematically in the following form: let $v$ be the velocity of recession of a galaxy, and let $x$ be its distance from us; then

$$v = Hx$$

$H$ is a constant of proportionality, usually called the *Hubble Constant*. It has units of velocity over distance and is usually expressed in kilometers per second per million light-years.

The Hubble Law is illustrated by Figure 11.4, showing a number of galaxies together with their spectra. In each case the spectrum of the galaxy is the tapering band of light in the middle, with laboratory spectral lines of known wavelength above and below for purposes of comparison. The clearest feature in the spectrum of each galaxy is a pair of calcium absorption lines that appear at the left in the top spectrum. The normal unshifted positions of these lines in the laboratory are marked by the two black lines at the top of the diagram. In the spectrum of a relatively close galaxy, 63 million light-years away in the direction of the constellation Virgo, these lines are shifted toward the red — which means to the right on this diagram — by a small and barely perceptible amount. A careful study of the spectrum shows that this small shift to the red corresponds to a speed of recession of 750 miles per second.

Next in the diagram is a galaxy 820 million light-years from us in the direction of the constellation Ursa Major. The calcium lines in the spectrum of this galaxy are seen to be shifted to the red by a considerably greater amount. Accurate measurements of the red shift of this galaxy indicate that it is receding from us at a velocity of 9300 miles per second.

The three remaining galaxies in the diagram are still more distant. The last galaxy, barely visible in this photograph, is 2.9 billion light-years away, at the very limit of the range of visibility with the 200-inch telescope. The pronounced red shift of the calcium lines in its spectrum corresponds to a velocity of recession of 38,000 miles per second, or more than 20 percent of the speed of light.

A plot of velocity versus distance for the five galaxies is shown in Figure 11.5. The points lie near a straight line. The departures from a straight line are due mainly to the errors and uncertainties in the measurement of

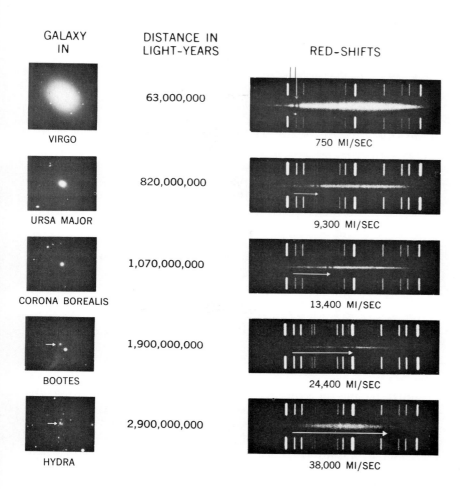

| GALAXY IN | DISTANCE IN LIGHT-YEARS | RED-SHIFTS |
|---|---|---|
| VIRGO | 63,000,000 | 750 MI/SEC |
| URSA MAJOR | 820,000,000 | 9,300 MI/SEC |
| CORONA BOREALIS | 1,070,000,000 | 13,400 MI/SEC |
| BOOTES | 1,900,000,000 | 24,400 MI/SEC |
| HYDRA | 2,900,000,000 | 38,000 MI/SEC |

*Figure 11.4   Galaxies at various distances and their spectra, showing the corresponding red shifts.*

their distances, the error in the measurement of the red shift being relatively small.

The slope of the line through the five points is the Hubble Constant *H*. This constant represents one of the most fundamental quantities in nature, because it tells how rapidly the Universe is expanding. To determine the critical Hubble Constant with greater accuracy, plots have been made of velocity against distance for all galaxies whose distances are known with a fairly high degree of accuracy. One of these plots is shown in Figure 11.6. The straight line that gives the best fit to the points plotted on this graph corresponds to a Hubble Constant of $17 Km/sec/10^6$ light-years.

In the remainder of this chapter we will use a value of 17 Km/sec per million light-years for the Hubble Constant, without further reference to the uncertainties in the distance measurements. You will see that the discussion depends in an essential way on the validity of the Hubble Law,

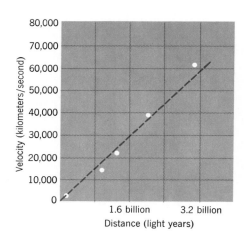

Figure 11.5 Velocity versus distance for the galaxies in Figure 11.4.

Figure 11.6 Velocity versus distance for 46 galaxies.

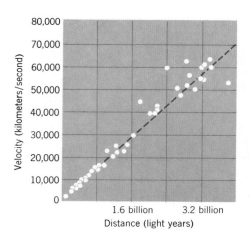

that is, on the straight-line relationship, but not on the precise value of the Hubble Constant in that law. If later measurements show that our value of 17 Km/sec/10⁶ light-years is inaccurate and another value should be used instead, our results for such quantities as the age of the Universe will be changed somewhat, but the basic cosmological ideas will not be affected.[2]

*The Hubble Constant.* The Hubble Constant is the keystone of cosmology. We have seen that its approximate value can be determined from measurements on distant galaxies. If we could find out whether that value has changed during the history of the Universe, and precisely how much it has changed, we would know the answer to the question: Which of the three cosmologies is correct?

The connection between changes in *H* and the validity of the various cosmologies can be perceived most clearly in the case of the Steady-State theory. If this theory is valid, then the rate of expansion of the Universe is the same at all times and, therefore, *H* must be the same at all times. If this were not the case, the Universe would be expanding at a different rate at an earlier time than it is today, and the basic idea of the Steady-State theory would be violated.

Thus, if we find evidence of a change in *H* with time, we know immediately that the Steady-State theory is wrong.

Now let us consider the other two cosmologies. Both assume that today's Universe is expanding as the consequence of a big bang that occurred billions of years ago. Immediately after the big bang, the internal pressures were enormous and the expansion was exceedingly rapid. In the course of time the expansion slowed down because the attraction of gravity acts continually to pull back all of the receding elements of the Universe. Therefore *H* must have a smaller value today than it had billions of years ago. According to the same reasoning, *H* will have a still smaller value billions of years hence when the Universe will be expanding more slowly than it is at the present time.

Thus, according to both the Big-Bang and the Oscillating theories, *H* must decrease with time, while according to the Steady-State theory, *H* is the same throughout all time.

Suppose now that we succeed in finding out what *H* was a billion years ago, and discover that it has, in fact, decreased appreciably in the last billion years. We will then know that the Steady-State theory is wrong, and one of the other two theories—the Big-Bang Universe or the Oscillating

[2] The value given for H was reported in 1972 by Alan Sandage, astronomer at the Hale Observatory. For many years prior to Sandage's announcement, the best value of H was considered to be 25 Km/sec/10⁶ ly. Sandage's reduction in H resulted from a re-calibration of the extragalactic distance indicators described in the Appendix.

Universe—is correct. If the Steady-State theory is eliminated, we will then be faced with the question of deciding between the Big-Bang and the Oscillating theories. Can an accurate determination of $H$ do this?

The answer is yes, because the Oscillating theory and the Big-Bang theory offer different predictions as to the *rate* at which $H$ decreases. You will recall that in both theories, $H$ must decrease with time because the expansion of the Universe is slowing down. However, the slowdown must be more pronounced in the Oscillating theory because in this theory the expansion comes to a dead halt, preparatory to its subsequent collapse. In fact, the theory predicts that $H$ will actually become equal to zero when the expansion is halted. In the Big-Bang theory, on the other hand, the expansion continues forever, and $H$ never becomes zero; it only drifts toward zero slowly in the course of time. Thus, $H$ decreases at a faster rate in the Oscillating theory than it does in the Big-Bang theory.

If we could measure $H$ at an earlier time and compare it with the value it has today, we would know the rate of change. Comparing the rate of change with the calculated value for each theory would tell us which cosmology is correct.

## A Cosmological Test

The changes in $H$ during a year, or even a lifetime, are too small to be detected. But if we could go back a billion years or more, we might detect a large enough change in $H$ to settle the question. But how could we measure the value that $H$ had a billion years ago? On the face of things that would seem to be an impossible task, since modern astronomical records go back no farther than a few hundred years. But consider the following facts: the light that reaches the earth from the Andromeda galaxy left that galaxy 2 million years ago. When an astronomer photographs Andromeda through a telescope, he sees that galaxy as it was two million years earlier, and not as it is today. Similarly, the light that reaches the earth today from the Virgo galaxy left that galaxy 63 million years ago. A photograph of the galaxy shows it as it was 63 million years in the past, and not as it is today.

Now we see how to obtain a picture of the Universe as it was a billion years ago: First, photograph galaxies that are within a distance of 100 million light-years. These galaxies will yield a picture of the expanding Universe as it has been during the last 100 million years. Since 100 million years is a relatively short time on a cosmic time scale, we can consider this picture to represent the Universe as it is today, and can regard the Hubble Constant derived from these measurements as the value of the Hubble Constant at the present time. Next, extend the measurements farther out into space, to galaxies whose distances from us are around 500 million light-years. The receding motions of these galaxies will give us another value of the Hubble Constant, representing the rate of expansion of the Universe during a time approximately 500 million years ago. If the accuracy of our measurements permits us to go still farther out into space, we can measure galaxies at a distance of one billion light-years, and then 2 billion light-years, and so on. The farther out we look in space the farther back we see in time. In this way, we can uncover the state of the expanding Universe in earlier epochs.

The idea behind this measurement is very simple, but the measurement is very hard to carry out in practice because it is difficult to measure the distances to remote galaxies with the necessary accuracy. The most complete study made thus far has been carried out at Mount Palomar by Sandage. He compiled information on 42 galaxies, ranging out in space as far as 6 billion light-years from us. His preliminary results indicate that the Hubble Constant was larger at an earlier time than it is today. That is, the Universe was expanding more rapidly at an earlier time. This conclusion indicates that the Steady-State theory is incorrect.

The Sandage measurements should also shed light on the question of the Big-Bang versus the Oscillating Universe. Taken at face value, the data indicate that the Hubble Constant is changing more rapidly than the Big-Bang Universe would allow. Therefore, the Universe must be oscillating. Unfortunately, the uncertainty in Sandage's values for the Hubble Constant is large enough so that his conclusion on this score cannot yet be accepted with confidence. Sandage estimates that another 10 years of work will be required to settle this basic question.

## The Primordial Fireball Radiation

An independent set of facts also favors the Big-Bang and Oscillating theories, and indicates that the Steady-State theory cannot be correct. This line of evidence starts with the remark that if the Universe was once in a dense, hot state, it must have been filled with an intense and brilliant radiation at that time. In fact, the Universe would look like the fireball that forms when a hydrogen bomb explodes. The intensity of the fireball must have diminished as the Universe expanded, but a small remnant of the original fireball radiation should still be present today. This fireball radiation should be detectable with a sensitive radio antenna. The special characteristic of the fireball radiation, which should enable astronomers to distinguish it from all other kinds of radiation reaching the earth, is the fact that it fills the Universe uniformly, and consequently must bombard the earth with the same intensity from all sides. If an astronomer points his antenna in many directions, he should measure the same intensity of radiation in every case.

This feature of the Big-Bang and Oscillating theories was pointed out by a Princeton physicist named Robert Dicke in 1965.[2] Dicke saw that if

[2]Gamow proposed the same idea in 1948, but his suggestion drew no reaction at that time.

the fireball radiation were discovered, it would settle the controversy between these theories and the Steady-State cosmology. He set about constructing an apparatus to search for the remnant of the fireball radiation, unaware that two Bell Laboratory physicists—Drs. A. Penzias and R. Wilson—had already found it. They, too, were unaware that they had made the discovery, for they were not looking for fireball radiation; they were measuring the intensity of radio noise received in a large antenna that had been set up some time before in connection with the communications satellite program.

The events that followed were strikingly reminiscent of the events that led to the discovery of radio astronomy in 1931 by Karl Jansky, another Bell Telephone scientist. Penzias and Wilson, like Jansky before them, noticed a puzzling radiation that could not be easily accounted for. Although Jansky's radiation was centralized in the direction of the Milky Way, the Penzias—Wilson radiation had precisely the same intensity no matter what direction their antenna was facing. In other words, it seemed to fill the heavens uniformly, just as Dicke had predicted for the remnant of the primordial fireball.

Penzias and Wilson were unable to explain the source of this sky-filling radiation, until a friend told them of Dicke's work. Then, they and Dicke realized that they had stumbled on the evidence for the primordial fireball. The rest is scientific history.

Subsequently, other physicists and astronomers confirmed the existence of the primordial fireball radiation. Their measurements constitute strong evidence for the Big-Bang and Oscillating cosmologies, for no satisfactory explanation for this radiation has been provided by the Steady-State cosmologists.

## Big-Bang or Oscillating Universe?

With the Steady-State Universe almost certainly excluded by two independent sets of facts, the choice now rests between the Big-Bang and Oscillating theories. The Big-Bang theory implies that the Universe had a definite beginning and is slowly approaching an end. The Oscillating theory implies that the Universe has neither beginning nor end, but repeats its cycle of alternate expansion and contraction throughout eternity. The choice between the two theories brings us back to the questions with which we opened the chapter: Was there a beginning? Will there be an end? Is the Universe eternal?

The particular kind of eternal Universe represented by the Steady-State theory has been more or less eliminated by observations. On the other hand, the more complicated sort of eternity described by the Oscillating theory seems to be supported by Sandage's observations. But, as we have pointed out, the uncertainty in Sandage's observations is such that the support for the Oscillating Universe is only very tentative. In 10 years or so, Sandage's results may become precise enough to settle the question. In the interim, can we satisfy our curiosity in some other way? Is there any other source of information that can settle this most important of all cosmological issues?

## Density of Matter in the Universe

The answer is straightforward. If the density of matter in the Universe is sufficiently great, the gravitational attraction of the different parts of the Universe on one another will be strong enough to bring the expansion to a halt, and reverse it to commence a renewed contraction. That is, the Universe will be in an oscillating state. On the other hand, if the density of matter in the Universe is not great, the force of gravity will not be sufficient to halt the expansion, and the Universe will continue to expand indefinitely into the future, as predicted by the Big-Bang theory.

In other words, the density of matter in the Universe is a critical factor in deciding between the two cosmologies. What is the critical density of matter required to slow down and reverse the expansion? A calculation shows that the present expansion of the Universe will be halted if the density of matter is $5 \times 10^{-30}$ gm/cm³ or greater. This density corresponds to one proton or hydrogen atom in a volume of 10 cubic feet.

How does the critical value of the density compare with the observed density of matter in the Universe? The matter whose density can be most readily estimated is that which is present in the galaxies in a visible form, as stars and dense concentrations of gas. If we were to smear out the matter of the galaxies into a uniform distribution filling the entire Universe, the density of this smeared-out distribution of matter would be $5 \times 10^{-33}$ gm/cm³. That is, the quantity of matter contained in all the galaxies of the Universe is too small by a factor of 1000 to halt the expansion.

Since energy and radiation are equivalent to matter by Einstein's formula, $E = mc^2$, we must add to the above figure the contribution from various types of radiant energy in the Universe, such as starlight and the primordial fireball radiation. These forms of energy turn out to increase the average density of matter by one or two percent, which is not enough to affect the outcome.

What about matter that is unobservable because it is not luminous? For example, this matter could exist in the galaxies in the form of dead stars, or stars of very low mass and negligible luminosity. It could also be present in the form of gas in the space between the galaxies.

The invisible matter is very difficult to detect, but its amount can be estimated by an indirect method. Galaxies usually are grouped in clusters, the galaxies in a cluster being held together by the force of their mutual gravitational attraction (p. 16). In such a cluster, the individual galaxies revolve around one another in a swarming motion, like bees in a hive. The more matter a cluster of galaxies contains—in any form, visible or invisible—the stronger the pull of its gravity, and the faster the swarming motions of the galaxies. If the velocities of the galaxies in a cluster can be measured, the total mass of the cluster can be calculated.

This idea has been applied to the large cluster in the constellation Coma Berenices. The results are surprising. On the basis of the motions of the galaxies in that cluster, the amount of matter it contains in an invisible form is thirty times greater than the amount present in the form of luminous stars and other directly observable objects.

Although the estimated density of matter in the Universe is greatly increased as a result of this determination, it is still thirty times too small to bring the expansion of the Universe to a halt. Thus, the facts seem to favor the Big-Bang cosmology. If this theory is correct, the Universe began suddenly, some 13 billion years ago. But what are we to make of such a picture? The Universe is the totality of matter; if there was a beginning, what came before? When all the stars go out, what comes after? On philosophical grounds, the concept of an eternal Universe seems more acceptable than the concept of a transient Universe that springs into being suddenly and then fades slowly into darkness.

Astronomers try not to be influenced by philosophical considerations. However, their curiosity regarding the scientific workings of the Universe has led them into numerous efforts to see whether the Oscillating theory might be saved by the discovery of some previously unsuspected source of energy or matter, that might add to the known matter in the Universe and slow down the current expansion. Estimates have been made of the number of neutrinos in the Universe, and the number of invisible dark objects—for instance, burned-out stars and black holes in space—that might exist. The result is always the same: none of these additional sources is sufficient to increase the density of matter in the Universe by more than a few percent. The total is still too small.

## THE BOUNDARIES OF THE KNOWABLE

Although two independent sets of observations seem to exclude the Steady-State Universe, no clear-cut evidence has yet been found to distinguish between the Big-Bang and Oscillating theories. The observations discussed above, which seem to support the Big-Bang Universe, and may be reversed at any time by new or improved measurements. Because of the receding motions of the galaxies, we can be sure that the ultimate cosmological theory will resemble one of the two forms: Big-Bang or Oscillating. But which one? Until the choice between these two cosmologies is settled, we will be unable to answer the questions with which this chapter opened. Was there a beginning? Will there be an end? Is the physical Universe eternal?

There the matter rests for the moment. Astronomers have exposed very interesting details in the history of the Universe — the birth of stars, the assemblage of the elements within the stars out of the three basic particles, and their dispersal to space in supernova explosions — but science has been unable to solve the fundamental problems of beginning and end.

## Questions

1. What is the cosmological significance in the life story of the stars?
2. Give a one-paragraph summary of each of the three cosmologies.
3. Describe the evidence which indicates that the Universe is expanding.
4. The observed red shift for Galaxy 3C295 is 36 per cent. What is its velocity relative to us? What is its distance? Describe the condition of the materials of the solar system when the light we now receive from this galaxy set out on its journey.
5. What is the Hubble law? Define the Hubble Constant. Explain how changes in the Hubble Constant with time can be measured. How can the rate of change of the Hubble Constant with time be used to prove which of the three cosmologies is correct?
6. What evidence indicates that an explosion took place at the birth of the Universe? What additional evidence seems to exclude the Steady-State theory of cosmology?
7. Which basic force may halt the expansion of the Universe? Why not the other two basic forces?
8. As we look outward in space, we look backward in time. As an example, 3C273 is 2 billion light-years away; hence, we see this quasar as it was 2 billion years ago. Suppose that the light from a quasar at a distance of 8 billion light-years showed a red shift, while the light from a quasar at a distance of 5 billion light-years showed no Doppler shift, and the light from quasars within 5 billion light-years showed a blue shift. Could this pattern of Doppler shifts be explained by any of the three cosmologies? If so, which one? Explain your reasoning.
9. Plot a graph of velocity versus distance for each of the following four sets of hypothetical observations of galaxies. (Minus and plus signs signify velocity toward and away from us, respectively. Draw a smooth curve through each set of points.)

| Distance | Velocity (km/sec) | | | |
| (lt.-yrs.) | (a) | (b) | (c) | (d) |
|---|---|---|---|---|
| 500 million | 12,500 | 13,000 | 1,000 | −1,000 |
| 1 billion | 25,000 | 28,000 | 3,000 | −2,000 |
| 2 billion | 50,000 | 58,000 | 10,000 | 2,000 |
| 4 billion | 100,000 | 136,000 | 40,000 | 10,000 |
| 6 billion | 150,000 | 260,000 | 120,000 | 25,000 |

10. If the radius of the observable Universe is 10 billion light-years and the average distance between galaxies is 2 million light-years, how many galaxies are contained in the observable Universe? If each galaxy contains 100 billion stars and the average mass of a star is $10^{33}$ grams, what is the total mass of the Universe? What is the average mass density of the Universe? Compare your result with the mass density required to halt the expansion of the Universe. What is your conclusion?

11. List the sources of matter and energy in the Universe. Summarize recent findings that may substantially increase current estimates of the total amount of mass and energy in the Universe. What is the cosmological significance of these findings?

12. If you were located in a galaxy at a distance of 10 billion light-years from the Milky Way Galaxy, what would be the appearance of the sky in all directions?

13. If the Steady-State cosmology did not appear to be excluded by scientific evidence, and all three major cosmologies were scientifically admissible, what would be your personal preference among the three theories? What are the reasons for your preference?

**Part Three    The Planets**

# 12　The Sun

The sun was formed 4.6 billion years ago, is nearly half-way through its life, and will not change its properties appreciably until it moves off the Main Sequence in five or six billion years to become a red giant. It is an ordinary body in the cosmic hierarchy, similar to countless other G2 stars on the Main Sequence in its general characteristics; but it has one unique feature: it is 300,000 times closer to us than the next nearest star.

The closeness of the sun gives it a considerable astrophysical interest. The sun may be one point among many on the H-R diagram from the viewpoint of the stellar evolutionist, but the solar astronomer, who studies its properties in detail, finds it to be a complicated and interesting object. The sun's surface is a tempestuous region, marked by violent outbursts known as flares whose origin remains largely unexplained. These solar eruptions produce effects in the earth's atmosphere that have major consequences for the inhabitants of this planet including radio blackouts and possibly changes in climate. The sun also provides us with our only opportunity to take a close look at a stellar atmosphere. The precise measurement of the intensities and detailed shapes of the lines in the solar spectrum, made possible by the sun's proximity to the earth, provides an observational checkpoint for calculations of absorption lines in stellar atmospheres. The information yielded by these calculations is the basis for the interpretation of all stellar spectra. Thus, the sun is the indirect source of a large body of astrophysical knowledge.

*A Solar Prominence*

Table 12.1
**Properties of the Sun**

| Quantity | Value | Method of Measurement |
|---|---|---|
| 1. Average sun–earth distance | 92,956,000 miles<br>149,598,000 kilometers | Radar reflection from Venus |
| 2. Angular diameter | 32' | |
| 3. Radius | 432,000 miles<br>696,000 kilometers | Angular size and distance to the earth |
| 4. Mass | $1.99 \times 10^{33}$ g | Orbits of the planets |
| 5. Average density | 1.41 g/cm³ | $\rho = \dfrac{\text{Mass}}{\text{Volume}}$ |
| 6. Solar constant | 1.947 cal/min/cm²<br>$1.358 \times 10^{6}$ ergs/sec/cm² | high-altitude aircraft measurements |
| 7. Luminosity | $3.90 \times 10^{33}$ ergs/sec | Solar constant and sun-earth |
| 8. Surface temperature | 5800° K | Luminosity and radius $(L = 4\pi R^2 \sigma T^4)$ |
| 9. Spectral type | G2 | |
| 10. Apparent magnitude | −26.8 | Photometer |
| 11. Absolute visual magnitude ($M_v$) | 4.79 | Apparent magnitude and sun-earth distance |
| 12. Bolometric correction (B.C.) | 0.07 | Spectral type |
| 13. Bolometric magnitude | 4.72 | $M_v$ − B.C. |
| 14. Rotation Period | Sunspots   Photosphere | Motion of sunspots |
| Equator | 25.0 (days)   26.0 | Doppler shift in Photo- |
| 30° | 26.4        27.3 | sphere spectrum |
| 60° | −        32.5 | |
| 80° | −        ∼ 35 | |
| 15. Magnetic field | ∼ 1 gauss averaged over surface, fluctuating and irregular. Hundreds of gauss over disturbed areas. Thousands of gauss in sunspots. | Zeeman effect |
| 16. Composition: most abundant elements in decreasing order of mass fraction (percent) | H    ∼ 75<br>He   ∼ 25<br>O     0.8<br>C     0.3<br>N     0.2<br>Ne    0.2<br>Si    0.06<br>Fe    0.04 | Solar absorption spectrum |

THE SOLAR SYSTEM

# HISTORY OF THE SUN

The history of a star with the mass of the sun was described in Chapter 7 as a case study in stellar evolution. Recapitulating the beginning of the history, condensed pockets of gas are believed to form and dissolve repeatedly in the course of the random movements of interstellar matter in the Galaxy. Sometimes these temporary condensations become permanent, because the atoms and molecules of the pocket of gas are held together by the attraction of their own gravity. As soon as a gravitationally bound cloud of particles forms, it commences to contract, falling inward on itself under the continuing attraction of gravity. The energy released by the collapse of the cloud is converted into heat, and the temperature at its center rises. The contracting, self-heating cloud is a protostar.

The protostar that became the sun is called the protosun. In the beginning the prostosun contracted steadily, but after roughly 20 million years proton-proton reactions flared up at its center, the resultant release of nuclear energy slowed the contraction to a halt, and the hot, dense cloud of gas, now the sun, settled down to a stable existence on the Main Sequence. According to theoretical studies of solar evolution, at that time the sun had a surface temperature very close to the present value of 5800° K, and the same yellow-white color with which we are familiar today. However, the young sun was roughly half as luminous as it is today.

## Changes in the Composition of the Sun

Has the sun changed in any other way during its lifetime? One major change has taken place as a consequence of the nuclear reactions that have been going on steadily at its center. It is likely that the sun was uniform in composition initially, with the same mixture of elements and the same relative abundances throughout its interior. Its main ingredients were primordial hydrogen and helium plus a small amount of heavier elements.[1] Hydrogen made up about 75 percent of the mass of the primitive sun, helium made up most of the remaining 25[2] percent,

---

[1] In addition to primordial helium, dating back to the Big Bang, the sun must also contain helium that was manufactured out of hydrogen in other stars and added to the interstellar medium later. The relative contributions from Big Bang helium and helium manufactured subsequently are a point of controversy in cosmology (see Chapter 11). However, the available evidence appears to favor the view that most of the helium in the Cosmos, and presumably in the sun, is primordial.

[2] This value is an average between an estimate of 30 percent deduced from the intensities of helium lines in the chromosphere (p. 302) and 20 percent deduced from the abundance of helium in solar cosmic rays.

and elements heavier than helium constituted one or two percent.[3]

Once nuclear burning began, helium started to build up in the center of the sun and the concentration of hydrogen began to diminish, as a

Figure 12.1 Comparison between initial concentrations of hydrogen and helium and their concentrations after 4.6 billion years, based on theoretical studies of solar evolution. The curves show the depletion of hydrogen within a zone of roughly 200,000 km around the center, and the corresponding build-up of helium in this zone.

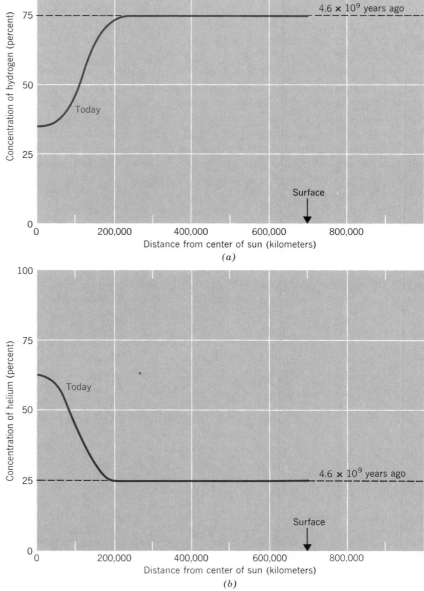

consequence of the steady conversion of protons to helium nuclei in fusion reactions. Gradually, a zone of helium-rich material spread outward from the center. Today, after four and a half billion years of steady burning, the concentration of hydrogen at the center has been depleted twofold, from 75 percent to approximately 35 percent. At the same time, the concentration of helium at the center has risen from 25 to 65 percent.

Figure 12.1 shows the variation in the concentrations of hydrogen and helium from the center of the sun to its edge. The changes are confined to a region extending out about 200,000 kilometers from the center, or roughly one-third of the sun's radius. Beyond this distance the helium/hydrogen ratio is the same as it was 4.6 billion years ago.

In the lifetime of the sun thus far, only five percent of the sun's total mass has been converted from hydrogen to helium. The reason is that although the change in composition near the center is drastic, it has occurred in a space that makes up only a small fraction—about one fiftieth—of the sun's total volume.

## THE SUN'S INTERIOR

Theoretical studies of stars of one solar mass have been carried out by many theoretical astrophysicists under a variety of assumptions, and agreement has been reached regarding the general conditions that exist in the interior of the sun. Typical results of these calculations are shown in Figures 12.2 and 12.3, which represent the temperature and density of the sun at various points between the center and the surface. Figure 12.2 shows that the temperature decreases from a central value of approximately 15 million degrees to a value that appears to be zero at the surface. In reality, the surface temperature is 5800 °K, but this value would be less than the thickness of a pencil line if represented on the million-degree scale of the graph on Figure 12.2.

Figure 12.3 indicates that the density within the sun falls off very sharply with increasing distance from the center. The central density is about 150 g/cm³, or 13 times the density of lead. Halfway from the center to the surface, the density has decreased to 1 g/cm³, which is the density of water. At the surface, the density is $10^{-7}$ g/cm³, or approximately one ten-thousandth of the density of air at the earth's surface.

As a result of the rapid falloff in the density of the sun, most of its mass is concentrated in a relatively small volume, approximately 90 percent of the sun's mass being contained in the inner half of its radius. The average density of the sun is 1.4 g/cm³, or somewhat greater than the density of water.

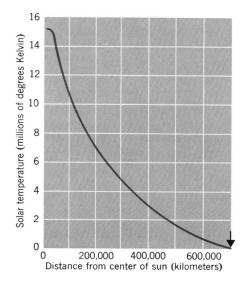

Figure 12.2 *Temperature at various depths in the sun's interior. The arrow indicates the surface.*

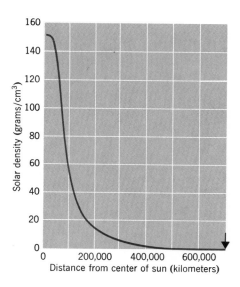

Figure 12.3 *Density at various depths in the sun's interior.*

[3] Expressed in numbers of atoms, hydrogen and helium constitute roughly 90 and 10 percent, respectively.

*The Solar Neutrino Experiment.* Although these temperature and density values are considered to be close to the true conditions within a star of solar mass, confidence in their detailed accuracy has been diminished somewhat by the results of a recent experiment, in which an attempt was made to measure the number of neutrinos emitted from the sun's interior. These massless, chargeless particles are created in the core of the sun in nuclear reactions such as those on p. 152, and escape to space directly because of their negligible interaction with matter. In the experiment, a tank-car-sized neutrino detector consisting of 600 tons of liquid tetrachlorethylene ($C_2Cl_4$), was buried in a mine in order to screen out interfering cosmic rays. This substance, similar to cleaning fluid but less poisonous, is relatively effective in recording the presence of neutrinos, which are captured by the chlorine atoms in the fluid.

After running the experiment for an extended period, physicists reported that the number of neutrinos emitted from the sun was 10 times smaller than the number predicted. The predictions were based on laboratory measurements of nuclear reaction rates, combined with theoretical studies of the temperature and density at the center of the sun.

In order to bring the calculations on the number of solar neutrinos into agreement with the results of the experiment, it would be necessary to reduce the computed temperature at the center of the sun by about 1.5 million degrees. This reduction represents a change of only 10 per cent in the temperature values shown in Figure 12.2; however, a 10 per cent correction is far greater than the uncertainty that the astronomers had previously attached to their calculations of conditions in the interior of stars of solar mass.

The solar neutrino experiment poses a serious problem in astronomy. Although the experiment is difficult, most astronomers and physicists consider that it has been done carefully and its result must be taken at face value. A possible explanation of the experiment is that the neutrino has unsuspected properties that have not been revealed previously. Another possibility is that the internal structure of the sun is more complex than previously assumed. One theoretical study suggests that occasionally, perhaps every few hundred million years, the core of the sun becomes unstable and expands, the temperature at the center drops, and the nuclear reaction rate and neutrino production decrease accordingly. The sun's surface temperature and luminosity diminish also. This means that for a time the central temperature of the sun, the neutrino production rate and the solar luminosity all are below their "normal" values. That could be the condition that prevails today, and would explain the result of the neutrino experiment.

Later, according to the studies, conditions in the sun's interior and on the surface return to normal. After perhaps 200 million years the sun's core becomes unstable again, and the entire cycle repeats. It has been suggested that the periodic changes in solar energy output that take place during this cycle might be the cause of the major ice ages on the earth.

Although a generally accepted explanation of the solar neutrino experiment has not yet been provided, the experiment and its possible theoretical interpretations are of great interest, partly because they have forced astronomers to reopen a nearly closed chapter in stellar evolution, and partly because they may have interesting implications for the history of the earth.

## "Experiments" In the Sun's Interior

How can the astronomer state with assurance that conditions in the interior of the sun are even within 10 percent of the values in Figures 12.2 and 12.3? The answer concerns a fundamental difference between astronomy and many other fields of scientific enquiry. Most branches of science are based largely on laboratory experiments, in which the scientist studies the behavior of an object under carefully controlled conditions. The object might be an electron, a proton, or a virus. The clear-cut relationships that make up the body of science have been obtained mainly as a result of such tightly controlled experiments in the laboratory. But in astronomy, this kind of controlled experimentation is not possible. The objects under study are usually galaxies, stars, or planets, which are too large to be brought into the laboratory or explored in their natural state.

"Experiments" are carried out in astronomy in a completely different way, involving the use of high-speed electronic computers. These computers have created a new mode of research in science, called the "numerical experiment." A numerical experiment starts with a set of laws or formulas—such as Newton's laws of gravity—that describe the object under study, and how it reacts to natural forces. For example, Newton's law of gravity tells us how the force of gravity acts on the atoms of a cloud of matter in space. Newton's law of motion—which is a separate law—tells us how rapidly these atoms move toward one another under the action of this force. When an atronomer conducts a numerical experiment on a star, he starts out with the assumption that the star is a spherical distribution of matter held together by gravity. In his first step, he writes a computer program, usually on punched cards, containing all the basic formulas relating to the star, such as the law of gravity, the law of motion, or the formula that gives the rate at which nuclear energy is released by the fusion of hydrogen into helium. Then the astronomer enters the computer program into the memory of the computer, together with numbers representing the mass of the star, the value of the universal constant of gravitation, the laboratory-measured rates of the various nuclear reactions that take place, and so on.

Next, he enters special instructions into the computer memory to describe the particular conditions under which the hypothetical experiment is to be conducted. For example, these instructions might specify the forces that are acting on the atoms of the star in addition to gravity, or the nuclear reactions that are important. The final instructions tell the computer to commence the sequence of numerical steps that simulate the behavior of the star. The steps consist in substituting the numbers that describe the initial properties of the object, and the magnitudes of the applied forces, changes in composition, and so on, into the equations representing the basic laws of physics.

The computer does all the necessary arithmetic very rapidly, at the rate of several million additions, subtractions, multiplications, and divisions per second. Finally, it produces a complete description of the star, including the temperature, density, and composition in its interior, and the luminosity and temperature on its surface.

If the object of the numerical experiment is to determine how the properties of the star change with time, that is, to describe stellar evolution, it is necessary to allow for the fact that hydrogen is being used up steadily at the center of the star. This is the main change that occurs in the star's lifetime. Since hydrogen burning is the source of the star's energy, when the amount of hydrogen changes all the other properties of the star also change.

In principle, conditions in the star's interior change every second because the amount of hydrogen goes down continuously. However, during most of the star's life its structure changes so slowly that a second-by-second description is not necessary. The time spent on the Main Sequence, for example, is divided up into, say, 200 time intervals. In the case of the sun, whose life span on the Main Sequence is 10 billion years, each time interval is 50 million years. Later on in the star's life, as it passes into the red-giant stage and beyond, it evolves more rapidly, and shorter time intervals become necessary. In calculations on the evolution of the sun, the time interval shrinks to about a million years at the beginning of the red-giant stage and becomes still shorter as the sun grows into a fully developed red giant. For a brief moment during the helium flash, the changes must be computed every few seconds. However, this period of intensely rapid change lasts for less than a minute.

Let us suppose that the star is in a stage in which a 50-million-year time interval is required. The astronomer proceeds as follows: first, he calculates conditions in the star assuming that its hydrogen content remains constant for 50 million years; second, he calculates the amount of hydrogen that would have been burned up in 50 million years, under the conditions of temperature and density that he has just calculated for the center of the star; and third, he *recalculates* the conditions in the star, allowing for the fact that less hydrogen is present. As a result of his calculation, he knows approximately how the conditions in the star have changed in that particular 50-million-year interval. In this way, 50 million years at a time, he traces the evolution of the star through the hydrogen-burning stage of its life. The modern description of stellar evolution depends nearly as much on the use of the computer in such studies as it does on the telescope and the spectroscope.

## The Zone of Convection

In the deep interior of the sun the temperatures range up to many millions of degrees. In this range of temperatures, collisions between atoms are sufficiently violent to eject many electrons from their orbits. Light atoms are completely stripped of their electrons, and heavy atoms lose their outer electrons, retaining only the tightly bound inner electrons. These inner electrons cannot be dislodged easily by absorption of a photon. Consequently, photons pass readily through the inner part of the sun.

Nearer to the surface of the sun the temperature falls, and the heavier atoms, such as iron, begin to recapture their outer electrons. The outer electrons in an atom are bound to the nucleus by relatively small forces and, therefore, can be easily separated from the nucleus by the absorption of a photon. For this reason, photons are strongly absorbed by atoms that possess their outer electrons. The appearance of these absorbing atoms in appreciable numbers below the sun's surface tends to block the flow of photons coming from the interior.

If photons are the only means of carrying energy up to the surface of the sun, the blocking of these photons will cause the temperature to drop sharply at some depth below the surface. In this situation, the outer region of the sun now consists of a layer of relatively cool gas resting on a hotter interior. This layer of cool gas reacts in the same way as a pot of water placed on a hot stove. The water at the bottom of the pot, heated by contact with the stove, expands, and rises to the surface. At the surface, the water loses some of its heat to space, cools, and descends to the bottom of the pot. There it is reheated and rises again. The result is a circulating current of water, which carries heat from the bottom of the pot to the surface (Figure 12.4).

In the same way, the gas at the bottom of the cool outer layer of the sun is heated by its contact with the hot gas in the interior, expands, and rises toward the surface. At the surface, the hot gas loses its heat to space, cools, and descends again into the interior. As a consequence the entire outer layer of the sun breaks up into ascending columns of heated gas and descending columns of cooler gas.[4] As with the pot of water on the stove, these circulating currents of gas carry heat or energy upward to the surface from the interior of the sun.

The transport of energy by circulating currents of gas or fluid is called convection. The currents that carry the heat upward are called convection currents, and the region of the sun in which this large-scale upward

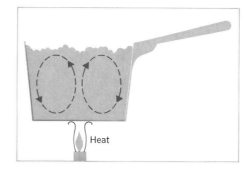

*Figure 12.4  Rising currents in a heated pot of water carry heat to the surface.*

---

[4] In the case of a pot of water on a hot stove, if the stove is very hot bubbles of water vapor, that is, steam, form at the bottom of the pot. These bubbles rise rapidly to the surface because of their buoyancy. That is, the water begins to boil. In the case of the sun, the materials are already in the gaseous state, and boiling does not occur.

and downward movement of gases occurs is called the _zone of convection_. It extends from a depth of about 150,000 kilometers upward to the surface of the sun.

At depths greater than 150,000 kilometers, energy is transported within the sun by radiation, that is, by the flow of photons. Near the surface the outward flow of radiant energy is blocked to a substantial degree by the absorption of photons, and convection sets in. From that depth out to the surface, energy is transported partly by convection and partly by radiation. Above the surface, radiation again becomes the sole means of energy transport.

_The Effect of Convection on the Temperature Profile._ The upward flow of hot gas in the zone of convection increases the temperature in the outer layers of the sun, erasing the sharp drop in temperature that would occur if the flow of radiation were the sole way in which energy could reach the surface. The temperature curve in Figure 12.2 was obtained from a theoretical calculation that added the energy carried to the surface by convection to the energy transported by radiation.

_Tiers of Convection._ The currents in the zone of convection are shown in Figure 12.5a. The diagram simplifies the structure of the zone by showing single columns of gas rising without a break from the lower boundary of the zone to the surface. The actual situation is believed to be closer to that shown in Figure 12.5b, in which several tiers of convection currents carry heat successively upward. Opinions vary regarding the number of tiers in the zone, but three is considered most probable. The lowest tier contains massive convective currents, perhaps 200 to 300 thousand kilometers in diameter. At an intermediate depth, these large convection currents give way to a second tier of currents. The convection currents in the second tier are about 30,000 kilometers in diameter and 15,000 kilometers deep. Above the second tier lies a third tier. The currents in the third tier are roughly 1000 kilometers across and 1000 to 2000 kilometers deep. The tops of the upward-moving columns of gas in the third tier make up the visible surface of the sun.[5]

## The Photosphere

The visible surface of the sun is called the _photosphere_. The photosphere is the sun's disk as observed visually or with a telescope. It has a uniform appearance when viewed with the eye or a small telescope, but inspection with a large telescope under good visual conditions reveals that it has a granulated texture. The granules are relatively small—about

[5] Evidence based on the Doppler-shift suggests that the large-scale, slow movements in the first and second tiers extend all the way to the surface. The small-scale, relatively rapid movements that make up the solar granules in the topmost tier are superimposed on these deeper-seated currents (pages 291–292).

1500 kilometers in diameter—and are difficult to observe under ordinary conditions because of the blurring effect of the earth's atmosphere. However, they appear clearly in photographs taken from instruments carried above the atmosphere in balloons (Figure 12.6), or from ground-based telescopes under good seeing conditions.

*Figure 12.5 (a) Upward-and-downward-moving columns of material in the zone of convection. Gas heated at the lower boundary ascends, loses a part of its heat to space, and descends.*

*Figure 12.5 (b) Tiers of convection.*

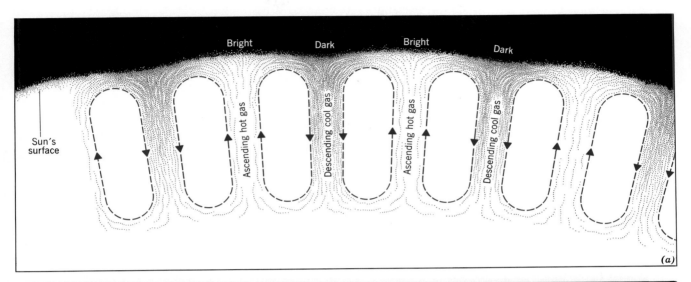

Bright    Dark    Bright    Dark

Sun's surface

Ascending hot gas

Descending cool gas

Ascending hot gas

Descending cool gas

*(a)*

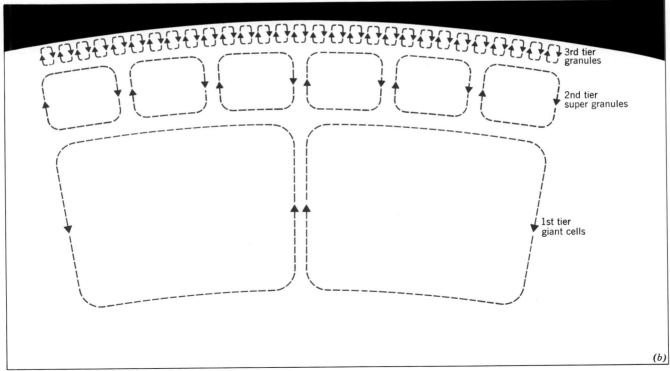

3rd tier granules

2nd tier super granules

1st tier giant cells

*(b)*

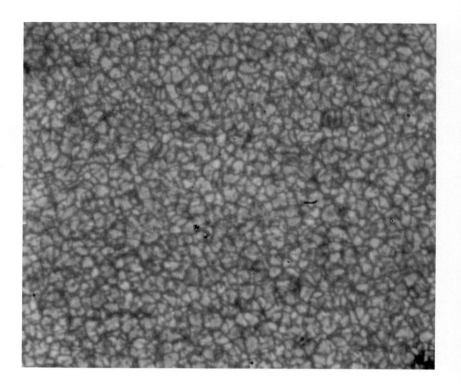

*Figure 12.6   Solar granules photographed from a balloon at 80,000 feet.*

*Solar Granules and Solar Convection.* Doppler-shift measurements indicate that in the bright center of a granule the gas moves upward, and at the dark boundary it moves downward. Apparently the granules are the tops of the ascending columns of hot gas in the uppermost tier of the zone of convection. The Doppler-shift measurements, combined with photographs such as that in Figure 12.6, are proof that the zone of convection actually exists in the sun.

The photograph in Figure 12.6 conveys the impression of irregularly shaped regions of glowing gas, separated by dark, relatively empty spaces. Actually, the density of the gas varies very little from the bright to the dark areas in the photograph. The center of each granule seems bright because it consists of relatively hot gas that has just risen from the interior. The spaces between the granules contain gas that has cooled and is descending again into the convective zone. Being cool, the gas in these regions radiates less strongly and appears darker than the gas at the tops of the upward-moving columns.

The formation of rising columns of gas above a hot surface is a familiar phenomenon in the earth's atmosphere. Frequently the sun's radiation heats the surface of the earth to a higher temperature than the air immediately above. The air, warmed by contact with the ground, expands, becomes buoyant, and rises in columns of heated gas, forming a zone of convection in the atmosphere. The top of the zone of convection is usually at an altitude of about 30,000 feet. As the columns of warm air ascend their temperature drops, and the moisture they have carried with them condenses into droplets of water, forming clouds and rain.

Sometimes, because of conditions in the atmosphere, condensation into clouds occurs only at the top of each upward-moving column of air. When this happens, the top of a column can be seen clearly as an isolated puff of cloud. Figure 12.7a shows a photograph of a field of cloud puffs formed in this way, marking a zone of convection similar to the zone of convection in the sun. Each cloud is analogous to a brightly glowing granule in the photosphere. Figures 12.7b and c compare the terrestrial clouds with solar granules and with convection cells produced in the laboratory by placing a pan of fluid over a uniformly heated surface.

*Supergranules.* Measurements of the Doppler shift in the solar spectrum reveal large-scale movements in the gas at the surface of the sun, similar to the movements of the gas in solar granules, but extending over much greater distances and persisting for longer times. The entire surface of the sun is broken up into a pattern of cells by these movements (Figure 12.8). The Doppler-shift results suggest that the gas flows from the center of each cell outward to its boundary. A recent refinement of the Doppler-shift technique indicates a small vertical velocity in each cell, in addi-

(a)

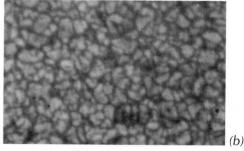

(b)

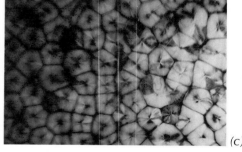

(c)

Figure 12.7 Comparison of convection cells in (a) the earth's atmosphere showing an overhead view of altocumulus clouds, (b) the sun and (c) a laboratory experiment with a pan of fluid heated at the bottom.

Figure 12.8 A photograph utilizing the Doppler shift to reveal large-scale movements of material at the surface of the sun. Light regions are material moving toward the observer and dark regions are material moving away. The alternating pattern of light and dark indicates that the sun's surface is broken up into cells of moving material about 30,000 km in diameter.

tion to the horizontal movements. As in granules, the vertical velocity is upward at the center of the cell and downward at its boundary.

This pattern of velocities suggests that the large-scale movements are the surface manifestation of a deep-seated network of convection cells, extending into the sun's interior to a depth approximately as great as their width, i.e., some tens of thousands of kilometers. Presumably, the large convection cells are the middle tier in the convection zone shown in Figure 12.5b.

The large cells are called *supergranules*. Each supergranule is about 30,000 kilometers in diameter, includes roughly 300 granules within its boundaries, and lasts for about one day.

Doppler-shift measurements also provide evidence for very large currents of moving material, called giant cells, which may be surface manifestations of the third tier of convection cells in Figure 12.5b.

*The Sharpness of the Sun's Disk.* Because the sun's density decreases smoothly and continuously as we pass through its outer layer, we might expect that the brightness of the solar disk would fade gradually into the blackness of space. However, inspection of the sun with the eye or a small telescope reveals that the edge is exceedingly sharp. The explanation for the sharpness of the edge comes out of a careful examination of the circumstances that control the depth in the sun from which photons can escape to space.

In the deep interior of the sun copious numbers of photons are emitted because the temperature of the gas is very high, and also because there are many radiating atoms per cubic centimeter. However, most of these photons are absorbed during their passage through the overlying layers of the sun. Very few reach the surface and escape to space. That is, very little of the sun's light comes from the deep interior.

At very great distances from the center of the sun the density of the solar gas diminishes to exceedingly small values, and the number of radiating atoms becomes very small. Therefore, very few of the photons that we see in the sun's radiation come from these outermost regions.

Most of the photons in the sun's radiation come from an intermediate level, at which the density of the solar gas is great enough to emit many photons, but the amount of overlying material is not so great as to prevent these photons from escaping to space.

The intermediate level is the sun's visible surface, or photosphere. It is a zone of finite thickness and not a sharply defined boundary. However, the sun's density decreases so rapidly with increasing distance from the center that the transformation from nearly full brightness to nearly complete transparency occurs in the relatively narrow distance of about 500 kilometers. Because 500 kilometers is a very small fraction of the sun's 1.4 million kilometer diameter, the solar disk appears to have a knife-edged definition when viewed with the naked eye.

*Limb-darkening.* The preceding discussion of the fate of a photon on its way out of the sun leads to an explanation of the phenomenon of *limb-darkening*. Although the disk of the sun appears uniformly bright to the eye, a measurement of the intensity of the light coming from different parts of the disk reveals that the sun's brightness falls off by roughly 70 percent from the center to the edge. This is the limb-darkening effect. It appears clearly in the photograph of the sun in Figure 12.9. The magnitude of the effect varies with wavelength, being greater at blue than at red wavelengths.

Figure 12.10 illustrates the origin of limb-darkening. Suppose the observer looks first at the center of the sun's disk along the line of sight *a*. Most of the photons he sees emerge from a narrow zone about 500 kilometers thick. This zone is the photosphere. It is shown as a shaded area in Figure 12.10. For definiteness, the region from which the photons emerge is shown as a single point *A* at the midpoint of the zone.

The brightness of the sun's disk at its center depends on the temperature at point *A*. The higher the temperature at *A*, the more copious the emission of photons, and the greater the brightness of the disk at this point.

Suppose now that we choose a point *A'* located on the limb of the sun at precisely the same depth below the surface as *A*. That is, *A'* is also located at the midpoint of the photosphere. The line of sight from *A'* to the observer is labeled *a'* in the figure. The photons moving toward

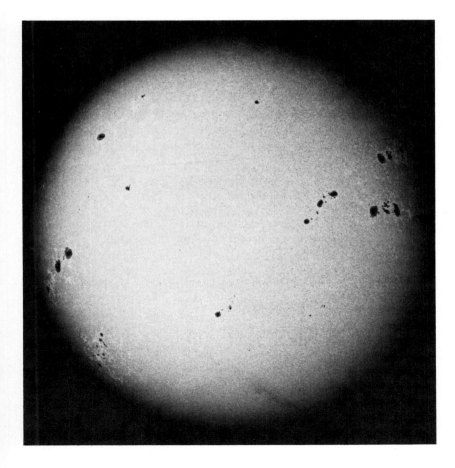

Figure 12.9 A photograph of the sun showing the decrease in brightness toward the edge of the solar disk, or limb-darkening. The photograph was taken at the peak of the sunspot cycle in 1957 and shows many sunspot groups.

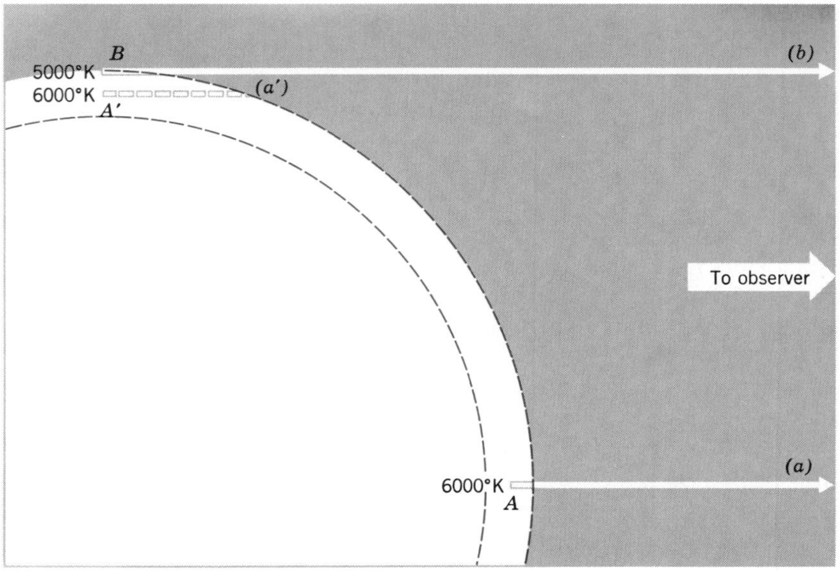

Figure 12.10 The limb-darkening effect. The zone between the dashed lines represents the photosphere, shown in exaggerated thickness. Radiation from the photosphere at the center of the sun's disk is emitted from a depth A. Radiation from the same depth at the limb of the sun (A') fails to penetrate the greater thickness of solar atmosphere. Radiation escaping from the limb must originate at a higher level B.

the observer along this line of sight travel obliquely through the sun's outer layers, and traverse a much greater thickness of gaseous matter than the photons originating from A. Consequently, most of these photons will be absorbed, and few will reach the observer.

The photons that the observer sees must originate from a point farther out, where the total amount of gaseous matter traversed is approximately the same as for photons originating at point A. The point from which these photons originate is marked B, and their line of sight is shown as b.

Since the temperature of the sun falls rapidly with increasing distance from the center, point B, which is farther away from the sun's center than A, must be at a lower temperature. Therefore, the rate of emission of photons from B will be less than from A. This means that the brightness of the sun's disk at the limb, viewed along the line of sight b, is less than brightness at the center of the disk, viewed along the line of sight a.

The limb-darkening effect occurs because the temperature of the sun decreases with increasing distance from its center. Applying the same reasoning in the opposite direction, astronomers have calculated the changes in temperature and density in the outer layers of the sun by measuring the limb-darkening effect for a range of wavelengths.

*The Temperature of the Photosphere.* Theoretical curves of radiated energy versus wavelength were shown in Chapter 5 for a range of temperatures. If these curves are matched to the measured spectrum of energy radiated by the sun at various wavelenghts, the best agreement is obtained for a temperature of approximately 6000°K. An appreciable variation in temperature occurs across the 500-kilometer-thick zone that consititutes the source of the sun's visible radiation. Some of the photons that we observe in the solar radiation actually come from a level a few hundred kilometers below the midpoint of the photosphere, where the temperature is about 8000°K, and some come from a level a few hundred kilometers above the midpoint, where the temperature is about 4000°K. The 6000°K temperature of the sun's visible surface is an average over this range of temperatures.

Another way of calculating the temperature of the photosphere depends on the observed luminosity and radius of the sun. Suppose that the sun were a solid sphere with a sharply defined surface, instead of being a sphere of compressible gas of varying density. In that case all the solar photons would come from the surface of the sphere, instead of being emitted from various depths within the sphere. If $T$ is the temperature at the surface of the sphere, and $R$ is its radius, the energy per second radiated to space by the sphere is

$$L = 4\pi R^2 \times \sigma T^4$$

where $\sigma$ is the Stefan-Boltzmann constant, which has the numerical value of $5.7 \times 10^{-5}$ in centimeter-gram-second units. $T$ can be calculated from this formula if the observed luminosity and radius of the sun are inserted. The calculation yields a value of 5800°K, which is called the *effective temperature* of the photosphere. This value is the temperature that a perfectly radiating sphere[6] would have if its radius were the same as the sun's radius, and its total energy output matched the sun's luminosity.

The effective temperature agrees closely with the temperature of 6000°K deduced from the curve of the variation of energy with wavelength in the sun's spectrum, and either value may be used as the temperature of the photosphere.

*Formation of Absorption Lines.* Moving upward from the photosphere, the temperature falls from 6000° to approximately 4000°K at an altitude of 500 kilometers, and remains at this relatively low temperature for a few thousand kilometers. This layer of relatively cool gas lying over the hotter gas beneath absorbs radiation at wavelengths characteristic of the atoms in the sun. The layer of cool gas is the region in which the solar absorption spectrum is formed (page 91)[7].

---

[6] The formula is valid for black bodies, that is, for objects that entirely absorb all light incident on their surfaces. According to a law of physics, a black body, or perfect absorber, is also a perfect radiatior. An object that absorbs part of the incident light and reflects part, such as a sphere painted dark gray for example, radiates uniformly less energy at every wavelength than this formula predicts. The sun is close to being a perfect absorber or black body and is, therefore, a nearly perfect radiator.

[7] The picture of a sharp distinction between the hot radiating region and the cool absorbing region is a simplified description. In reality, the continuum radiation and the absorption take place simultaneously over a range of levels in the outer part of the sun, with continuum radiation dominant at the lower levels and absorption dominant at the higher levels. The lower levels belong unambiguously to the photosphere and the upper levels are a transition zone between the photosphere and the chromosphere.

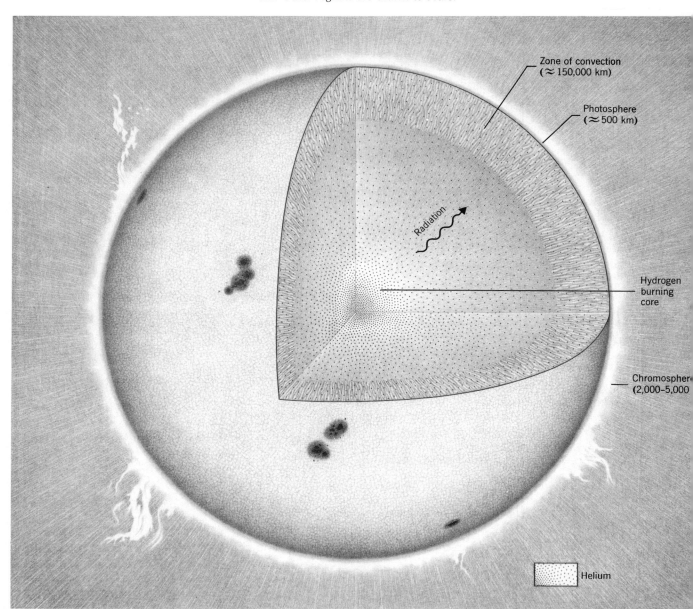

*Figure 12.11 A cross section of the sun showing the structure of the solar interior and atmosphere. The spacing of the dots represents the density of helium nuclei. Thicknesses of the photosphere and chromosphere are exaggerated by a factor of ten. Other regions are drawn to scale.*

Zone of convection
(≈ 150,000 km)

Photosphere
(≈ 500 km)

Radiation

Hydrogen
burning
core

Chromosphere
(2,000-5,000

Helium

Corona

## THE SOLAR ATMOSPHERE

The region of tenuous and essentially transparent solar gas lying above the photosphere is called the *solar atmosphere*. The outer boundary of the solar atmosphere is not clearly defined. The atmosphere extends out to a distance of about 5 million kilometers from the sun, if its limit is considered to be the point at which the density of the solar gas has decreased to the density of the gas in the space between the planets, but the sun's influence on the interplanetary gas can be detected as far away as the orbit of the Earth. In a sense, the solar atmosphere extends throughout much of the solar system.

The solar atmosphere is divided into two regions called the *chromosphere* and the *corona*. Both regions are invisible under ordinary conditions because their faint luminosity is masked by sunlight that has been

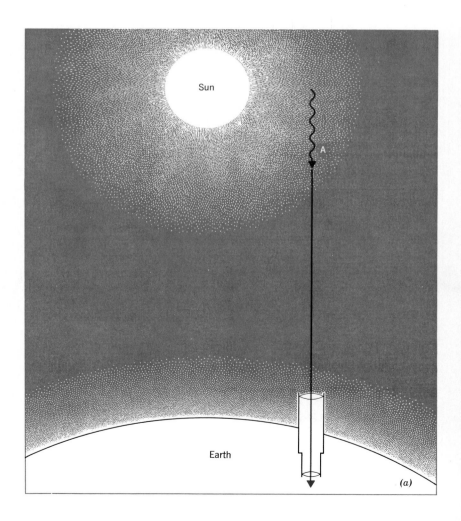

(a)

scattered in the earth's atmosphere or in the telescope itself. Figure 12.12 illustrates the way in which the photons emitted from the face of the sun can be scattered and changed in direction so that they appear to come from the solar atmosphere.

These scattered photons create an apparent halo of light around the sun that is enormously brighter than the true solar atmosphere. A street lamp, viewed on a foggy night, possesses a similar halo of light scattered by the water droplets making up the fog.

During the brief moments in a total eclipse when the face of the sun is completely covered by the moon, the halo of scattered light disappears, and the solar atmosphere becomes visible as a luminous aureole surrounding the moon's black disk and extending out into the space around the sun to a distance of as much as 10 solar diameters or 14 million kilometers. The sudden appearance of this pearl-white luminescence, radiating as much light as the full moon and covering 100

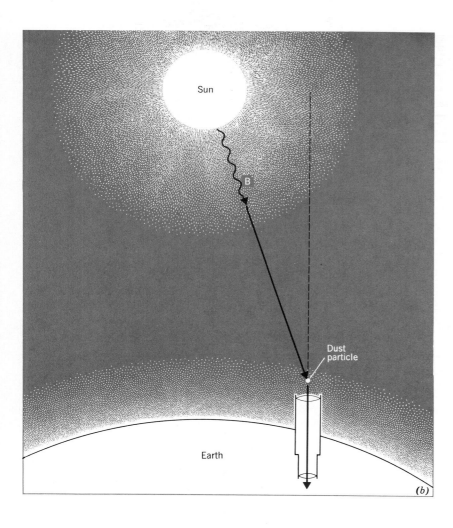

Figure 12.12 Masking of radiation from the solar atmosphere by sunlight scattered in the earth's atmosphere or in the telescope. (a) Photon A is emitted from the chromosphere or corona and received directly by the observer. (b) Photon B is emitted from the photosphere and scattered into the observer's telescope by a collision with a particle in the earth's atmosphere so that it appears to originate in the solar atmosphere.

times the moon's area, has a greater impact on the observer than any other visual display created by the movement of the celestial bodies.

## The Chromosphere

During a total eclipse the trained observer may see a very thin red crescent of light flash into view on the sun's eastern limb in the beginning of totality, at the precise instant when the moon's black disk first covers the photosphere. At the same time the corona appears as a surrounding halo extending far out into the space around the sun. The crescent is a part of the chromosphere of the sun. The corona persists throughout the period of totality, which may be several minutes, but the red flash of the chromosphere disappears in one or two seconds, and is easily missed.

The chromosphere is visible for a very short time because it is a relatively narrow zone in the solar atmosphere, and its rays are cut off quickly by the apparent motion of the moon. Most of the light from the chromosphere originates in the region of the solar atmosphere extending from the photosphere to a height of approximately 2000 kilometers. This narrow zone, only one four-hundredth of the sun's diameter, is exposed at the eastern limb of the moon at the beginning of totality (Figure 12.13a). As the moon continues its apparent motion to the east, it conceals successively higher levels in the chromosphere. The zone of concealment sweeps upward through the chromosphere at a rate of 300 kilometers per second, cutting off the red light from the lower chromosphere in a few seconds, and concealing the entire chromosphere after 15 or 20 seconds (Figures 12.13b and c). The sequence is repeated in reverse order on the western limb of the sun at the end of totality.

*The Flash Spectrum.* If the spectrum of the sun is observed during a total eclipse, the familiar absorption lines in the solar spectrum remain visible as long as a part of the photosphere is exposed. At the moment that the photosphere is entirely covered, the dark-line absorption spectrum suddenly changes to a bright-line emission spectrum. This is the spectrum of the chromosphere. It is called the *flash spectrum* because it flashes into view briefly at the beginning and end of the period of totality of the eclipse.

Many of the bright lines in the flash spectrum are identical with dark lines in the normal absorption spectrum of the sun's disk. The reason is that the two spectra are produced by two successive stages in the same sequence of events. When we observe the surface of the sun, we see a continuous spectrum of radiation from the photosphere after this radiation has passed through a cooler layer of absorbing gas. A photon from the photosphere is absorbed when its energy fits the difference in energy between two states of an atom in the chromosphere. The absorption of

*Figure 12.13 Visibility of the chromosphere during a total eclipse. (a) At the beginning of totality, the entire chromosphere is visible on the eastern limb of the sun. (b) A little later the moon's disk cuts off the light from the lower chromosphere. (c) Still later the entire chromosphere is concealed. The full sequence lasts 15 to 20 seconds.*

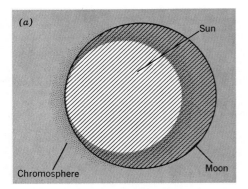

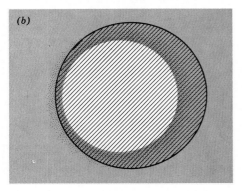

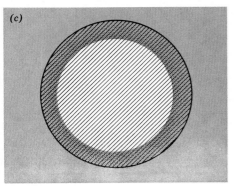

photons with this particular energy gives rise to a dark line in the sun's spectrum. This is the first event in the sequence.

When the atom absorbs a photon, it is raised to an excited state. After a brief interval, the excited atom collapses to a lower excited state or to the ground state. Photons are emitted in the collapse to the lower state, creating the emission spectrum of the chromosphere. This is the second event.

The lines in the emission spectrum are the same as the lines in the sun's absorption spectrum because the absorption spectrum is produced by the initial transition of atoms from a low-energy to a high-energy state, while the emission spectrum results from the subsequent transition of the same atoms in the reverse direction. Since the energy differences are the same for both spectra, the lines have the same wavelengths.

To see the emission spectrum of the chromosphere, we must look at the region just beyond the limb of the sun. The atoms in this region absorb photons from the photosphere below and reemit them in all directions. Some photons are emitted along the line of sight from the limb of the sun to the observer, who detects a glow of radiation coming from the part of the solar atmosphere immediately outside the disk of the sun (Figure 12.14, page 302). If this radiation is spread into its component wavelengths by a spectrograph, it forms the flash spectrum.

Figure 12.15 (page 303), shows the flash spectrum photographed in the 1973 eclipse, with the first two Balmer lines of hydrogen, a line of helium, and lines of several metals identified. The slit normally used at the entrance to a spectrograph to separate closely spaced lines is not needed in obtaining these spectra, because the source of light in the chromosphere is itself a very narrow zone. Each line in the flash spectrum is an image of the narrow zone at one particular wavelength. The lines are curved into the shape of narrow crescents because they are formed by the circular segment of the chromosphere that lies just outside the concealing disk of the moon, as shown in Figure 12.13a.

Figure 12.15 shows that the most intense radiation in the flash spectrum comes from the first Balmer line of hydrogen at 6563Å. The strong 6563Å line gives the chromosphere its characteristic red color.

If a line is produced by a transition at very great altitudes, the region of the solar atmosphere responsible for this line will be exposed throughout the period of totality, and the line will appear in the flash spectrum as a more complete segment of a circle. This is true for lines originating in the corona. The lines in Figure 12.15 at 5303Å and 6374Å appear as longer circular arcs than any other lines in the flash spectrum, demonstrating that these lines originate high in the solar atmosphere, and therefore are coronal lines. The faint green circle in Plate 18 is the same coronal line at 5303Å. It appears as a complete circle because it originates at altitudes that lie entirely outside the moon's eclipsing disk.

*The Temperature of the Chromosphere.* Although many lines are identical in the flash spectrum and the sun's absorption spectrum, the correspondence is not perfect. The flash spectrum has lines of neutral

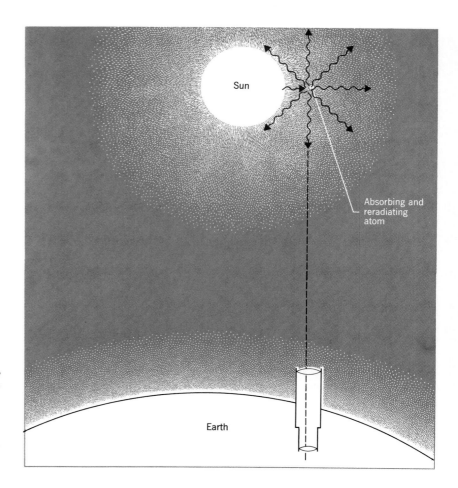

Figure 12.14 Photons emitted from the photosphere at the limb of the sun are absorbed by atoms in the chromosphere. The excited atoms collapse to lower states, emitting photons in all directions. Occasionally, a photon is emitted in the direction of the earth, giving rise to the observed emission spectrum of the chromosphere.

and ionized helium, as well as some lines of ionized metals that are missing or very weak in the solar absorption spectrum. The presence of these lines in the flash spectrum suggests that the temperature at the upper levels of the chromosphere is considerably higher than the temperature at the surface of the sun, since the electrons in the helium atom are tightly bound to the nucleus, and high temperatures and correspondingly violent collisions are required to raise these electrons to excited orbits or remove them from the atom entirely.

Neutral helium atoms, for example, can be excited in a gas only if the temperature is greater than 10,000°K, and the appearance of ionized helium atoms in appreciable numbers requires a temperature of at least 20,000°K. The presence of these lines suggests that the temperature at some level in the chromosphere is at least 20,000°K.[8]

[8] Recent studies indicate that ultraviolet radiation from the corona also contributes to the ionized helium lines in the flash spectrum.

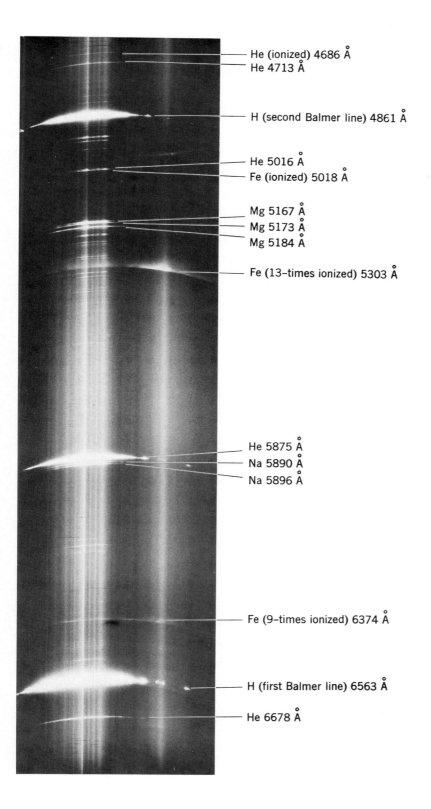

He (ionized) 4686 Å
He 4713 Å

H (second Balmer line) 4861 Å

He 5016 Å
Fe (ionized) 5018 Å

Mg 5167 Å
Mg 5173 Å
Mg 5184 Å

Fe (13–times ionized) 5303 Å

He 5875 Å
Na 5890 Å
Na 5896 Å

Fe (9–times ionized) 6374 Å

H (first Balmer line) 6563 Å

He 6678 Å

*Figure 12.15 A flash spectrum photographed from Mauritania in the 1973 eclipse at the conclusion of totality. The most intense line in the spectrum is the first Balmer line at 6563Å in the red region of the spectrum.*

These qualitative conclusions are confirmed by more precise studies in which many separate exposures of the flash spectrum are taken in succession, at the rate of one or two per second, during the 15- or 20-second period in which the chromosphere is exposed at the beginning or the end of totality. The flash spectrum changes character during the course of these brief periods because, as noted above, the moon's disk successively covers, and then uncovers, different regions in the chromosphere as it slides across the face of the sun. Thus, the sequence of flash spectra yields the contributions from different altitudes in the chromosphere. The structure of the chromosphere can be determined with a precision of a few hundred kilometers in this way.

The results of the flash spectrum studies are shown in Figure 12.16. The temperature of the chromosphere falls from 6000°K at the photosphere to a minimum of approximately 4000°K and stays in the range between 4000°K and 6000°K up to approximately 2000 kilometers. Above that height the temperature begins to rise very steeply, reaching the million-degree level at an altitude of roughly 5000 kilometers and remaining at that level throughout the inner corona.

At the high temperatures that prevail in the upper chromosphere, all

*Figure 12.16 Temperatures in the chromosphere. The transition zone is highly variable and inhomogeneous, with spicules and interspicular matter (see p. 307) contributing to the average conditions shown in the graph.*

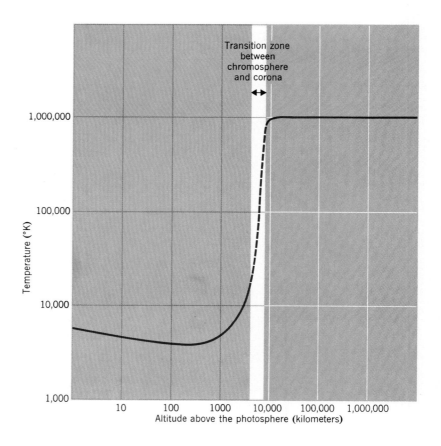

hydrogen and helium atoms are ionized, and the 6563Å line and other emission lines of neutral hydrogen and helium disappear. Elements heavier than hydrogen and helium, such as calcium and iron, also lose several electrons at this temperature, although they are not completely stripped as is true for hydrogen and helium. Thus, the lines of all these elements, which are prominent in the spectrum of the lower chromosphere, disappear gradually as the altitude increases and are entirely missing from the spectrum of the corona.

*Photographs of the Chromosphere.* The radiation from the chromosphere, summed over all wavelengths, is roughly 1000 times fainter than the radiation from the photosphere. For this reason the chromosphere is not visible in ordinary photographs of the sun taken in white light. However, if the light from the sun is passed through a filter that only transmits light in a limited band of wavelengths, a different situation may prevail. Suppose that the filter transmits light at the wavelength of the first Balmer line of hydrogen, at 6563Å, and blocks light at all other wavelengths. The 6563Å line of hydrogen is one of the strongest absorption lines in the solar spectrum. This means that most of the 6563Å photons coming from the lower levels in the photosphere are absorbed, and very few escape to space.

Each atom that absorbs a 6563Å photon subsequently collapses to its initial state, emitting another 6563Å photon, but this photon also is absorbed if the atom is at a depth where the density is substantial.

Only if the atom emitting the 6563Å radiation is located in the chromosphere, where the density is lower and the gas is relatively transparent, will a 6563Å photon be likely to escape from the sun. Therefore, an observer, viewing the sun through a filter transmitting only 6563Å radiation, does not see the photosphere; instead he sees the chromosphere.

It is possible to explore the structure of the photosphere and chromosphere at several levels by using slightly different wavelengths, all in the neighborhood of the 6563Å line. Figure 12.17 shows the variation of absorption with wavelength in the vicinity of this line. If the filter transmits light at the precise center of the line, where the absorption is strongest, the photographs will show details of chromospheric structure several thousand kilometers above the surface. If the filter is modified to transmit light half or three-quarters of an angstrom away from the center of the line, the absorption will be somewhat less than at the center, and the photograph will show the structure of the chromosphere at a somewhat lower altitude. If the transmitted light is, say, two angstroms from the line center, where the absorption is relatively small, the photograph will show some of the structure of the underlying photosphere.

In this way, by varying the wavelength in the vicinity of a strong absorption line, the observer can see into the chromosphere to different depths and obtain a picture of the way in which the properties of the chromosphere change with height. Plate 25 illustrates the effect on a photographic image of variations in wavelength in the vicinity of the 6563Å line.

Although 6563Å radiation is frequently chosen for photographs of the

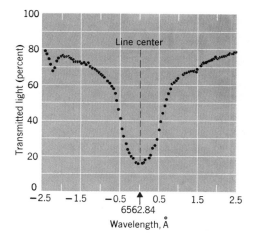

Figure 12.17 Measurement of the percentage of light transmitted through the chromosphere in the vicinity of the absorption line of hydrogen at 6563Å. The shallow depression 2.5Å to the short-wavelength side of the center of the 6563Å line is the combined result of absorption by silicon atoms in the sun and water molecules in the earth's atmosphere.

**305**

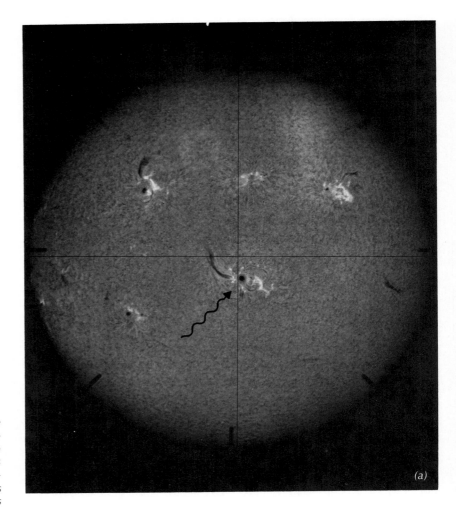

*Figure 12.18    The sun photographed in the light of (a) the hydrogen absorption line at 6563Å and (b) the calcium absorption line at 3934Å. The 6563Å photograph was taken from space during the Skylab mission on September 5, 1973, and shows several active regions with sunspots, plages and filaments (see pages 320–324), in addition to the network of large cells. The arrow points to an active region at the center of the sun's disk.*

chromosphere, other absorption lines can also be used. In particular, the very strong absorption line of calcium at 3934Å often is employed for this purpose.

Figures 12.18*a* and *b* show the disk of the sun photographed in the light of the 6563Å line of hydrogen and the 3934Å line of calcium. The photographs show patches of brightness, produced by high-temperature regions in the chromosphere. These bright patches are caused by strong magnetic fields on the sun's surface (page 320). Both photographs, but particularly the image taken in 3934Å calcium light, suggest a network of cells covering the entire solar disk, giving the sun's image a mottled or granulated appearance like the skin of an orange. These cells are approximately the same size as the supergranules that appear in the Doppler-shift image of the photosphere (Figure 12.8 and page 291). It is usually assumed that the chromospheric network of cells visible in Figure

(b)

12.18b is associated with the supergranule network in the underlying photosphere.

Figure 12.19, also taken in 6563Å light, shows the detailed structure of a small region in the chromosphere. The photograph reveals many short, dark lines, resembling blades of grass, that seem to outline or partially outline a dozen or so irregularly-shaped fields on the sun's surface. Each field is about 30,000 kilometers across. These fields have the same size as the supergranules in Figure 12.8 and the cells in Figure 12.18b, and are believed to be identical with them. The blade-like features outlining the cells are thought to be jets of matter that are squeezed out of the sun's surface at the boundary between two adjoining supergranules.

*Spicules.* Figure 12.20 shows a photograph of the limb of the sun, also taken near the center of the 6563Å line. The photograph reveals

numerous spears or tongues of luminous material that rise out of the base of the chromosphere. These jets, called *spicules* (from the Latin *spiculum* for javelin), originate in the lower and middle chromosphere at an altitude of 1000 to 2000 kilometers, and generally reach heights of several thousand kilometers, although some rise to altitudes greater than 10,000 kilometers. They vary in diameter from a few hundred to a thousand kilometers. Each spicule lasts about 10 minutes on the average, and new ones continually form, grow, and disappear. Only a small fraction of the sun's surface is believed to be covered by the spicules at any given time, but when viewed in profile, as in Figure 12.20, they seem to be quite dense and give the chromosphere the appearance of a burning prairie.

Spicules are suspected to be identical with the dark blade-like features in Figure 12.19, although this association has not been established conclusively. Calculations show that a jet of gas forming a spicule could appear bright when viewed against the blackness of space at the limb of the sun, while the same jet could absorb enough 6563Å radiation to appear dark, as in Figure 12.20, when viewed against the photosphere beneath.

Figure 12.19 (opposite) A photograph of an area on the sun about 100,000 kilometers across, showing several large areas approximately 30,000 kilometers in diameter, whose boundaries are partially outlined by dark blade-like jets of matter. The jets may be spicules. The photograph was taken at a wavelength 7/8Å from the center of the first Balmer line at 6563Å. The light forming an image at this wavelength comes from an intermediate level in the chromosphere, several thousand kilometers above the photosphere.

Figure 12.20 Spicules at the limb, photographed 1Å off the center of the 6563Å line of hydrogen.

## The Corona

The chromospheric spicules generally rise to a height of about 5000 kilometers. This altitude may be called the upper boundary of the chromosphere. Above it lies the region of the solar atmosphere called the *corona*. As seen during an eclipse, the visible corona extends out from the edge of the solar disk many millions of kilometers (Figure 12.21). When viewed from the ground, the luminosity of the corona fades into the background of scattered light from the sky at a distance of roughly 10 million kilometers from the sun, but photographs taken from a balloon at high altitudes, where the sky is darker, show a visible corona out to 30 solar radii or 20 million kilometers, and measurements of the influence of the corona on radio waves show a detectable effect halfway out from the sun to the earth. Other measurements made from satellites and space probes suggest that the corona has no outer boundary. A stream of gas called the *solar wind* flows out of the corona and into the solar system at all times, continuously immersing the earth and its sister planets in the tenuous gases of the solar atmosphere.

*The Coronagraph.* The chromosphere and corona can be studied at leisure, without limiting the period of observation to the few minutes of

Figure 12.21 *The solar corona 30 seconds after the start of totality during the eclipse of March, 1970. Features are visible at a distance of 4.5 solar radii or 3 million kilometers. The sun's axis of rotation is about 40° counterclockwise from the vertical in the photograph.*

totality of a solar eclipse, by blocking out the interfering light from the disk of the sun artificially. A telescope can be modified for this purpose by placing an opaque disk, whose diameter is precisely equal to the apparent diameter of the sun, at the focus of the objective lens where the first image of the sun is formed. Telescopes constructed in this way are called *coronagraphs*.

Even with the opaque disk in place, it is still essential to minimize the effect of sunlight that has been scattered in the atmosphere prior to entering the telescope, as in Figure 12.10*b*, or has been scattered on passing through the objective lens prior to the formation of the sun's image. The first requirement is satisfied by placing the coronagraph at a high altitude where the air is thin and relatively free from dust. To satisfy the second requirement, the coronagraph is constructed with a simple convex lens as the objective. Most telescope objectives are constructed out of several separate lenses, in order to correct color fringing and other distortions in the image, but this is undesirable in a coronagraph, since each lens in the combination would be a source of scattered light. The corrective lenses are introduced into the coronagraph after the image has been formed and the sun's disk has been blocked out, when scattered light is no longer a problem.

Great care also is taken to eliminate blemishes in the objective or dust on its surface, which are residual sources of scattered light. A filter admitting a narrow band of wavelengths is used to minimize color fringing.

The coronagraph has greatly increased the amount of information available to the solar astronomer. However, under the best conditions coronagraphs still are unable to detect the weak radiation from the

THE SOLAR SYSTEM

outer corona, which is too faint to be seen except during a total solar eclipse. Because a total eclipse provides unique conditions for observing the outer corona, solar astronomers travel to remote and nearly inaccessible places, if necessary, to set up their instruments in a region in which the period of totality is greatest. Calculations based on the orbits of the earth and moon reveal that the longest possible period of totality in an eclipse is seven minutes and 40 seconds. Eclipses with periods of totality exceeding seven minutes are exceedingly rare, although three have occurred in this century, in 1937, 1955 and 1973. The next seven-minute eclipse will occur in 2150.

*The Spectrum of the Corona.* The Balmer line and other strong emission lines of hydrogen and helium gradually disappear from the spectrum of the solar atmosphere with increasing altitude, and are replaced by the continuous spectrum of white light characteristic of the corona. The disappearance of these emission lines marks the transition from the chromosphere to the corona. The corona also contains emission lines, but they are weak in comparison to its continuous spectrum.

The continuous spectrum of radiation from the corona consists mainly of visible light from the sun's surface that has been scattered in the direction of the earth by collisions with free electrons (Figure 12.22, page 312). Although the density of these coronal electrons is low, the intensity of sunlight is great enough to make the scattered light detectable. Because the light from the sun has a continuous spectrum and a white color, the scattered light from the corona has the same properties.

The spectrum of the corona also contains many emission lines. The coronal lines are relatively faint compared to typical lines in the spectrum of the chromosphere, the most intense being the green line of ionized iron seen in Color Plate 18. Most of the lines cannot be identified with the spectrum of any familiar chemical element that has been studied in the laboratory, and for a time it was thought that the corona might contain a previously unknown rare gas, called "coronium." Since the rare gas helium was discovered by the presence of its lines in the sun's absorption spectrum before it was known on the earth, this explanation seemed plausible. Around 1940, astronomers realized that the strange lines are produced by familiar elements, but in a very highly ionized state in which the atoms of these elements have lost as many as a dozen or more electrons. The green coronal line in Color Plate 18, for example, is produced by a transition in atoms of iron that have lost 13 electrons out of their normal complement of 26. Atoms that have lost so many electrons cannot be produced in the laboratory in sufficient numbers to permit a study of their spectra. This was the reason why the lines in the spectrum of the corona had never been detected on the earth.

*The Temperature in the Corona.* An enormous amount of energy is required to remove 13 electrons from an iron atom. The outermost electrons can be dislodged with relative ease, but the inner electrons are bound very tightly to the nucleus, and collisions of exceedingly great violence are required to eject them from their orbits. Collisions with the necessary degree of violence occur only at a temperature of

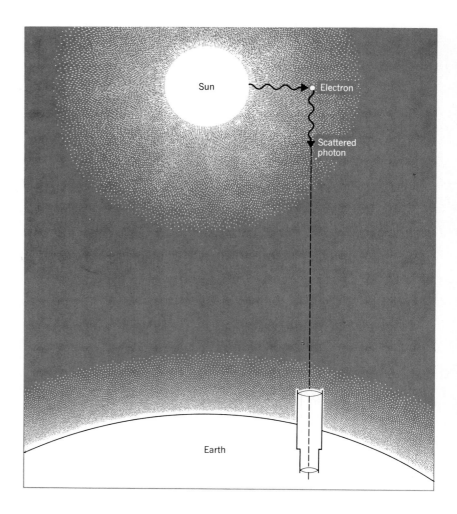

Figure 12.22 Light from the sun is scattered by free electrons to produce the corona's visible luminosity.

~~500,000°K or more.~~ Other lines in the spectrum of the corona, such as the lines produced by transitions in calcium atoms that have lost 14 electrons, require a temperature of at least one million degrees.

The presence of these lines indicates that the temperature in large regions of the corona is close to one million degrees. This discovery led to one of the greatest problems in the study of the sun. What is the source of the energy that heats the corona to a temperature of a million degrees? The tenuous gas of the corona is too rarified to absorb an appreciable amount of energy from the sun's radiation as it passes through, and even if most of this energy were absorbed, it would only heat the corona to the temperature of the photosphere, which is only 6000°K. Some other agent must be at work to produce the million-degree temperatures in the corona.

No completely satisfying resolution of the difficulty has been obtained thus far, but one factor that is believed to be important is the continual agitation of the gases of the corona by shock waves rising out of the turbulent, boiling surface of the sun. These shock waves are sharply defined sound waves, like a thunderclap or the boom produced by a supersonic aircraft. Each shock wave is a zone in the gas that has been compressed and heated. The compressed, heated wave travels through the atmosphere at approximately the speed of sound. The degree of heating produced by a single shock wave is very small, but if innumerable shock waves follow one another in rapid succession, the heating effect can accumulate and raise the temperature of the corona considerably. Intense magnetic fields on the surface of the sun may also contribute to the heating of the corona, as described on page 320.

## SOLAR SURFACE ACTIVITY

Up to this point the sun has been treated as a typical Main-Sequence star, whose internal structure and evolution were discussed in detail in Chapter 7 because the sun's position on the Main Sequence, midway between the largest and smallest stars, made it particularly well suited to serve as an example of an average star. Throughout the discussion the questions of primary interest were the release of nuclear energy within the sun by the fusion of hydrogen, the transport of this energy to the surface, and its radiation into space. Theoretical studies of these processes, based on experiments in the nuclear physics laboratory combined with an enormous body of knowledge derived from stellar spectra, lead to predictions of the evolutionary tracks of stars of various masses that constitute one of the great syntheses of theory and observation in science.

These predictions seem to include all the basic properties of the sun as a star. They predict that a star with one solar mass, and roughly 5 billion years old, will have a luminosity of $4 \times 10^{33}$ ergs/sec and a surface temperature of 6000°K. They also predict the details of the internal structure of a sun-sized star, including the amounts of hydrogen and helium at various distances from the center, the values of density and temperature at all depths from the surface to the center, and the nature of the zone of convection.

But they completely fail to predict the occurrence of solar "weather." They fail to predict violent storms that rage across the face of the sun in an 11-year cycle, geysers of hot gas that rise hundreds of thousands of miles into the solar atmosphere, and explosive outbursts of gamma rays, x-rays, ultraviolet radiation, and energetic protons that erupt sporadically from the surface.

We see the tempestuous solar "weather" only because of the closeness of the sun. If our star were thousands of light-years away, most of the activity on its surface would be undetectable. The reason is that the surface storms, violent as they are, contain only a millionth part of the sun's total output of energy. Consequently, they play a limited role in the broad scheme of evolution of Main-Sequence stars. However, they produce a great variety of displays that are scientifically interesting, strikingly beautiful, and of great practical importance to the inhabitants of the earth. A major branch of solar physics is devoted to the study of the pyrotechnics that play across the face of the sun, waxing and waning with the 11-year sunspot cycle.

### Sunspots

Sunspots are the only sign of solar surface activity that can be detected with the naked eye under ordinary circumstances. When a large spot or group of spots is present on the face of the sun, it can be seen easily at sunset, or through a thin haze of clouds or a filter.

References to sunspots go back to a Greek observer in the fourth century B.C. There was little further mention of them during the long period in which western astronomy languished, the commonly held belief being that the sun was a perfect, unblemished sphere. The invention of the telescope in 1609 destroyed this illusion immediately. By 1611, four observers, among them Galileo, had independently studied the spots with the aid of the newly invented instrument.

When sunspots are photographed with high resolution under good

*Figure 12.23 A sunspot photographed with a balloon-borne telescope. The dark umbra and bordering penumbra of the sunspot and the granules in the surrounding photosphere are visible.*

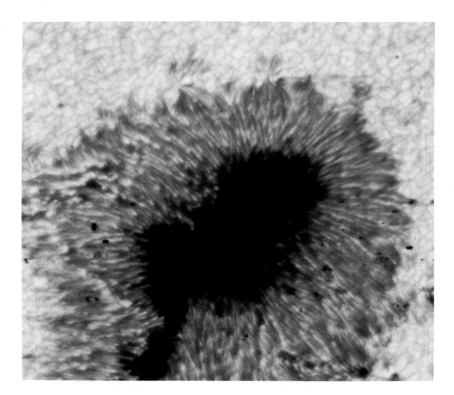

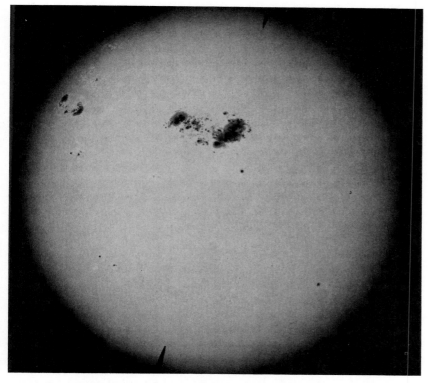

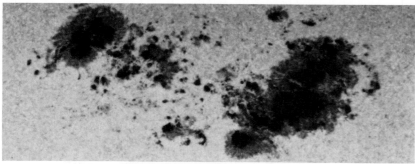

*Figure 12.24  A large cluster of sunspots photographed with the 100 inch telescope in 1947 at sunspot maximum. The lower photograph is an enlarged view.*

seeing conditions, they look like irregularly shaped holes or craters in the sun's surface. Figure 12.23 is an example. It shows a photograph of a sunspot about 30,000 kilometers in diameter, taken through a telescope suspended from a balloon at an altitude of 80,000 feet. Atmospheric blurring of the image is reduced at that altitude, and the details of sunspot structure emerge more clearly than they can be seen from the ground. The black inner region of the spot is called the *umbra* and the surrounding fringe is called the *penumbra.* The granulation in the photosphere is also visible in this photograph.

The average size of sunspots is about 10,000 kilometers, but on rare occasions spots appear that extend across more than 150,000 kilometers

**315**

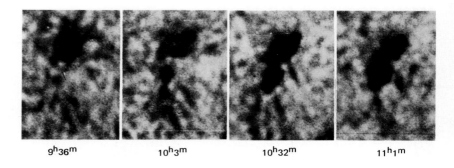

Figure 12.25 A sequence of photographs showing the birth of a pore. The new pore appears beneath an existing pore and coalesces with it during the 85-minute sequence.

9ʰ36ᵐ     10ʰ3ᵐ     10ʰ32ᵐ     11ʰ1ᵐ

of the sun's surface. The spots usually occur in pairs or in complex groups. Figure 12.24 shows a large sunspot group photographed with the 100-inch telescope on Mount Wilson that includes one enormous spot about 100,000 kilometers across.

The smallest sunspots, called *pores*, range in size from a few thousand kilometers down to the limit of telescopic resolution. Still smaller pores, too minute to be resolved in a telescope, may also exist. Large sunspots in groups usually are accompanied by considerable numbers of pores. Figure 12.25 shows the birth of a pore in a sequence of photographs taken on February 4, 1958, during an interval of an hour and 25 minutes. Pores frequently appear and disappear in a few hours, and generally do not last longer than a day.

Small sunspots persist for several days or a week, and the largest spots may last for many weeks, long enough to be carried across the entire face of the sun during the course of its rotation, and reappear about a month later on the opposite limb.

What are sunspots? Early observers of the spots through telescopes offered widely varying explanations. Some thought they were small planets circling the sun within the orbit of Mercury; others said they were mountains projecting above luminous clouds that covered the sun's surface; Galileo thought they were clouds drifting in the sun's atmosphere.

The answer turned out to be simpler than any of the theories proposed in early times. Sunspots are regions of the sun that are a few thousand degrees cooler than the gas surrounding them. Consequently, they radiate less energy to space, and appear darker. The average temperature in a

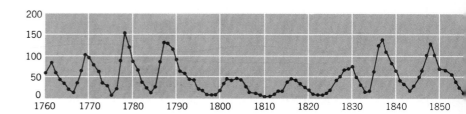

sunspot is about 4000°K, compared to 6000°K for the surface of the sun as a whole. Although the temperature difference is great enough to make sunspots appear black, the spots are intrinsically bright, the luminosity of a typical sunspot being hundreds of times greater than the light of the full moon. If a sunspot could be separated from the sun and viewed by itself in the sky, it would appear as an object of dazzling brightness.

*The Sunspot Cycle.* For about 200 years, astronomers have recorded the number of spots appearing every day and their position on the face of the sun. Figure 12.26 shows the variation in the number of sunspots from 1760 to 1969. The method of compiling the sunspot numbers is based on arbitrary rules relating to the size of a spot, and the records are considered to be unreliable for the period prior to 1850. In spite of these defects, the graph clearly displays a cyclic rise and fall with maxima and minima recurring approximately every 11 years. The periodic change in sunspot number is called the *sunspot cycle.*

Although the average length of the sunspot cycle is about 11 years, the interval from one maximum in the cycle to the next has been as short as seven or as long as 17 years. All the other manifestations of solar activity discussed later in this chapter, including flares, plages, and prominences, are keyed to the pattern of rise and fall in the sunspot numbers. In the years of sunspot maximum the surface of the sun is violently disturbed, and outbursts of particles and radiation of all wavelengths are a common occurrence. In the years of sunspot minimum, these outbursts are far less frequent.

The records of the positions of the spots also reveal the important fact that sunspots are almost entirely confined to the zone of latitudes between 40° and the sun's equator, and never appear near the poles. The spots are found at their highest latitudes at the start of a new sunspot cycle, immediately following the last minimum. Later in the cycle they tend to appear at lower latitudes, and the last spots of a given cycle usually lie close to the equator. No fully satisfactory explanation has been given either for the 11-year cycle or for the drift of the sunspots toward the equator during each cycle.

*Magnetic Fields in Sunspots.* If the light from a sunspot is passed through a spectrograph, some of the lines in the sunspot spectrum are found to be similar to the lines in the normal absorption spectrum of the sun, but others are strikingly different. The special lines are distinguished

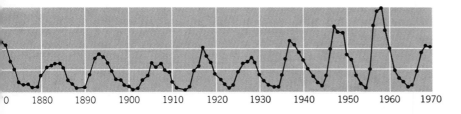

0    1880    1890    1900    1910    1920    1930    1940    1950    1960    1970

*Figure 12.26   The sunspot cycle.*

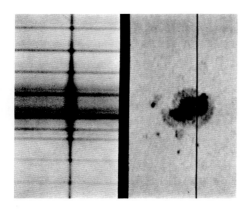

Figure 12.27 The splitting of a line into three separate lines (left) by the magnetic field in a sunspot (right). The black line on the right shows the position of the spectrograph slit with respect to the spot.

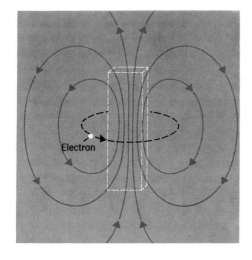

Figure 12.28 The magnetic force produced by an orbiting electron.

by the fact that each one is split into two or more closely spaced components. Figure 12.27 shows how a single line divides into three components in the interior of a sunspot.

The splitting of a spectral line into several separate lines, called the Zeeman effect, is a well-known phenomenon in the laboratory. It is produced by a magnetic field acting on the atoms of gas that radiate the spectrum, in the following way. Each atom consists of electrons orbiting around a central nucleus. The orbiting electrons are moving charges of electricity, equivalent to rings of electric current. According to the laws of electromagnetism, a ring of electric current generates the same magnetic force as a bar magnet (Figure 12.28). The magnetic field in the sunspot, acting on the innumerable atom-sized bar magnets, distorts their structure and changes the energy levels of each atom. The shift in energy changes the wavelengths of the lines in the spectrum. The change in energy is small and, therefore, the change in wavelength is also small.

Why is each line split into several separate lines, instead of being just shifted in wavelength? The reason is that every tiny atom-magnet tends to line up along the direction of the magnetic field but, according to the laws of the atom, it can only line up in certain allowed directions, just as electrons in the atom can only occupy certain allowed orbits (see pages 80–81). Each allowed direction corresponds to a different energy for the atom, and therefore a different wavelength for the photon emitted in a transition. In the case of the particular element producing the line shown in Figure 12.27, the atom-magnets are allowed to line up *with* the magnetic field, *opposite* to the field, or at *right angles* to it. The three directions give rise to three energy levels and three separate lines.

A theoretical study yields a formula connecting the degree of splitting in the Zeeman effect, that is, the separation between the lines, and the strength of the magnetic field. In this way, the measured separation between the lines yields the information that magnetic fields in sunspots are as high as several thousand gauss.

A magnetic field of several thousand gauss is a very strong field. By comparison, the strength of the earth's magnetic field is less than one gauss.

*The Origin of Sunspots.* Why is a sunspot cooler than its surroundings? Heat normally flows from hotter to cooler regions; why doesn't it flow into the sunspot from the surrounding high-temperature gas, and eliminate the difference in temperature?

The answer is connected with the very strong magnetic fields that exist in the interiors of sunspots. The critical point in the explanation of sunspots is that these strong magnetic fields bend the paths of electrically charged particles. If the particle is inside the region of the magnetic field, its path is bent into a circle (page 243). If the particle is outside the field and its motion carries it toward the field, it is deflected at the boundary and prevented from entering (Figure 12.29).

All the particles in the sun's interior are electrically charged. These particles carry heat from the interior of the sun to the surface in the form of convection currents. However, charged particles cannot readily enter regions where strong magnetic fields are present. Consequently the convection currents are partly suppressed, and heat is prevented from reaching the surface.

This explanation accounts for the fact that a sunspot is colder than the surrounding surface of the sun when it is first formed. However, it does not explain why the spot *remains* cold. Heat should leak into the region under the spot in the form of photons, which are not electrically charged and, therefore, cannot be affected by the magnetic field. Theoretical studies suggest that a sunspot should warm up to a normal temperature in a few days as a result of the heat carried into its interior by the flow of photons. But large sunspots live for many weeks. What keeps them cold? The answer is not clear. A complete theory of sunspots is one of the major challenges in solar astronomy.

*Sunspot Pairs.* Sunspots often appear on the surface of the sun in pairs, aligned in approximately an east-west direction. Magnetic field measurements using the Zeeman effect show that the two sunspots often have opposite magnetic polarities, and are connected by lines of magnetic force that emerge from the surface at the position of one spot and re-enter at the position of the other, as in Figure 12.30.

Frequently, a complex group composed of many sunspots appears. Magnetic field measurements show that in this case also the region containing the group of sunspots often is divided into two adjacent regions of opposite magnetic polarity. Presumably, magnetic lines of force run from one region to the other. Figure 12.31a (page 320) shows a group of sunspots photographed on July 6, 1965. Regions of positive and negative magnetic polarity are indicated by blue and grey shading, respectively. Figure 12.31b, taken in the center of the 6563Å line, shows the structure of the overlying chromosphere. Lines running from one sunspot group to the other suggest magnetic lines of force connecting the two regions of opposite polarity.

For a given sunspot cycle, and a given hemisphere on the sun, the polarity of the leading, or westernmost, sunspot is always the same. In the latest sunspot cycle, the polarity of the leading sunspot invariably is negative in the northern hemisphere and positive in the southern hemisphere. With each new sunspot cycle, the polarities reverse. A possible explanation of this pattern is given on pages 328–329.

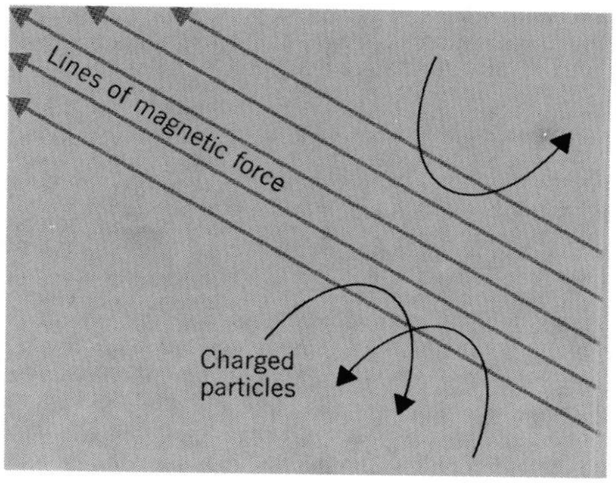

Figure 12.29 Deflection of electrically charged particles at the boundary of a magnetic field.

Figure 12.30 The magnetic field over a pair of sunspots.

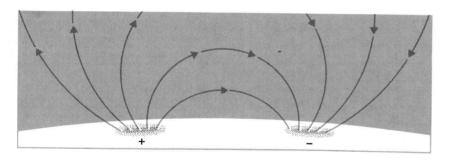

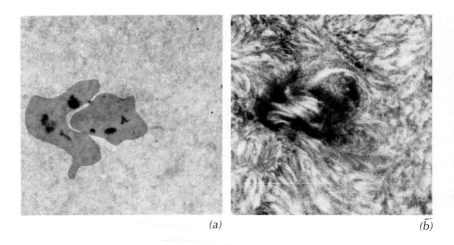

(a)                                      (b)

Sometimes one group of sunspots, all having the same magnetic polarity, is observed, but no sunspot or group of sunspots of opposite magnetic
polarity is visible nearby. However, magnetic measurements always
reveal a strong magnetic field of opposite polarity in an adjacent region
(Plate 21). Generally this is a region in which sunspots have formed
and disappeared, leaving behind a less concentrated but still relatively
intense magnetic field.

### Plages, Flares, Prominences and Filaments

*Plage.* Figure 12.32a shows a sunspot group photographed near the
limb of the sun on October 8, 1964. Figure 12.32b shows the same area
photographed in 6563Å radiation, revealing the structure of the
chromosphere. The bright glow in the 6563Å photograph indicates a
highly disturbed condition in the gas above the sunspots. These bright
patches are called *plages* (French: beaches). They are nearly always
found above regions in the photosphere in which a strong magnetic
field exists, regardless of the presence of sunspots.

A plage usually precedes the appearance of sunspots in a high-magnetic
field region on the sun. It remains while the sunspots are visible, and persists for several weeks or more after the spots disappear.

A plage is a region in the chromosphere that apparently has been
heated to incandescence by a local concentration of magnetic fields. A
satisfactory explanation of the heating action of the fields has not been
provided. It is possible that the lines of magnetic force guide energetic
charged particles to the region, forming small volumes of hot, dense
gas; or the field lines may act as guides for travelling vibrations, similar
to the sound waves caused by a thunderclap or a supersonic boom,
conducting the vibrations upward into the chromosphere and corona
(page 313).

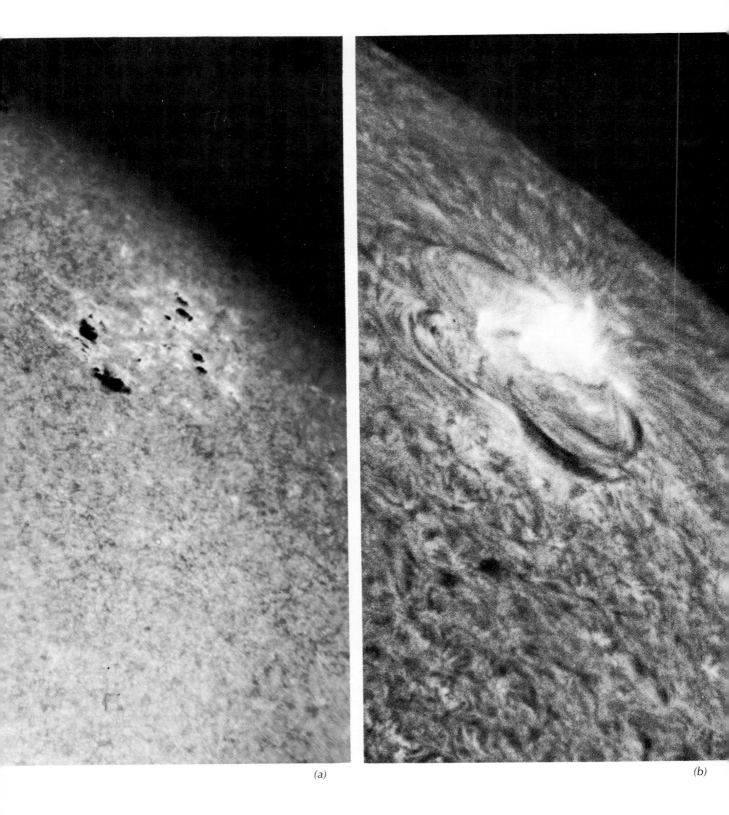

(a)

(b)

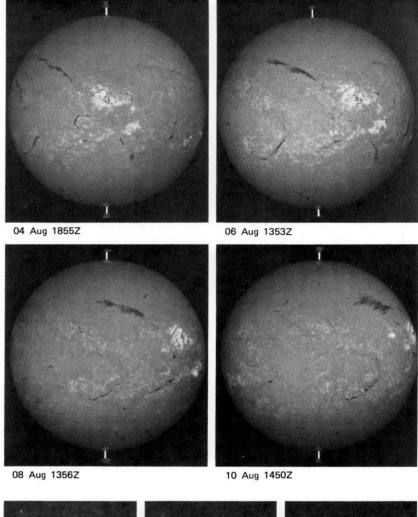

04 Aug 1855Z                    06 Aug 1353Z

08 Aug 1356Z                    10 Aug 1450Z

*Figure 12.33 A strongly disturbed region rotates from the center of the sun's disk to the limb in August, 1972. Intense flares occurred on August 4 and 7.*

*Figure 12.34 Buildup of an intense flare on August 4, photographed in 6563Å light. (a) 0620; (b) 0638; (c) 0738 (Greenwich Mean Time).*

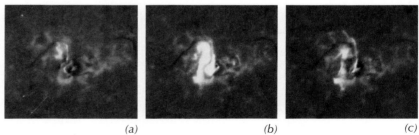

(a)                  (b)                  (c)

*Flares.* Frequently during the peak periods of the sunspot cycle, and less often at other times, the surface of the sun is marred by explosive outbursts of energy that hurl particles and radiation into the solar system. These outbursts, called *solar flares*, usually are observed in the vicinity of large, complex groups of sunspots such as the group in Figure 12.24.

Flares build up to peak intensity in a few minutes, and disappear in a period varying from 10 to 15 minutes to several hours, depending on the size of the flare. The burst of radiation and energetic particles produced by a large flare may play havoc with radio communications and cause substantial changes in the normal magnetic field of the earth. Large flares also create the luminous draperies and filaments of the aurora at latitudes as low as the southern United States and Mexico, which normally do not witness these spectacular atmospheric displays.

*The August 1972 Flares.* An exceptional series of solar flares occurred in August 1972. The flares originated in the disturbed region of the sun's surface that had first appeared a month earlier as a small and innocuous cluster of sunspots. The cluster disappeared on the west limb in mid-July in the course of the sun's 27-day rotation, and reappeared on the east limb on July 29. On August 2 the first of a number of flares erupted from the region. On August 4 and August 7 enormous flares occurred. These were among the largest ever recorded.

Photographs of the sun on August 4, 6, 8 and 10 are shown in Figure 12.33 (opposite). The flare region is in the dead center of the disk on August 4. On August 8 it has rotated more than halfway to the east limb, and on August 10, the region, still very active, has reached the east limb.

Figure 12.34 is a detailed view of a part of the disturbed region, showing the rapid buildup of the large flare that occurred on that day. The visible luminosity of the flare grew to full intensity in 18 minutes, and disappeared an hour later.

A spectacular eruption of luminous gas, shown in Figure 12.35, occurred on August 11, when the disturbed region had rotated beyond the limb. The arrow indicates an ejected mass of material moving rapidly away from the sun. This material rose to a height of 250,000 kilometers above the surface in approximately 20 minutes.

High-energy protons ejected in the flare of August 4 ionized atoms and molecules in the earth's upper atmosphere and disrupted radio communications at high latitudes for several days. Slower-moving particles travelling outward from the sun strengthened the normal force of the solar wind (page 309) on arriving at the earth two days later. Some of the charged particles in the solar wind penetrated the earth's magnetic field, while others were turned aside by it. Both effects combined to create a major disturbance in the geomagnetic field, triggering strong currents of electricity in the ground and tripping circuit breakers in power-lines in several places in the United States and Canada.

*Prominences.* Prominences are masses of luminous gas that appear in the corona far above the sun's surface. Prominences consist of gas that is cooler and denser than the surrounding corona. They are luminous because at the relatively low temperatures and high densities that prevail in a prominence, its ions recapture electrons and emit photons.

Sometimes the material in a prominence seems to rise upward from the chromosphere in surges and eruptions; the cloud of ejected matter in Figure 12.35 is an example of an eruptive prominence. At other times, the material in the prominence streams downward from great

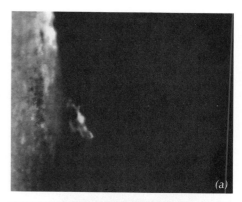

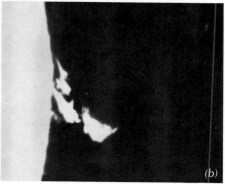

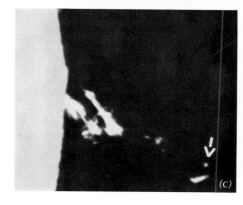

Figure 12.35 Ejection of a huge mass of material into space from the disturbed region at the limb of the sun on August 11. The arrow points to ejected matter 250,000 kilometers above the sun's surface. The times are (a) 2029; (b) 2040; (c) 2049 (Greenwich Mean Time).

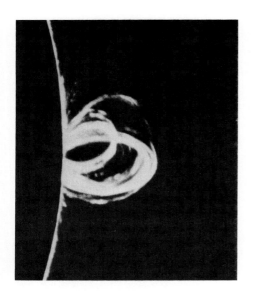

heights, like luminescent rain. Often the glowing streams of gas form graceful curves that appear to be shaped by lines of magnetic force looping upward out of the chromosphere (Figure 12.36). Many prominences are relatively stable and quiescent, and seem to float for hours or days above the solar surface. A striking example of a quiescent prominence appears on page 278.

*Filaments.* Photographs of the chromosphere above a region of high magnetic field often show long dark streaks called *filaments*. These are regions of relatively cool and dense gas that appear dark against the solar disk in 6563Å photographs, because they are cooler than their surroundings and absorb 6563Å radiation more effectively. Several filaments are visible on the sun's face in Plate 24.

Filaments and prominences are the same objects. If a filament is followed to the edge of the solar disk during the sun's rotation, and photographed silhouetted against space, it appears as a luminous prominence (Figure 12.37).

*Figure 12.36 A giant loop prominence photographed in 6563Å light. Although the appearance of the loop suggests that the material rises out of the surface along one branch of the loop and returns along the other, motion picutres show, surprisingly, that luminous regions condense out of the corona at the top of the loop and move downward along both branches.*

## The Sun's Rotation and the Solar Magnetic Field

The blackness of sunspots has been explained in terms of intense magnetic fields of thousands of gauss that exist in the interior of a typical sunspot. Less intense but still relatively large magnetic fields are also observed outside the sunspots, both in the photosphere and in the chromosphere above. In fact, sunspots, flares, prominences and all other disturbances on the surface of the sun seem to be connected with the presence of magnetic fields that are concentrated in relatively small areas of the photosphere and chromosphere.

The origin of these strong, localized fields is one of the major mysteries in solar astronomy. The sun has a weak, irregular surface field with a strength of roughly one gauss, which is too little to provide a simple explanation of the strong fields in sunspots and disturbed areas.

An explanation that has gained wide acceptance among astronomers connects the intense magnetic fields observed in sunspots to a peculiar property of the sun's rotation. The sun rotates on its axis approximately once a month, but the rotation is more rapid at the equator than at other latitudes. The period of rotation is roughly 26 days at the equator, 28 days at a latitude of 45°, and still longer at higher latitudes. A rigid body like the earth or the moon could not rotate in this way, but the sun, being a gaseous sphere, is capable of doing so. Important consequences for the solar magnetic field can be deduced from the fact that this variation in the sun's rate of rotation occurs.

The starting point in the discussion is the assumption that the sun's general field resembles a bar-magnet field at high latitudes (Figure 12.38, page 326). If the resemblance to a bar-magnet field were accurate, the lines of force would emerge from the north pole, curve around in space and re-enter at the south pole. At low latitudes, the lines of force would

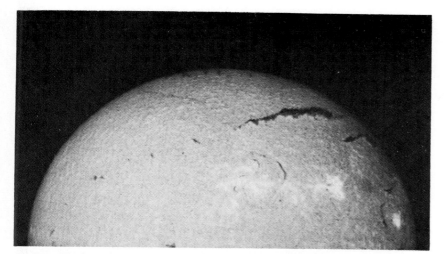

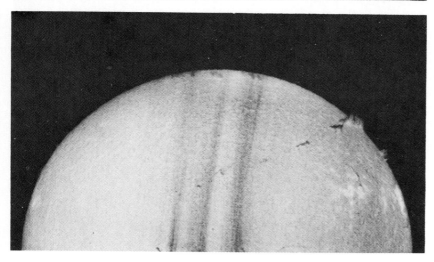

Figure 12.37 A filament on the face of the sun rotates to the limb to become a prominence.

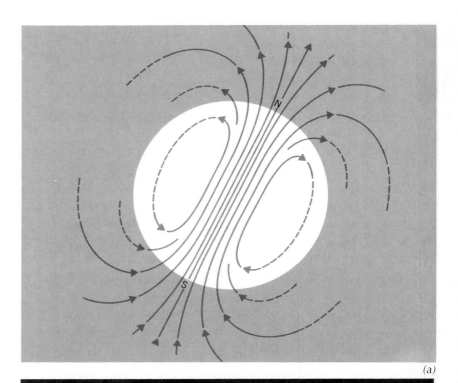

(a)

Figure 12.38 (a) The lines of force of a
bar-magnet. (b) A photograph of the corona
shows streamers suggesting the bar-magnet
field at high latitudes. At lower latitudes,
where the field is represented by dashed
lines in (a), the sun's field is highly irregular near the surface.

(b)

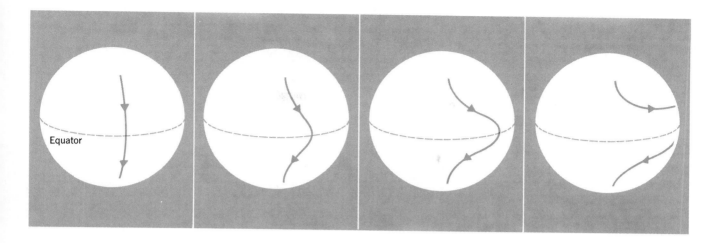

Equator

run largely in a north-south direction, as in the figure. We know that the magnetic force at the surface is highly irregular and variable at low latitudes, and does not bear any resemblance to the uniform north-south field shown in the figure, but it is possible that the irregularities in the field only exist near the surface, and that the magnetic field in the sun's interior has a more uniform appearance.

Now consider the effect of the sun's differential rotation on a bar-magnet field in the interior. The interior of the sun contains electrically charged particles—mainly electrons and protons—at a relatively high density. Studies of the behavior of a magnetic field in a dense, gaseous mixture of electrons and protons show that the lines of magnetic forces are carried along with the flow of the particles in the gas. Thus, lines of force that would normally run from north to south in the sun are stretched out into long loops by the more rapid rotation of the gas at the equator, as in Figure 12.39. After the sun has rotated a number of times, the direction of these lines of force is changed from north-south to east-west. When several years have elapsed and many rotations have occurred, the lines of force become very tightly wrapped around the sun at low latitudes, creating an intense east-west magnetic field just below the sun's surface, as in Figure 12.40 (page 328).[9]

Theoretical studies of magnetism also show that two adjacent lines of magnetic force tend to move apart, as if they had a physical reality and were acted on by a force of mutual repulsion. The repulsion increases in strength when the field is strong. If the field is many thousands of gauss, this repulsive force can become larger than the force of the sun's gravity. The studies indicate that when this happens, a loop of magnetic force may burst out of the sun's surface (Figure 12.41, page 329).

*Figure 12.39 A line of magnetic force running from north to south is stretched into a long loop by the more rapid rotation of the sun at the equator.*

[9] There is also evidence that the rate of rotation changes with depth, enhancing the effect of the differential rotation.

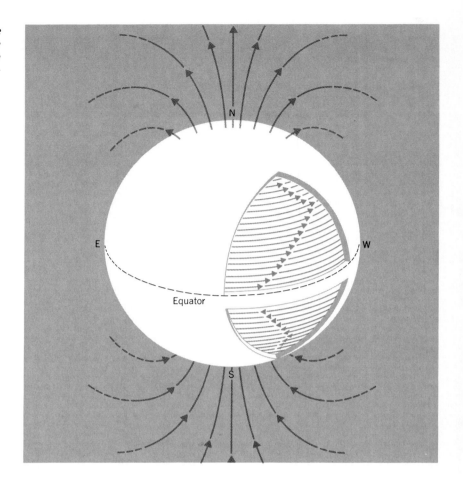

*Figure 12.40 The magnetic field in the interior of the sun after many rotations. The field lines are stretched into an intense east-west field under the surface as a result of the sun's differential rotation.*

In addition, because a magnetic field deflects charged particles the gas in a strong-field region has a lower density than the surrounding gas and becomes buoyant. The buoyancy adds to the upward force propelling the material to the surface.

Where the lines of force leave the sun's surface, there is an intense magnetic field directed vertically upward, i.e., with positive polarity. Where the lines of force return to the sun at the other end of the loop, they enter the surface proceeding vertically downward, creating a strong magnetic field with negative polarity (Figure 12.42). This description corresponds to the appearance of pairs of sunspots with magnetic fields of opposite polarity, that are commonly observed in disturbed regions on the sun (pages 319–320).

Figure 12.43 (page 330) shows the sequence of events in a disturbed region. Prior to the commencement of the disturbance, the lines of magnetic force run evenly beneath the surface in an east-west direction (Figure 12.43*a*). In the first phase of the disturbance, lines of force rise

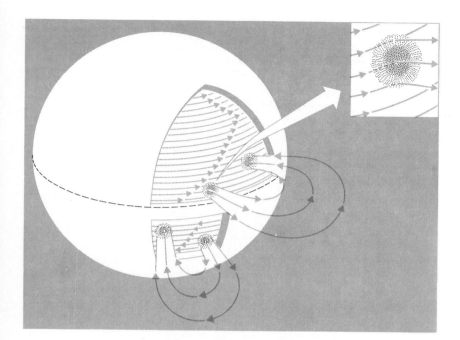

Figure 12.41 A loop of magnetic force bursts out of the sun's surface.

to the surface and break through, forming a magnetic arch connecting two regions of moderately strong magnetic field and opposite polarity (Figure 12.43b). Charged particles are constrained to follow the direction of the magnetic arch, forming a pattern of moving gas in the chromosphere that reveals the direction of the lines of force when the region is

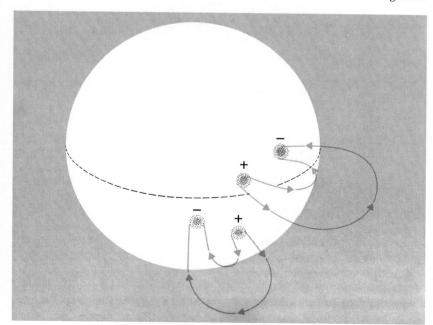

Figure 12.42 A pair of sunspots with opposite magnetic polarity remains where the loop of force broke through the surface. In the present sunspot cycle, the polarity of the leading (westernmost) spot is negative in the northern hemisphere and positive in the southern hemisphere. The polarities will reverse in the next sunspot cycle when the general magnetic field of the sun reverses.

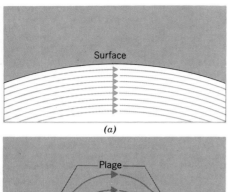

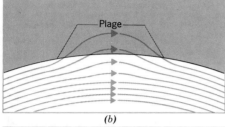

(a)

(b)

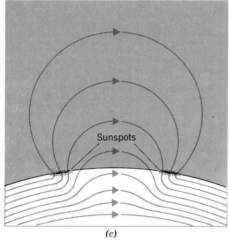

(c)

Figure 12.43 Sequence of events in a disturbed region on the sun's surface. The lines indicate the local magnetic field.

photographed from above in 6563Å light, as in Plate 26. Particles and sound waves are guided upward along the lines of force, heating the chromospheric gas to incandescence and forming plages.

In the second phase, the lines of force, continuing their upward movement, leave the surface of the sun at a nearly vertical angle and loop far out into the corona, occasionally reaching heights of 100,000 kilometers or more (Figure 12.43c). The regions in which the lines of force go through the surface vertically, and where the field strength is greatest, form the interiors of sunspots. The lines of force looping upward high into the corona sometimes lead to the luminous veils, streamers and loops classified as prominences.

In the next phase of the storm, the magnetic field diminishes in strength in the disturbed region, the sunspots disappear, and the region regains the appearance of Figure 12.43b, with magnetic arches and overlying plage.

Normally, an interval of 10 to 15 days elapses between the initial appearance of plages and the peak level of activity, with a gradual buildup of sunspots, prominences and flares. Most of the sunspots are gone after about four weeks or one solar rotation. Filaments often form and erupt during this period. After two rotations the brightness of the plage is considerably diminished, and after four rotations it usually disappears.

In the final phase the magnetic field grows still weaker and spreads in space, sometimes persisting for more than a year. The superposition of many such old disturbed regions seems to be the explanation for the weak irregular field on the sun's surface.

## Questions

1. How has the sun changed during its history? What is the principal reason for the change?
2. Why does the density of the sun increase with depth? Why does the temperature increase with depth?
3. Describe the process of convection. Give examples of convection on the earth. Where does convection occur in the sun? Why? How is convection in the interior of the sun manifested on the surface?
4. The surface temperature of the sun is approximately 6000°K. The radius of the sun is $7.0 \times 10^{10}$ cm. What is the luminosity of the sun?
5. What features of the sun are revealed during an eclipse that are normally not visible. Describe an instrument for making the same features visible under non-eclipse conditions.
6. How is the flash spectrum used to determine the temperature at various levels in the solar atmosphere? Describe the temperature variations in the solar atmosphere.

# THE SOLAR ATMOSPHERE

The solar atmosphere is the gaseous material lying above the visible surface of the sun. It is divided into the *chromosphere*, predominantly red in color, reaching from the surface up to a height of approximately 5000 kilometers, and the tenuous *corona*, extending from that height out to interplanetary space.

PLATE 18 (top). The flash spectrum photographed at the completion of totality during the eclipse of March 1970. A slit was not used in obtaining this spectrum because the source of light for most of the lines is a very narrow zone in the chromosphere immediately above the surface, less than 5000 kilometers, or 7 seconds of arc, in width.

Each curved line is an image of the chromosphere in one wavelength. The complete circle in the green region at 5303Å is produced by ionized iron atoms that have lost 13 electrons in the million-degree temperature of the corona. Because the region emitting the 5303Å radiation is located at high altitudes in the solar atmosphere, it is visible around the entire rim of the moon's concealing disk.

PLATE 19. The sun photographed during the eclipse of March 1970. The red light appearing along one segment of the moon's concealing disk has a wavelength of 6563Å, corresponding to the first Balmer line of hydrogen. This light comes from the chromosphere. The prominence near the middle of the red segment of the disk is a mass of relatively cool and dense gas extending up into the corona to a height of about 50,000 kilometers.

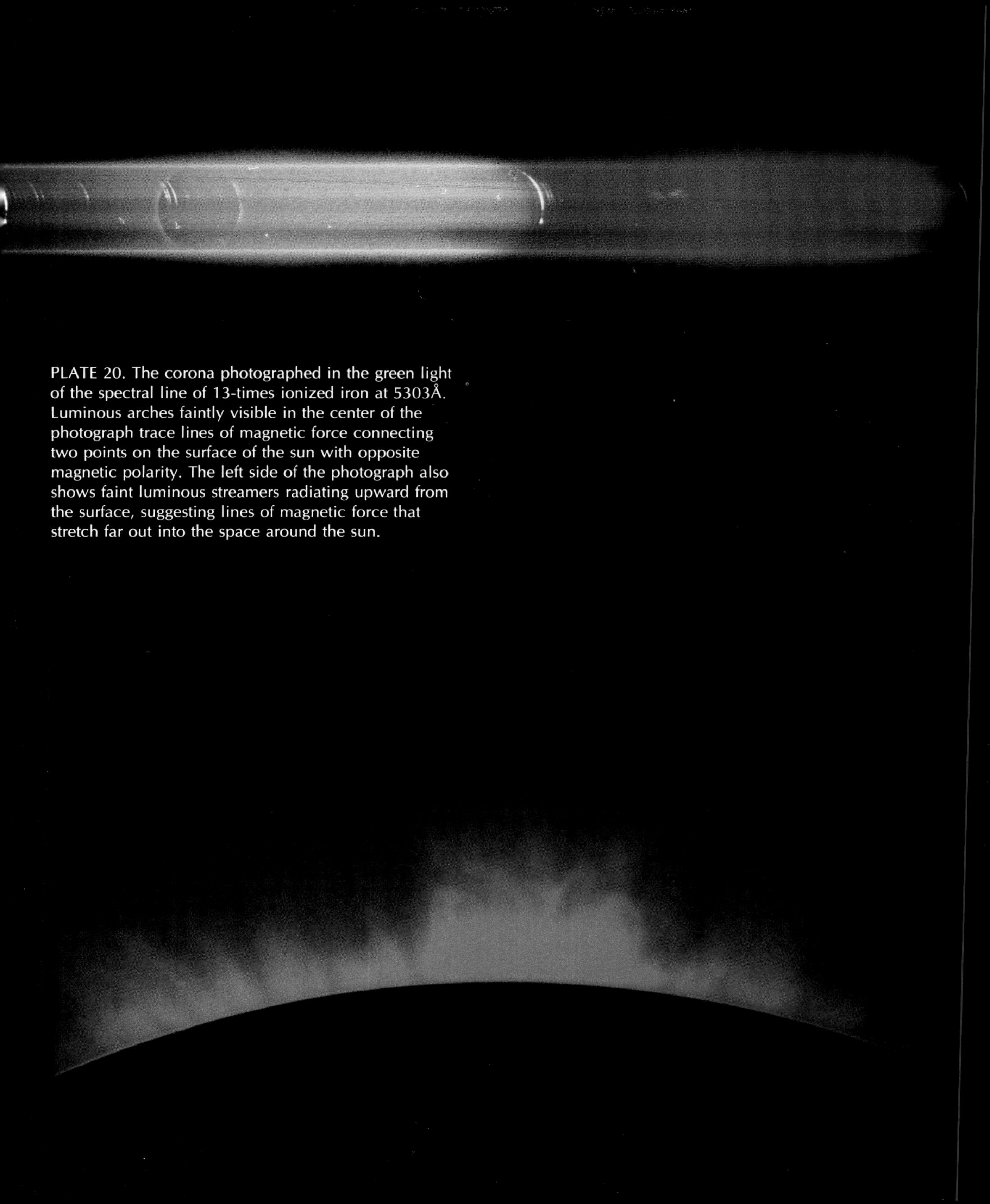

PLATE 20. The corona photographed in the green light of the spectral line of 13-times ionized iron at 5303Å. Luminous arches faintly visible in the center of the photograph trace lines of magnetic force connecting two points on the surface of the sun with opposite magnetic polarity. The left side of the photograph also shows faint luminous streamers radiating upward from the surface, suggesting lines of magnetic force that stretch far out into the space around the sun.

# SOLAR SURFACE ACTIVITY

Disturbed areas on the face of the sun are always the sites of intense magnetic fields, ranging up to hundreds of gauss in disturbed areas in general and thousands of gauss in the interiors of sunspots. In a disturbed region the lines of magnetic force usually emerge from a region of one magnetic polarity and re-enter the surface in another region of opposite polarity located nearby. Disturbed regions may or may not contain sunspots.

PLATE 21 (far right). A color representation of the magnetic field in the photosphere, in a disturbed region about 50,000 kilometers across. The color image is superimposed on a photograph of the same region taken in 6563Å light, showing chromospheric structure. Luminous areas are plages. Red-orange indicates strong magnetic fields of negative polarity. Blue-green indicates strong magnetic fields of positive polarity. Lines of magnetic force probably emerge from the sun's surface in the sunspot group (bottom right) and re-enter the surface in the orange area (center). A sharply defined boundary separates the regions of positive and negative magnetic polarity. The color-coded photograph illustrates the fact that disturbed regions in the chromosphere are associated with strong magnetic fields of opposite polarity in the underlying photosphere.

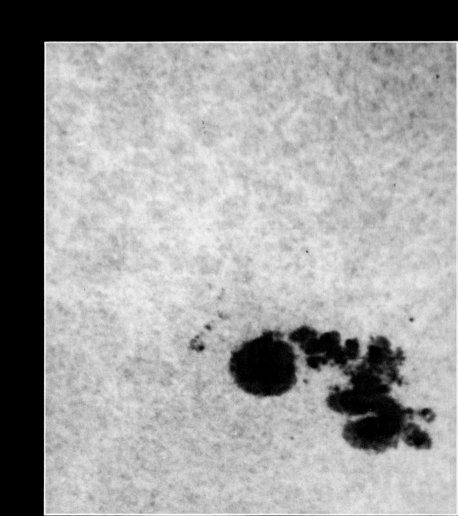

PLATE 22 (right). The same region in white light, showing the underlying photosphere and sunspot group.

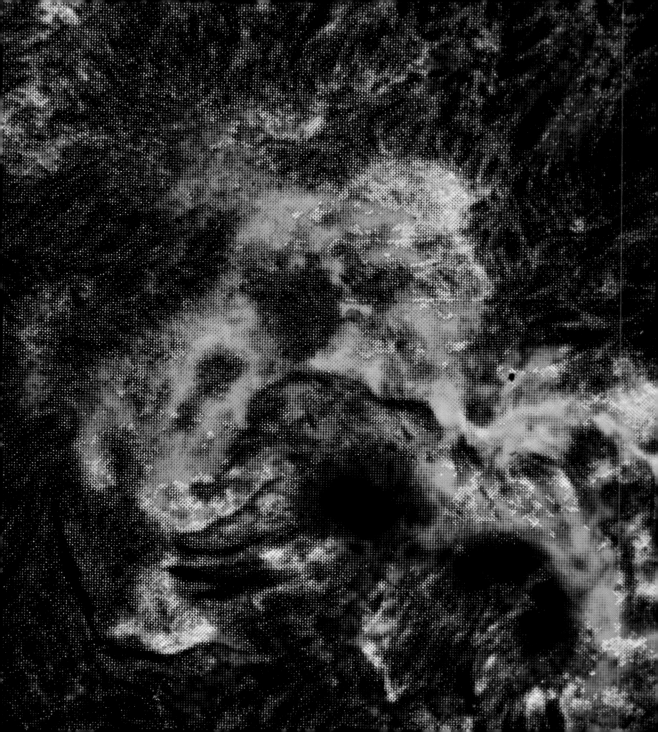

PLATES 23 AND 24. The Disk of the Sun Photographed in White Light and at 6563Å.

PLATE 23 (right). A white-light photograph taken on May 18, 1970, and showing the photosphere, reveals three sunspot groups at lower left, slightly above and to the left of center, and at upper right.

PLATE 24 (far right). This photograph, taken at 6563Å, shows the structure of the sun's chromosphere at the same time.

The bright areas in the 6563Å photograph are disturbed, hot regions in the chromosphere called plages. Bright plages appear near the three sunspot groups in the white-light photograph.

The plages without sunspots presumably are areas in which the magnetic field is intensified and sunspots either will appear, or were present and disappeared.

The large sunspot group at upper left consists of two distinct groups separated by a distance of about 100,000 kilometers. Curved lines, visible in the 6563Å photograph in the vicinity of this double sunspot group, suggest that lines of magnetic force emerge from one group of sunspots and return to the other group, in a pattern resembling the magnetic field around a bar magnet ( pages 319–320). The 6563Å photograph also shows the dark wisps and streamers called filaments. These are relatively cool masses of gas far above the sun's surface, which absorb 6563Å radiation strongly because of their lower temperature, and therefore appear black against the solar disk when the sun is photographed at this wavelength (page 325). If viewed

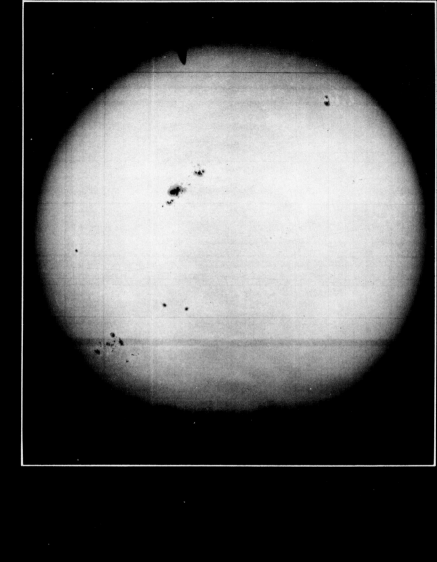

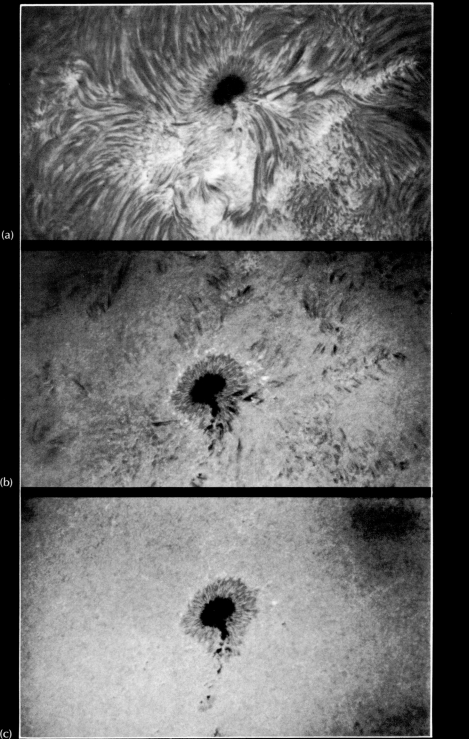

(a)

(b)

(c)

PLATE 25 (left). A Sunspot Seen at Three Wavelengths.

Photographs (a), (b) and (c) show a sunspot observed through filters transmitting light at three wavelengths near the first Balmer line of hydrogen. The wavelengths were (a) 6563Å, in the center of the absorption line, (b) $^7/_8$Å off center and (c) 2Å off center. In each case the width of the transmitted wavelength was $^1/_4$Å.

The photographs illustrate that we see into the chromosphere to a depth that depends on the wavelength of the light used in forming the image. In (a), because of the strong absorption in the center of the first Balmer line, the light from the photosphere is screened out, and only the structure of the chromosphere appears. The dark streaks reveal the direction and sometimes the changes in polarity of the magnetic fields in the chromosphere.

In (b) the absorption is less pronounced and the structure of the photosphere emerges, together with features of the chromosphere, such as the dark jets of matter identified with spicules (page 307).

In (c) the wavelength is sufficiently far from the center of the line so that the absorption is substantially reduced, spicules disappear, and further details of the photosphere appear including granulation.

PLATE 26 — (opposite). Chromospheric Structure Near a Sunspot Group. This 6563Å photograph shows two sunspot groups (lower right and upper left) The bright areas are plages. The dark streak AA' is a filament (page 324 and Plate 23). Numerous streaks and threads of material, dark because they absorb 6563Å radiation strongly, originate in the plages. The threads indicate the magnetic field pattern in the chromosphere. An example is seen to the right of the larger sunspot group, where dark threads emerge from one plage (B) and re-enter the surface at another plage (B'). The regions B and B' clearly have opposite polarities.

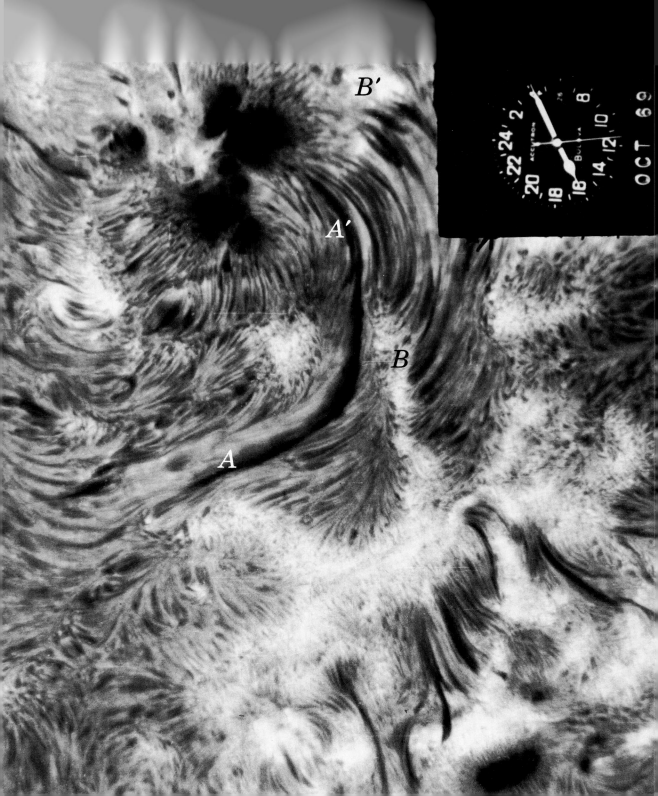

In regions in the corona that are cooler and denser than average, hydrogen atoms recapture electrons, emitting photons in wavelengths corresponding to transitions between states of the neutral atom, such as the first Balmer line at 6563Å. The emitting regions appear bright in photographs taken in 6563Å light, forming luminous streamers called prominences that extend high into the corona.

The neutral atoms that emit the 6563Å radiation are a small part of the gas in a prominence. Much of the material in the prominence consists of ions and electrons, which, because they are charged particles, are constrained to follow the direction of the lines of magnetic force in the corona, leading to graceful curving forms.

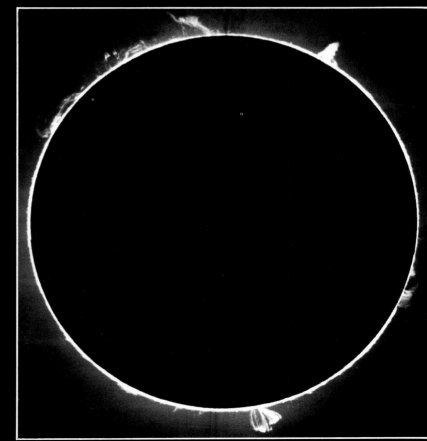

PLATE 27. A photograph of the sun taken in the light of the calcium absorption line, during the eclipse of December 9, 1929. The narrow ring of luminosity is the chromosphere. Prominences rise into the corona to heights of as much as 150,000 kilometers at several places on the limb. The striking feature at bottom is a loop prominence originating in a disturbed region on the sun containing a strong localized magnetic field. Its shape traces lines of magnetic force that curve upward into the corona and down again.

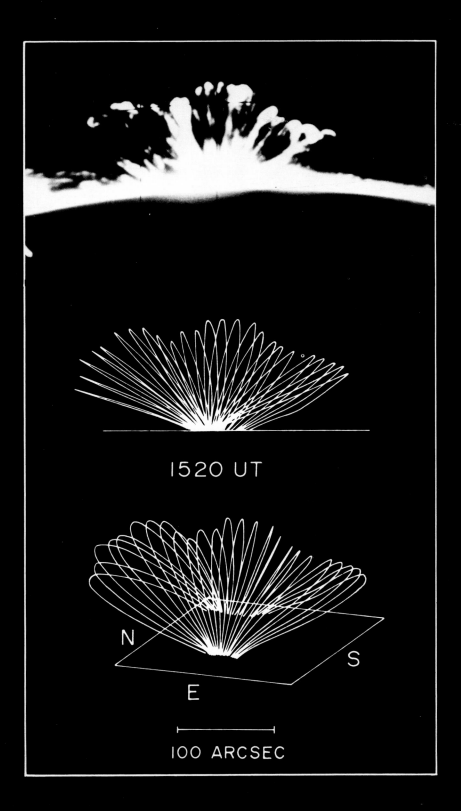

1520 UT

N

E

S

100 ARCSEC

PLATE 28. A loop prominence photographed in 6563Å light on November 18, 1968 at the limb of the sun. Below, a side view and an oblique overhead view of magnetic field lines in the corona, calculated from the observed pattern of the magnetic field at the surface of the sun underneath the prominence. The observed structure of the prominence closely follows the computed pattern of the magnetic field lines.

PLATE 29 (following page). Two stages in the eruption of one of the greatest solar prominences ever photographed, on June 4, 1946. The eruption had commenced prior to sunrise. Below, shortly after sunrise, the prominence was already 200,000 kilometers above the sun's surface. Above, forty minutes after, it had risen to a height of 500,000 kilometers, expanding at a speed of about 400,000 kilometers (250,000 miles) per hour.

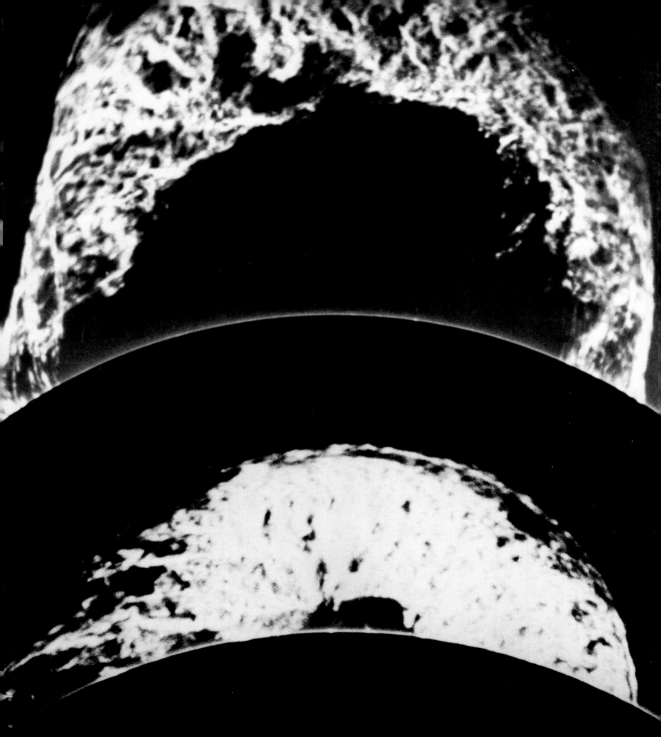

7. The chromosphere radiates about a thousandth of the energy radiated by the photosphere. How can the structure of the chromosphere be observed against the much brighter background of the photosphere? Explain.

8. Describe the sun's corona. What produces the radiation from the corona? Why is the corona normally invisible?

9. Describe a sunspot. Why are sunspots darker than the surrounding surface of the sun? Summarize the life history of a sunspot.

10. Describe the following: plage; flare; filament; prominence. Do these phenomena have a common cause? Explain.

11. What effects are produced on the earth by a large solar flare?

12. According to current theories of solar surface activity, two key properties of the sun give rise to sunspots, flares and other disturbances. Describe these properties and the role they play in generating surface disturbances.

# 13　The Solar System

According to the available evidence, the Universe began its existence 13 billion years ago as a dense, hot cloud. The cloud cooled as it expanded, and after a time, stars began to form. Within each star the manufacture of heavier elements out of hydrogen and helium began through a series of nuclear reactions. In this way, the elements of the Periodic Table were assembled out of the basic building blocks of matter.

In the smaller stars, these elements were locked within the star permanently by gravity, but the larger stars ended their lives in the cataclysmic explosion of the supernova, radiating the heat of billions of suns, and spraying out to space the elements that had been manufactured within a star during its lifetime. There these elements mingled with the primordial hydrogen and helium to form an enriched mixture of gas. Later, new stars were born out of the mixture.

Four and a half billion years ago, after innumerable supernova explosions, the concentration of the heavier elements amounted to about one per cent. Around that time, the sun and its family of planets condensed out of a cloud of gaseous matter located in one of the spiral arms of the

*Arend-Roland Comet.*

Milky Way Galaxy. How did the planets form? The first task in our study of the solar system is the investigation of the formation of the planets in this astrophysical context.

## THE ORIGIN OF THE SOLAR SYSTEM

The origin of the solar system is less of a mystery than the origin of the Universe, but it is still not a clearly understood event. When the authors were in high school, the commonly taught theory held that the planets came into being as by-products of a catastrophic event in which the sun collided with a passing star. The force of gravity tore huge streamers of flaming gas out of the bodies of the two stars during this encounter. As the intruding star receded into the distance, some of these streamers of gaseous material were attracted by the sun's gravity and captured into orbits circling around it. The earth condensed out of one of these streams of hot gas to form a molten mass, on whose surface a crust formed and gradually hardened with the passage of time.

It is easy to calculate the probability that the earth and other planets originated in a collision between two stars. The likelihood of a collision between the sun and another star depends on the size of the sun, and on the distance between it and its neighbors. Stars, large though they are, are minute in comparison with the average distances that separate them. The sun, for example, is one million miles in diameter, but 24 *trillion* miles from its nearest neighbor. A calculation shows that because the stars are so far apart, the possibility of a stellar collision is extremely small. In fact, a calculation shows that throughout the 10 billion years in which our Galaxy has existed, there have probably been no more than one or two stellar collisions.

Thus, the collision theory of the origin of planets implies that we are unique and alone in this corner of the Universe.

### The Condensation Theory

A very different prediction comes out of the modern theory of the origin of the solar system. This theory asserts that planets are formed as a natural accompaniment to the birth of stars. As the gas cloud of the star-to-be contracted under the inward force of its own gravity, the density of the gas increased, and atoms and molecules collided and occasionally stuck together. Gradually, small fragments of solid material accumulated in the cloud as a result of these collisions. What happened next is not well understood. According to one theory of the birth of the planets, when the fragments of solid matter were still quite small they were swept along with the surrounding gas in its motion around the center of the cloud. However, when the fragments became large enough—roughly basketball-sized or larger—they were too mas-

*Figure 13.1   Atoms and larger fragments in orbit around the protosun.*

sive to be carried with the gas. They began to "fall" out of the gas cloud toward the central plane of the solar nebula, where the density of the gas was greatest and the pull of gravity was strongest. In the course of time, the solid bodies became concentrated in a thin disk of matter in the midplane of the nebula.

The situation was analagous to that in a cloud of moist air in which water droplets are condensing; when the droplets are small, they move with the air currents, but when they have grown to a sufficiently large size, they fall out of the cloud in the form of rain.

Thus, according to our picture, at this early stage the solar system consisted of a relatively dense, contracting cloud—which eventually formed the sun—surrounded by a large number of solid fragments and planetesimals of various sizes, confined to a disk of matter circling the sun (Figure 13.1).

With the further passage of time, the central cloud continued to contract, until finally its temperature reached the critical level of 10 million degrees needed for the ignition of thermonuclear reactions. This point in the contraction marked the true birth of the sun. In our solar system, that occurred about 4.5 billion years ago (Figure 13.2a).

Throughout the period in which the sun was forming, the aggregation of small bodies into larger ones continued until the planets were completed (Figure 13.2b and c). The last stage in the formation of the planets may have occurred somewhat before the sun was born, or later. We are not sure which came first, the sun or the planets.

The 32 moons in the solar system could have been formed in the same manner, as still smaller condensations around their parent planets. Once again, the satellites may have condensed either a little earlier or a little later than the planets.

## Other Solar Systems

The condensation theory of the origin of planets implies that planets are formed nearly every time a star is formed, and must, therefore, be

Figure 13.2 Formation of the planets. Only the inner planets are shown in the final stage; if the giant planets and Pluto were shown also, the diagram would be four feet in diameter.

very common objects in the Universe. In fact, there must be literally billions of them in our Galaxy alone. Unfortunately, this important prediction cannot be tested directly by astronomers, because a planet circling a neighboring star, and shining only by the reflection of that star's light, is far too faint an object to be seen in the largest telescopes on the earth.

If a 200-inch telescope were placed in an orbiting satellite or in an observatory on the moon, far from the obscuring effects of the earth's atmosphere, it would be just barely able to detect a Jupiter-sized planet circling one of the stars near the sun. However, telescopes of this size will not be placed in orbit, or on the moon, for many years to come.

*Detection of Planets Circling Other Stars.* It might seem at first that we will have to wait a long time before we can find out if planets are common objects, as the condensation theory predicts. However, the situation is less discouraging than this. Although it is impossible at present to see a planet directly, it is possible to find large planets on neighboring stars—if these planets exist— by an indirect method.

The method consists in observing the motion of each star very carefully, and looking for signs of the effect of a planet's gravity on the star's motion. If the star has no planets, its motion across the sky is a smooth curve—very close to a straight line. But if a planet is circling around the star, the pull of its gravity disturbs the star's motion and produces a wiggle in the path of the star. The wiggle repeats itself once in every revolution of the planet (Figure 13.3).

As an example, if the earth were the only planet in the solar system, its pull on the sun would produce a series of wiggles in the sun's path across space, each wiggle lasting for one year.

Theoretically, an observer on another star could detect the existence of the earth in our solar system by the effect that it would have on the motion of the sun. In practice, however, the observer could only notice the effect of Jupiter, whose gravitational pull on the sun far outweighs that of any other planet in the solar system. Thus, the observer would see a wiggle in the sun's motion that repeated itself every 12 years, since 12 years is the time it takes Jupiter to revolve about the sun.

The deviation in the sun's motion produced by Jupiter is about one million miles in the course of 12 years. However, in 12 years the sun, which is moving through the heavens at 60 miles per second, covers a distance of 20 billion miles. Thus, the deviation of one million miles caused by Jupiter, although it seems large, is relatively small in comparison to the main part of the sun's motion during that period, and it could only be detected by the distant observer if he had taken pains to observe the sun's position with great care. If he knew the mass of the sun, this distant observer on another star could use his observations of the wiggle in the sun's motion, combined with the laws of gravity, to deduce both the *mass* of the unseen planet, Jupiter, and the *diameter* of its orbit.

*Discovery of Another Solar System.* Evidence has been reported for planets circling stars relatively close to the sun, using the method of looking for wiggles in their path. All the apparent planets discovered in this way have been relatively large, with masses ranging from slightly less than the mass of Jupiter to ten times Jupiter's mass. This is so because only a planet about as large as Jupiter, or larger, will exert a sufficiently strong tug on its parent star to produce a detectable wiggle in the star's motion. Smaller planets may also exist around these stars, in addition to the large ones, but their effect on their suns would be much harder to detect.

But just as our solar system contains eight other planets in addition to Jupiter, and five of the eight are earthlike, we can assume that there is a fairly good chance that these other solar systems also contain several other planets, some of which will be earth-sized.

Motivated by these ideas, a Swarthmore College astronomer named Peter van de Kamp has made a careful study of the motions of all the nearby stars in the sky, in the hope of discovering the wiggling motions that would betray the presence of planets. Starting in 1938, he commenced to take a series of photographs, at regular intervals of time, of about 40 of the stars closest to us. Then, he plotted the paths of the stars from these photographs and looked for the characteristic wiggles that would betray the presence of planets.

In 1963, after 25 years of patient effort, van de Kamp announced that Barnard's Star shows an apparent wiggle in its motion. The effect seems to be due to the presence of a planet orbiting Barnard's Star with a

Figure 13.3  *Influence of a planet on the motion of its parent star.*

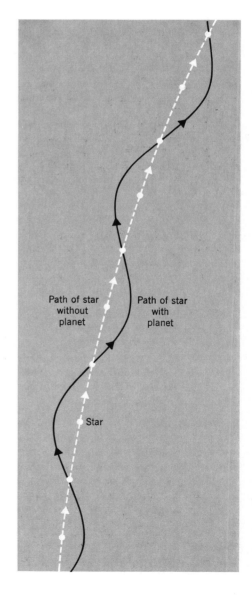

Path of star without planet

Path of star with planet

Star

mass approximately equal to the mass of Jupiter.

Why were 25 years of observation required to reveal the existence of this planet? The reason is that in spite of the closeness of Barnard's Star to us, and the large size of the planet van de Kamp discovered, the relative change in the star's position, caused by this planet, is still exceedingly minute. Its minuteness is demonstrated by the fact that the shift in angle, that is, the width of the wiggle, is equivalent to the width of a human hair when seen from a distance of one mile.

The final chapter in the story of Barnard's Star is still being written. In April 1969 van de Kamp reported evidence for a *second* planet orbiting Barnard's Star, with about four-fifths the mass of Jupiter. If these reports are confirmed, this body would be the smallest object yet detected outside our solar system.

### Solar Systems Versus Multiple Stars

Let us return to the question of the probability of formation of other solar systems in our Galaxy. The tentative indication of the second planet on Barnard's Star gives some support to the suggestion that where one planet exists in orbit around a star, it is likely that several exist, making up a family of planets just as in our solar system.

It is true that no evidence has yet been obtained for the existence of planets as small as the earth, circling other stars than the sun, but the continuing refinement of the observations of the motions of nearby stars, leading to the report of a second planet on Barnard's Star, suggests that someday a relatively small planet may also be detected circling one of these close stars.

Going beyond Barnard's Star, it has been found that at least one other star in our neighborhood shows the characteristic wiggle that may indicate the presence of a dark companion. This star, known as Lalande-21185 in the astronomical catalogues, is 8 light-years from the sun. The object circling Lalande-21185 is ten times as massive as Jupiter.

Although this object is more massive than Jupiter, it is still below the critical mass for a star; that is, it is not large enough to generate the temperature needed for nuclear burning. If it were somewhat more massive, it would be a faint companion star to Lalande-21185. The latter would then be a double star, rather than a single star like the sun, accompanied by planets.

These remarks show that it is easy to erase the distinction between a small star and a large planet. Thus, in our attempt to obtain information on the likelihood that planets are attached to stars, we can also look for the number of double stars, triple stars, and so on, that are in the sky. If multiple stars are common, stars with planets probably are also.

Observations indicate that multiple stars, usually consisting of one large star and one or two smaller companion stars, are very common in the sky. In fact, they make up half the population of the Galaxy. This must mean that whenever a cloud of gas and dust starts to condense in space, the condensing cloud tends to break up into a number of separate clouds of various sizes. The biggest cloud forms a star, while the other clouds form smaller stars or planets of various sizes. This combination of theory and observation indicates that solar systems like ours almost certainly are very common objects in our Galaxy, as well as in other galaxies.

*Planets Without Stars.* The condensation theory leads to still another interesting possibility. Suppose that a cloud condenses in space and forms one or more bodies, but none of the bodies—whether one or several have been formed does not matter—is massive enough to become a star. Such an object, or a group of objects if several have been formed out of one condensation, would resemble the planets of the solar system; yet, they cannot properly be called a family of planets, or a solar system, because a solar system is defined as a family of planets circling around a star, and there is no object in this group that is large enough to be a star.

Such relatively cold planet-sized objects would wander through space as dark, frozen worlds, free of the gravitational influence of any star. They may exist around us in considerable numbers, quite near to our solar system. Since they do not shine by their own light, and are too far away from any star to be visible to us by its reflected light, we have no way of detecting these "free planets" at present. We may stumble across one of them when we begin to explore with space vehicles the regions lying beyond the boundary of the solar system.

## Arguments For and Against the Theory

These ideas are derived from the condensation theory of the origin of the solar system, for which no definite proof has been supplied. However, there are good reasons for believing in the theory:

*First,* it fits naturally into the latest ideas on the birth of stars.

*Second,* it predicts that the planets are a natural accompaniment to the birth of stars, and the prediction is borne out by the detection of planets circling around stars in the neighborhood of the sun.

*Third,* multiple stars are very common objects in the sky, and they are not very different from stars accompanied by massive planets.

*Objections to the Theory.* Although nearly all astronomers believe in the condensation theory of the origin of the solar system, this theory still presents some serious difficulties. One of the difficulties concerns the rotation of the sun. The sun spins on its axis once every 27 days, just as the earth rotates once in 24 hours. It is easy to understand why the sun should be spinning in this way if we remember that it was originally formed out of swirling masses of interstellar hydrogen gas. Some of the swirling motion must have been retained in the cloud out of which the sun formed. When the sun-cloud was newly formed, and its dimensions were large, it probably rotated very slowly. But as it contracted it must have spun more and more rapidly, just as an ice skater spins faster when he pulls in his arms and draws his skates together, in contracting from a sweeping turn to a small circle. As a result of this effect, the sun should now be spinning on its axis at the rate of once every few hours. Actually, it turns at a far slower rate, 100 times less rapidly. What has slowed the sun down? Charged particles, blown off the surface of the sun in the so-called solar wind, may have exerted a braking effect, but a thoroughly satisfactory explanation of this effect on the sun has never been provided.

Another difficulty involves the earthlike planets only. According to the theory, these planets, as well as the giant planets, must be the result of a gravitational condensation of matter that took place in the parent cloud of the solar system, around the time the sun itself was forming. This idea works well for a planet such as Jupiter, which is massive enough so that the force of its gravity would unquestionably have been adequate to cause it to condense at the beginning of the solar system, and to remain condensed ever since that time. However, the idea does not work very well for the earthlike planets, because they are relatively small objects, and the force of their gravity could not have been sufficient to cause them to condense out of the parent cloud.

Therefore, we are forced to conclude that we can explain the giant planets easily, but we do not have a good theory for explaining how the earthlike planets came into existence.

A third problem arises out of the fact that the planets should, according to theory, contain a large fraction of the mass of the original solar gas cloud. A calculation by James Jeans leads to the conclusion that a third of the mass of the sun should have been left behind to form the planets. Actually, the planets have roughly one thousandth of the mass of the sun. What has happened to the missing material? Perhaps streams of energetic particles from the surface of the sun blasted it away, or it may have evaporated from the outer regions of the solar system. The answer is not known.

These difficulties of the condensation theory are not easily resolved. It can only be said that we have a strong suspicion that the earth was formed by condensation, along with the sun, 4.5 billion years ago, but no one yet has a clear understanding of the tangled complex of events that surrounded the genesis of the planets. It is possible that we may be in for a major surprise as we carry out our *in situ* explorations of the moon and the other planets in the coming decades.


## THE EARLY YEARS OF THE SOLAR SYSTEM

We believe that when the solar system came into being, it was only a cloud of gaseous hydrogen, mixed with small amounts of other substances. According to modern astronomy, this cloud contained the materials that are now in the bodies of the sun, the planets, and the creatures that walk on the surface of the earth. It was the parent cloud of us all. At its center existed a dense, hot nucleus that later formed the sun. The outer regions—cooler and less dense—gave birth to the planets.

Out of what materials were the planets formed? The bulk of the parent cloud must have been composed of the light gases, hydrogen and helium, because they are the most abundant elements in the Universe. Other elements relatively abundant in the Universe, although less so than hydrogen and helium, are carbon, nitrogen, and oxygen, metals such as iron, magnesium, and aluminum, and silicon. These substances must also have been present in relatively great abundance in the parent cloud of the planets. No doubt the remaining 80-odd elements were also represented, but in smaller amounts.

All the familiar chemical compounds of these substances would have formed in the cloud in a relatively short period of time. Hydrogen combines readily with oxygen to form molecules of water vapor; hydrogen also combines with nitrogen to form molecules of ammonia gas, and it combines with carbon to form methane, also called marsh gas, which is used extensively today for cooking. Carbon and oxygen combine to form carbon dioxide. Considerable amounts of each of these compounds must have formed in the parent cloud. However, they were probably not present in the form of gases, because of the low temperature—about 100 degrees Fahrenheit below zero—prevailing in the region of the cloud out of which the earth was formed. At this temperature they congealed into a slushy mixture of water, ammonia, and methane in liquid and solid form, plus solid carbon dioxide—dry ice. The other elements that were present in abundance—silicon, aluminum, magnesium, and iron—combined with oxygen to form grains of rocklike materials and metallic oxides.

These, then, are the substances out of which the planets condensed: a Neapolitan sherbet of frozen water, ammonia, and methane, plus various kinds of rocky substances—all immersed in a gaseous cloud of hydrogen and helium, plus smaller amounts of other vapors.

When the planets first condensed out of this mixture of gases and solid matter, the bulk of their mass should have consisted of hydrogen and helium. The giant planets—Jupiter, Saturn, Uranus, and Neptune—are, in fact, composed mostly of these very light elements, and are small-scale models of the sun and stars in their composition although, of course, they lack the nuclear energy resources of a star. There is little question but that these planets were formed in precisely the same way that a star is formed, by condensation out of gaseous matter under the force of their own gravity. As a result, they are very different from planets such as the earth. Our planet is a rocky ball of matter, almost entirely solid except for some molten material at the center.

Why hydrogen and helium are scarce on the earthlike planets is a mystery. Some students of the subject say that these planets condensed in the same way as Jupiter, as large spheres of gas composed mostly of hydrogen and helium. However, the force of gravity on Jupiter is so strong that it prevents even the lightest gases from escaping. Once a planet as massive as Jupiter has formed, the removal of its light gases becomes very difficult.

It seems more reasonable to assume that the light gases disappeared from the inner parts of the solar system—perhaps, blasted away during a temporary flareup of the newly-formed sun—before the earthlike planets condensed. These planets—Mercury, Venus, the earth, and Mars—were formed subsequently, out of the rocky materials left behind.

In that case, the materials out of which the earth condensed would have been particles of rock and carbon and small amounts of ice that circled the sun under the force of gravity, each a miniature planet in its own right.

This theory for the origin of the earthlike planets leads to a speculative picture of their formation. Occasionally, collisions would have occurred between neighboring particles in the course of their circling motion. Some collisions were gentle, and the particles stuck together. In this way, in the course of millions of years, small grains of rock gradually grew into larger ones. Some pieces of rock became large enough to exert a gravitational attraction on their neighbors. These were the nuclei of the modern planets. Once they had grown large enough to attract other particles by their own gravity, they quickly swept up all the materials in the space around them, and developed into full-sized planets in a short time.

The complete process of formation of the earthlike planets went on over a period of perhaps 10 million years, proceeding with extreme slowness at first, and then with rapidly increasing momentum in the final stages. At the end, the remaining matter in the solar system was gathered into the existing earthlike planets, and only a few atoms of gas remained in the space between. This is the situation in the solar system as it exists today.

## THE SOLAR SYSTEM TODAY

The solar system consists of a G2 star—the sun—circled by nine planets, 32 moons, asteroids, and a large number of comets.

The internal structure and evolution of the sun were discussed in Chapter 7 because the Sun's position on the Main Sequence, midway between the largest and smallest stars, makes it particularly well suited to serve as an example of the average star. In the remainder of the book we continue to treat the sun as a typical Main-Sequence star without unusual properties. The branch of astronomy known as solar physics is largely concerned with the sun's outer layers and atmosphere, and with disturbances in these outer layers, associated with the waxing and waning of the sunspot cycle, which create a variety of effects on the earth and in interplanetary space. These effects—including the solar wind, solar cosmic rays, the aurora, radio blackouts, magnetic storms, and the Van Allen radiation belt or magnetosphere—are of considerable scientific interest and practical importance. However, they depend on aspects of stellar structure that are surface details in the context of a general treatment of the properties of stars and planets, and are outside the confines of the present discussion.

The sun is 1000 times more massive than the largest planet—Jupiter—

and seven hundred times more massive than the rest of the solar system including Jupiter. The sun's gravity controls the motions of the other members of the solar system. Without it, they would drift away and wander into space. This fact—that a force of attraction is required to keep the planets in their orbits—was not appreciated until relatively late in the history of astronomy. As late as the time of Galileo, it was generally believed that the circle represented one of nature's ideal forms of motion, requiring no force for its indefinite continuation. Galileo shared this erroneous view, which probably prevented him from discovering the law of gravity, although he stood on the threshold of that discovery as a consequence of his penetrating analysis of the movements of objects on the earth. Newton was the first person to strip away the old misconceptions and to replace them with a clear-cut statement of the nature of motion.

### The Laws of the Solar System

In the third book of the *Principia* Newton included a figure intended to explain the orbital motion of the moon (Figure 13.4). The figure shows a

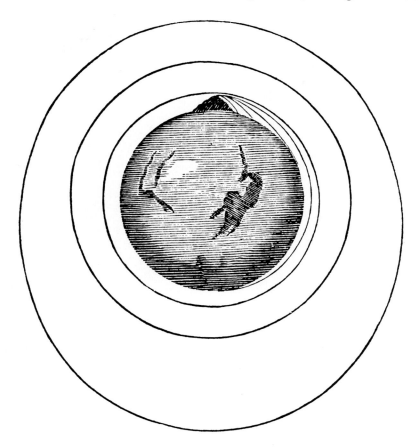

*Figure 13.4  Newton's diagram illustrating the motion of the moon, from the third volume of* Principles of Natural Philosophy. *The cannon placed on a mountaintop fires a shot that travels a curved path, compounded of its horizontal forward velocity and a downward motion produced by the gravitational attraction of the earth.*

mountain on whose top a cannon is mounted with its barrel directed horizontally. When a ball is fired from the cannon, it moves forward under the impetus of the pressure exerted by the hot gases in the barrel. At the same time it is subjected to the downward attraction of gravity, which pulls it toward the center of the earth. The combination of the forward motion and the motion downward under gravity is a curved path, which terminates when the projectile hits the ground.

If the charge of explosives is increased, the forward velocity increases and the ball traverses a greater distance before it is pulled to the ground by gravity. It is conceivable that a cannon could be constructed of such power that the ball would travel around the earth without striking the ground. The combination of the forward motion produced by the discharge of the cannon, and the downward deflection produced by gravity, would curve the path of the projectile into a circular orbit around the earth. The cannon ball would be a satellite.

Newton constructed this imaginary experiment to explain the motion of the moon around the earth, but it is also the explanation for the motion of a planet, or any other object, around the sun.

If the object has no forward momentum carrying it around in its orbit, then it will fall into the sun, drawn by its gravity. Suppose now that the object is given a small forward momentum. Then it will fall in toward the sun on a curved path, loop around it on a curved path at a close distance, and move outward again in an elliptical orbit (Figure 13.5a).

If the object has a very large forward momentum, it will escape the sun's gravity and leave the solar system entirely. If the forward momen-

*Figure 13.5   Types of planetary orbits.*

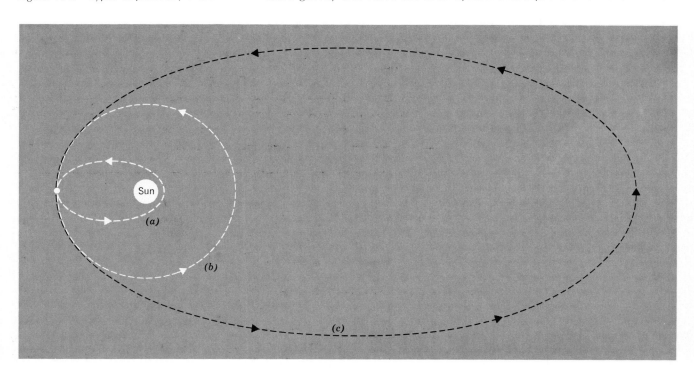

tum is large, but not large enough to carry the object out of the solar system, the object will follow a curved trajectory carrying it a considerable distance from the sun, but will eventually return on a path which is again an ellipse (Figure 13.5c).

Between these two cases must lie a third possibility, in which the path of the object will be bent into an approximate circle by the combination of its forward momentum and the pull of the sun's gravity. If the two motions are in the proper proportion, the path of the object will be a perfect circle (Figure 13.5b).

*The Law of Gravity.* The gravitational force between two objects grows weaker as the distance between them increases. Newton proved that this effect is governed by the following law: *The gravitational force between two objects varies as one over the square of their distance.* That is, if the distance between two objects is doubled, the force of gravity diminishes to one quarter of its former value. If the distance is tripled, the force of gravity decreases to one ninth of its previous value, and so on. This relationship is known as the *inverse-square law of gravity.*

Suppose that the masses of the two objects are $m_1$ and $m_2$, respectively, and the distance between them is $d$. Then the mathematical statement of Newton's law of gravity is

$$\text{Force} = G\,\frac{m_1 m_2}{d^2}$$

In this formula, $G$ is the constant of proportionality, called the *universal constant of gravitation.* In the cgs system, $G$ has the value of $6.7 \times 10^{-8}$.

A legend has it that Newton's thoughts on gravity were stimulated by an apple that struck him while he was resting under a tree. Newton speculated that the earth's gravity—the force that makes an apple fall to the ground—might be the same as the force that holds the moon in its orbit. Others had made similar speculations, but Newton went beyond his contemporaries. He calculated the strength of the earth's gravity at the distance of the moon by computing the pull that would be required to bend the path of the moon into a circular orbit around the earth.

Comparing his results with the force of gravity on the surface of the earth, Newton discovered that in the space around the earth extending out to the moon, gravity falls off as one over the square of the distance to the center of the earth.

Newton wondered whether the same law governed the motion of the planets around the sun. The clue to the answer appeared in a discovery made by the astronomer Johannes Kepler, many years before Newton was born. Kepler had discovered that the orbits of the planets are ellipses.[1] Newton proved that the ellipse is the path that must be followed by an object moving under an inverse-square law of attraction. Since the planets move in ellipses, he concluded that they are

[1] Kepler proved his result only for Mars, but in Newton's time it was generally accepted that all planets moved in elliptical orbits.

governed by a force obeying the inverse-square law. The force that held the planets in their orbits around the sun, the force that held the moon in its orbit around the earth, and the force that made objects fall to the earth's surface must be one and the same.

Thus Newton extended the law of gravity from the earth to the moon and the planets. Encouraged by his success, he conjectured that the law of gravity was valid throughout the Universe. The boldness of his step cannot be exaggerated. In one flight of thought, Newton united the heavens and the earth. This achievement laid the cornerstone for the structure of modern science.

## GENERAL PROPERTIES OF THE SOLAR SYSTEM

Important information regarding the planets may be derived from an examination of their general properties as a group of bodies circling the sun. The minor bodies of the solar system are also of considerable interest, partly because they can yield information regarding the primitive composition of solar system materials, and partly because the orbits of these objects occasionally take them close to the earth, with spectacular consequences for the observer of the night sky.

The basic properties of the nine planets are listed in Table 13.1.

Table 13.1 **Properties of the Planets.**

| Name | 1 Distance from Sun $10^6$ miles | 2 Distance from Sun Relative to Earth's Distance (A.U.)[a] | 3 Inclination to Ecliptic | 4 Eccentricity of Orbit[b] | 5 Revolution Period (earth–years) |
|---|---|---|---|---|---|
| Terrestrial Planets | | | | | |
| Mercury | 36 | 0.39 | 7°00′ | 0.21 | 0.24 |
| Venus | 67 | 0.72 | 3°24′ | 0.01 | 0.62 |
| Earth | 93 | 1.00 | 0°00′ | 0.02 | 1.00 |
| Mars | 142 | 1.52 | 1°51′ | 0.09 | 1.88 |
| Giant Planets | | | | | |
| Jupiter | 483 | 5.20 | 1°18′ | 0.05 | 11.86 |
| Saturn | 886 | 9.54 | 2°30′ | 0.06 | 29.46 |
| Uranus | 1780 | 19.18 | 0°46′ | 0.05 | 84.01 |
| Neptune | 2790 | 30.07 | 1°07′ | 0.01 | 164.97 |
| Pluto | 3670 | 39.44 | 17°19′ | 0.25 | 248.4 |

[a] The earth's mean distance from the sun—92,600,000 miles—is known as the astronomical unit (A.U.).
[b] Difference between the closest and farthest distance to the sun, expressed as a fraction of the average radius of the orbit.

## The Motions of the Planets

Columns 1 and 2 of Table 13.1 indicate that the planets revolve around the sun in more or less regularly spaced orbits, the distance from one planet to the next increasing by approximately a factor of one and one-half to two in most cases (Figure 13.6).

A conspicuous gap lies between the orbit of Mars and the orbit of Jupiter, which are located, respectively, at 1.5 and 5.2 times the distance from the earth to the sun. A planet should be located midway between Mars and Jupiter, at a distance of, perhaps, three times the earth's orbital radius, but none is found there. Instead, a swarm of rock and iron fragments known as the asteroid belt circles the sun in that neighborhood at an average distance of 2.9 earth-orbit radii.

The orbits of all the planets are within a few degrees of the plane of the orbit of the earth called the *ecliptic*. The orbits of the innermost planet, Mercury, and the outermost planet, Pluto, are exceptions (column 3 in Table 13.1). All orbits are close to perfect circles, with the exception, again, of Mercury, Pluto and to a lesser extent, Mars (column 4). The distance from Pluto to the sun varies by approximately one billion miles during the course of the planet's year. The farthest reach of Pluto's orbit marks the outer boundary of the solar system, except for the minor bodies known as comets. Some cometary orbits reach out one-fifth the distance to the next-nearest star.

| 6 | 7 | 8 | 9 | 10 | 11 | 12 |
|---|---|---|---|---|---|---|
| Rotation Period (hours or days) | Mass (earth's mass = 1) | Diameter (miles) | Density (water = 1) | Effective Temperature[c] °K | Incident Solar Energy (relative to earth)[d] | Number of Satellites |
| 59d | 0.05 | 3,000 | 5.4 | 450 | 6.7 | 0 |
| 243d | 0.82 | 7,600 | 5.1 | 283 | 1.9 | 0 |
| 23h 56m 4.1s | 1.00 | 7,930 | 5.52 | 240 | 1.00 | 1 |
| 24h 37m 22.6s | 0.12 | 4,270 | 3.97 | 220 | 0.43 | 2 |
| 9h 50.0m | 317.80 | 89,000 | 1.33 | 100 | 0.04 | 12 |
| 10h 14m | 95.2 | 75,000 | 0.68 | 75 | 0.01 | 10 |
| 10h 49m | 14.5 | 30,000 | 1.60 | 50 | 0.0031 | 5 |
| 15h | 17.2 | 28,000 | 2.25 | 40 | 0.001 | 2 |
| 6.39d | 0.1? | 4,000(?) | 4? | 40 | 0.0006 | 0 |

[c]Temperature at the surface if there were no atmosphere.
[d]The earth receives on the average 2 calories per square centimeter per minute.

In the space between the planets there is a tenuous cloud of gas, mostly hydrogen, with a density of 100 to 1000 atoms per cubic inch. Beyond the boundary of the solar system the density of matter drops to its interstellar value of 10 atoms per cubic inch.

*Length of Year.* The length of a planet's year—that is, the time required to complete one circle around the Sun—increases with distance from the sun, varying from 88 days for Mercury to two and a half centuries for Pluto (column 5). The relationship between a planet's distance to the sun (*R*) and the length of its year (*T*) is given by:

$$\frac{T \text{ (planet)}}{T \text{ (earth)}} = \sqrt{\frac{R^3 \text{ (planet)}}{R^3 \text{ (earth)}}}.$$

This formula was obtained by Kepler by trial and error at the beginning of the seventeenth century. Newton used Kepler's law to derive the inverse

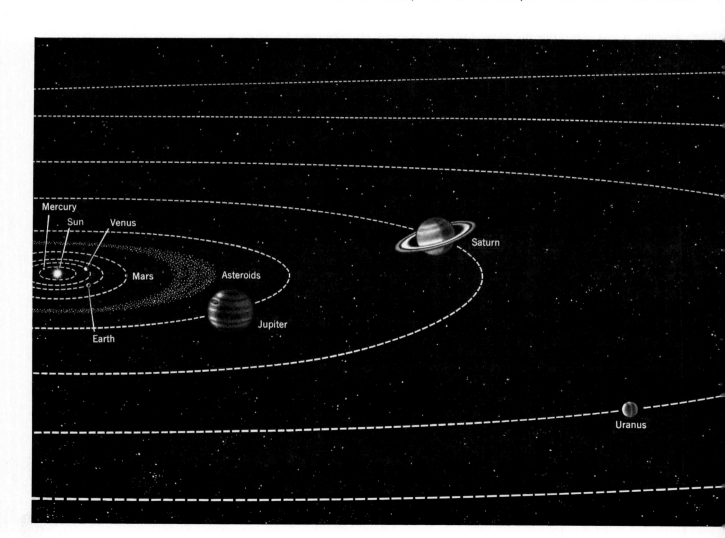

*Figure 13.6   The spacing of the orbits of the planets.*

square law of gravity and tested his result against the orbit of the moon.

*Length of Day.* The length of a planet's day — that is, the time required to complete one rotation about its axis — varies from 9 hours and 50 minutes for Jupiter to 243 days for Venus (column 6). The similarity in the lengths of the day for the earth and Mars probably is a coincidence. The significance of the long period of rotation of Venus is discussed in Chapter 15. All giant planets rotate very rapidly on their axes. The explanation for the rapid rotation is to be found in the "ice-skater" effect on page 340. The rotation produces a pronounced bulge at the equator which is particularly large for Saturn, amounting to three thousand miles or 9.5 percent of the radius of the planet. In contrast, the earth's equatorial bulge is 14 miles.

*Rotation of Mercury.* A surprising development occurred recently in astronomy, when a standard observation of Mercury, assumed correct

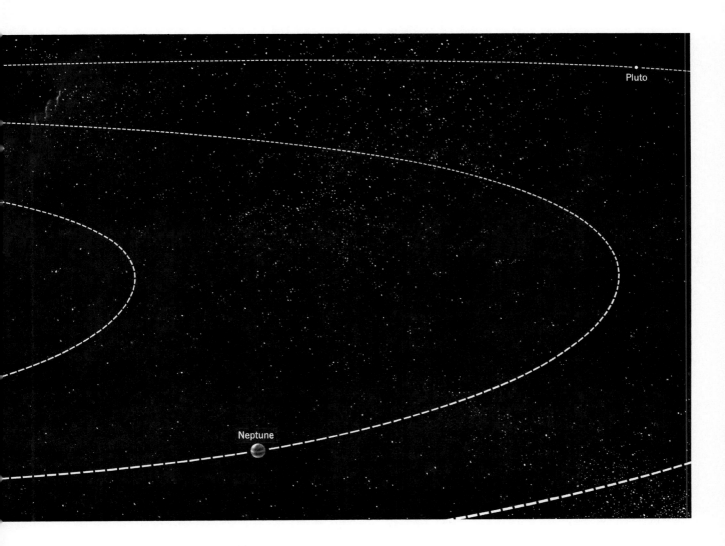

for many years, was discovered to be in error. Early measurements of the length of the day on Mercury, based on tracking the movements of visible markings across the planet's surface, indicated that the length of its day was 88 earth-days. That is, the length of the day and the length of the year were thought to be the same. This is known as *synchronous rotation.*

Synchronous rotation was expected for Mercury on the grounds that the powerful pull of the sun's gravity on its nearest planet would keep the same face of Mercury turned toward it at all times, just as the pull of the earth's gravity keeps one side of the moon facing our planet at all times. The visual tracking of the movements of markings on the surface seemed to support this theory. Around 1965, however, Doppler measurements of the planet's rotation, carried out by bouncing radar beams off its surface and analyzing the echoes, showed that the length of the Mercury day was 59 earth-days, or two-thirds of the length of its year.

One theory attributes the effect to the presence of two bulges on the face of the planet, located on diametrically opposite sides of the planet, that were pulled out of the planet by the sun's gravity at an earlier stage and were then "frozen" into the permanent shape of the planet. Each time Mercury passes through a point of closest approach to the sun in the course of its elliptical orbit, the sun's gravity, which is strongest at that point, must line the planet up so that one of these two bulges is pointing directly toward the sun. This condition is satisfied if the length of the day, divided by the length of the year, is equal to one-half, one, one and one-half, two, and so on. The third case, which makes the length of the day two-thirds the length of the year, is the mode of motion into which Mercury has settled, according to this explanation.

### The Terrestrial Planets

The nine planets divide into two groups differing greatly in their size, mass, and composition (Figure 13.7). Mercury, Venus, and Mars resemble the earth in being composed almost entirely of rocky materials and iron, and are known, together with the earth, as the *terrestrial planets.* The moon is only slightly smaller than Mercury and is sometimes included with the terrestrial planets, being composed of similar materials.

The average densities of the terrestrial planets vary considerably according to column 9 of Table 13.1, suggesting a substantial difference in composition. However, when allowance is made for the increase in density in the interior of each planet caused by the pressure of the overlying layers, the so-called "uncompressed density" derived in this way turns out to be approximately the same for all planets. For each terrestrial planet, the uncompressed density is somewhat greater than the density of surface rocks on the earth, as would be expected for a planet composed of rocky materials plus a substantial admixture of iron.

The density of Mercury is somewhat greater than you would expect for this composition. Probably its closeness to the sun and high temperature resulted in the loss of a fraction of the lighter, relatively volatile elements that were retained on the other terrestrial planets. A close view of Mercury was obtained in 1975 from the Mariner space craft (Figure 13.7a).

Table 13.1 lists hypothetical surface temperatures for each planet, calculated from the intensity of the solar radiation reaching the planet at its average distance from the sun (column 10). In the case of the earth and Venus, the actual surface temperatures are substantially higher because the relatively dense atmospheres of these planets act as insulating blankets. Mars has a very thin atmosphere, and Mercury and the Moon have substantially no atmospheres, hence the true surface temperatures for these bodies are close to the ones listed in the table.

Other properties of the terrestrial planets are discussed in later chapters.

*Moons of the Terrestrial Planets.* The terrestrial planets possess a meager supply of moons—three in all—compared with 29 satellites claimed by the giant planets. One of them—our Moon—is the subject of Chapter 15. The other two, known as Deimos and Phobos—Greek for Terror and Fear—belong to Mars, the God of War in mythology. Deimos and Phobos are small fragments of rock only a few miles in diameter. Phobos hurtles over Mars at an altitude of only thirty-seven-hundred miles. It completes an orbit in 7 hours, moving faster than the planet rotates beneath it, so that it rises in the west and sets in the east.

## The Giant Planets and Pluto

Five planets lie outside the orbit of Mars. They are the giant planets—Jupiter, Saturn, Uranus, and Neptune—and the small planet, Pluto.

*Pluto.* Pluto, found in 1930, was the ninth and last planet to be discovered in the solar system. Its orbit carries it farther from the sun than that of any other planet, and probably marks the outer boundary of the solar system. Because Pluto is so far away, we have been able to learn very little about it, except that it appears to be a body similar in composition to the terrestrial planets. It must be a frozen, silent world, far too cold to support any form of life.

The absence of hydrogen and helium from Pluto probably marks the end point of a trend discernible in the compositions of the giant planets. Progressively with increasing distance from the sun, larger amounts of primordial hydrogen and helium appear to be missing from these planets. The trend toward a decrease in the concentrations of hydrogen and helium probably is the result of the sun's weak force of gravity at great distances. The lightest gases—hydrogen and helium—would be the first elements to escape to interstellar space under these circumstances. Pluto may have accumulated out of a residue of heavier elements, achieving a composition similar to that of the earth, although for entirely different reasons.

*The Giant Planets.* More is known about the giant planets. They are five to ten times larger than the earth and far more massive, but considerably lower in density. In general, their density is about the same as that of water; Saturn, in fact, is less dense than water; it would float in the bathtub if you could get it in.

Figure 13.7a  *The cratered moon-like surface of Mercury photographed in March 1975 by Mariner 10 from an altitude of 11,800 miles.*

Figure 13.7b *The sun and the planets: relative sizes. The sun and planets, arranged in order of distance from the center of the solar system, are shown here in proportion to their actual sizes. The sun is one million miles in diameter, or 13 times the size of the largest planet, Jupiter (pp. 352-353).*

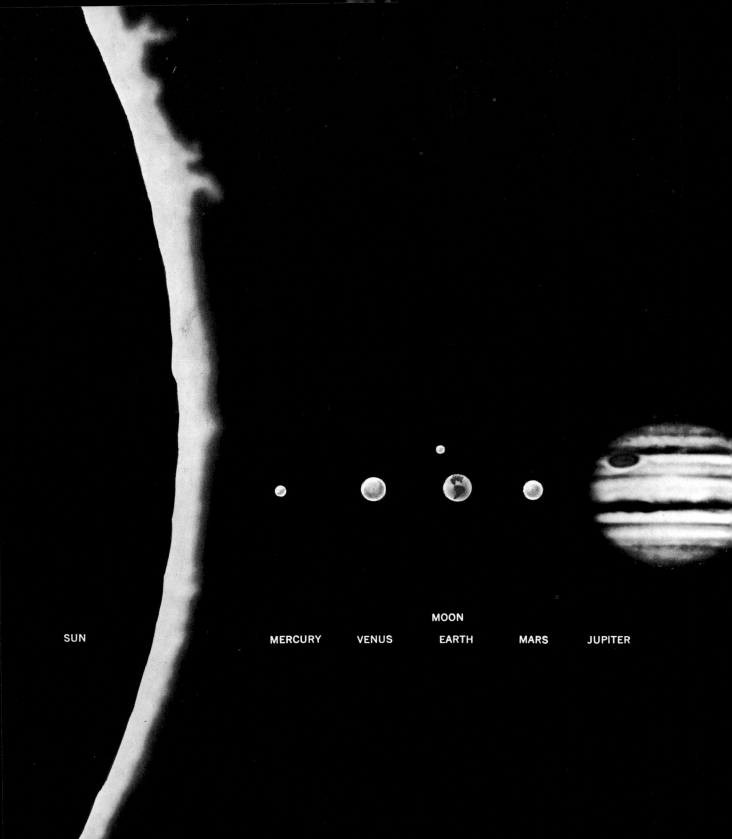

SUN  MERCURY  VENUS  MOON  EARTH  MARS  JUPITER

SATURN                                                 URANUS       NEPTUNE                PLUTO

As noted above, the giant planets are less dense than the earth and its neighbors because they contain large amounts of the lightest elements, hydrogen and helium. As a consequence, the structure of a giant planet is entirely different from that of a terrestrial planet. Probably the giant planets lack a well-defined surface distinct from the overlying atmosphere. Instead, the outer layers of a giant planet must consist of gases that become steadily more compressed as one descends deeper into the planet, but remain in the gaseous state.

In addition to their hydrogen and helium, the giant planets must contain the rocky materials, iron, and other metals that make up the bodies of the terrestrial planets. Each giant planet probably contains a sufficient abundance of earthlike materials to make up a terrestrial planet of respectable size. However, conditions in the interiors of the giant planets are so uncertain that it cannot be predicted whether these substances are collected at the centers of the giant planets in earthlike cores, or are dispersed throughout the interiors of the planets.

*The Moons of the Giant Planets.* Twenty-nine moons in all circle the giant planets. Jupiter has 12—more than any other planet—which orbit it like a solar system in miniature. Six moons of the giant planets—four around Jupiter and one each around Saturn and Neptune—are the size of the earth's moon or larger. One of these moons may serve as a base for the scientific exploration of Jupiter in the future, since direct manned reconnaissance of the largest planet is forbidden by the crushing force of its gravity.

The remaining satellites of the giant planets are very small, with diameters ranging from 70 miles down to 4 miles. These satellites resemble the asteroids in size, and some may be asteroids that were captured from their normal orbits by the gravitational forces of the giant planets.

*The Rings of Saturn.* The rings of Saturn are an extraordinary phenomenon familiar to every amateur astronomer. The four rings[2] occupy the region between 46,000 miles and 85,000 miles from the center of Saturn. The innermost ring is 9500 miles above the planet's surface. The outermost ring ends 1600 miles inside the orbit of Janus, the closest of Saturn's moons. The rings are paper-thin relative to their diameter, with estimates of their thickness ranging from 4 inches to 10 miles. Consequently, when the rings are viewed edge-on to the earth, they disappear, although when viewed face-on they are dazzlingly bright (Figure 13.8). Stars can be seen through the rings, indicating that they are not solid sheets. Probably they consist of crystals of ice and grains of ice-coated dust.

The origin of the rings lies in the gravitational force exerted by Saturn on its satellites. Since the force of gravity increases in strength with decreasing distance, the near side of a moon feels a somewhat stronger gravitational pull toward its parent planet than the far side. The stronger attraction on the near side tends to wrench the material on this side out

[2] The fourth ring was discovered in 1971.

of the body of the moon. The force of the moon's internal gravity, holding it together, resists this effect. If the moon is too close to its planet the excess pull on the near side becomes too great to be counteracted and the moon is torn apart. Each of the resulting fragments then circles in orbit around the planet as a miniature moon.

A moon must keep a minimum distance from its parent planet in order to stay intact. The minimum distance is called the *Roche limit*. A calculation shows that the rings of Saturn are inside the Roche limit for that planet. Either the rings are fragments of a moon that spiralled in too close and was torn apart; or, more likely, they are grains of material that were within the Roche limit around Saturn originally and, therefore, were prevented from collecting into a single object when the moons of Saturn were first forming.[3]

All planets and stars possess Roche limits. The Roche limit for the sun is one million miles from its center, or approximately 500,000 miles

[3] Artificial satellites orbiting the earth within the Roche limit are preserved by their rigid construction.

*Figure 13.8 Photographs showing various orientations of the rings of Saturn.*

above its surface. No planet could form within this distance. The Roche limit for the earth is 10,000 miles from its surface, well inside the orbit of the moon.

## The Asteroids

Between the orbits of Mars and Jupiter there is a gap in the distribution of the planets. We might expect to find a planetary body located outside the orbit of Mars, about three times the earth's distance from the sun; but instead we find only a large number of small bodies—planetesimals—circling in a ring. These are called *asteroids.* The largest of the known asteroids is Ceres with a diameter of 480 miles. Three other asteroids—Pallas, Vesta, and Hygiea—have diameters greater than 200 miles. The remaining asteroids—estimated to number tens of thousands—are far smaller.

Asteroids are discovered by taking a long exposure photograph of stars and noting any objects which move relative to the stars (Figure 13.9).

The orbits of several asteroids bring them dangerously close to the earth. Hermes went by at a distance of 400,000 miles in 1937, and Icarus approached within 4 million miles of the earth in 1968. The visit of Hermes was a near miss for earth inhabitants; if it ever strikes the earth—as it may in the future—the force of its impact will liberate the energy of 10 million hydrogen bombs, and may destroy a substantial fraction of the population of the earth. We know from the geological record that the land areas of the earth have not been struck by large

*Figure 13.9 Discovery of the asteroid Icarus by its motion against the background of the fixed stars.*

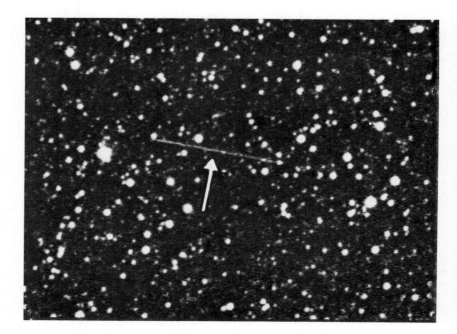

objects during the last few hundred million years. However, the first years of the earth's record have been wiped out, and during that early period such collisions may have been frequent. The great craters and circular maria of the moon show the marks of collisions with asteroids and planetesimals the size of Hermes and larger.

The unusual properties of asteroid orbits provide a clue to the origin of these bodies. Many orbits are inclined at very large angles to the plane of the ecliptic. (Figure 13.10). The orbits also tend to be highly elliptical in contrast to the near-circular orbits of the planets (Figure 13.11). The orbit of Icarus, for example, carries it to within a distance of 19 million miles from the sun and out to a distance of 180 million miles. These peculiarities can be explained by the powerful gravitational force of Jupiter. Jupiter's gravity affects all nearby objects, and occasionally pulls an asteroid out of its normal orbit, and may set it on a collision course with another asteroid. If a collision occurs, fragments of the two asteroids will leave the scene of the collision traveling through space in many

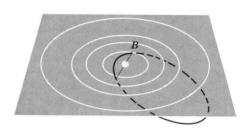

*Figure 13.10 Orbit of Icarus. Icarus is about one mile in diameter. Its orbit is inclined at 23 degrees to the ecliptic; the revolution period is 409 days; the orbital eccentricity is 0.83, and at perhelion the asteroid is only about 19 million miles from the Sun — within the orbit of Mercury.*

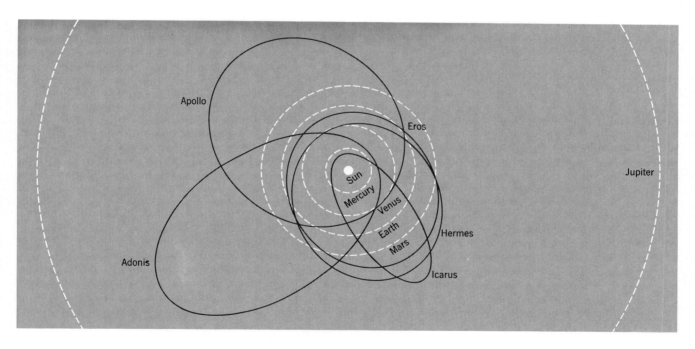

different directions and with different speeds and will fall into new orbits around the sun, including orbits that make a large angle to the plane of the ecliptic. Some may be slowed down by the collision and fall in toward the sun, as presumably was the case with Icarus, while others are hurled further out into the solar system. This collision theory would explain the great variety to be found in the asteroid orbits; it also explains the fact that they have never accumulated into a single large planetary body during the history of the solar system.

*Figure 13.11 The orbits of various asteroids.*

**357**

THE SOLAR SYSTEM

## Meteorites

Among the fragments ejected from the asteroid belt in a collision, some fall by chance into orbits crossing the orbit of the earth. It is believed that most of the meteorites that hit the earth have this origin. The examination of meteorites that survive the searing passage through the earth's atmosphere reveals that most are pieces of rock and iron with a rather complex physical and chemical history, suggesting that they were broken off from larger bodies during repeated collisions.

Most meteorites consist of rocky materials primarily, but fragments of pure iron and nickel are not uncommon. The meteorite that blasted out the Arizona meteorite crater was a large block of iron weighing approximately one million tons (Figure 13.12).

Meteorites range in size from blocks of material weighing many tons down to invisible grains of rock or iron dust called micrometeorites. If the meteorite is the size of a grain of sand or larger, it leaves a fiery trail of incandescent matter behind as it passes through the earth's atmosphere. When we see this trail in the night sky, we call it a shooting star. The great majority of the meteorites are so small that the heat of their passage through the atmosphere vaporizes them before they reach the ground. If the meteorite is the size of a basketball or larger, its entry into the atmosphere creates a spectacular sight called a fireball.

Prior to the landing on the moon, meteorites were the only samples of extraterrestrial matter available to man. Their examination in the laboratory has provided two results of great importance to students of the origin of the solar system. One relates to the age of the solar system. Measurements of the ages of meteorites by the technique of radioactive dating (Chapter 14) yield results ranging up to 4.6 billion years. The oldest rocks found on the earth are about 3.6 billion years old. Since

*Figure 13.12   An iron meteorite from the Arizona Crater.*

these 4.6-billion-year-old meteorites are the oldest objects known, their age is taken as a lower limit to the age of the solar system and, therefore, to the age of the earth. This conjecture is strengthened by the dating of the lunar rocks, which also yields ages up to 4.6 billion years (Chapter 15).

A second point of interest is the composition of meteorites. They show no signs of having been worked over by air and water erosion or subjected to a long history of chemical separation, as is the case for rocks on the surface of the earth. For this reason, they are believed to give a better indication of the original state of the materials in the solar nebula than can be obtained on the earth.

The chemical composition of one type of meteorite in particular, called the chondrite, is believed by some geologists to provide an accurate indication of the composition of the materials out of which the earth accumulated as a new planet. Chondrites are often used as a starting point in theoretical studies of the earth's history from the time of its formation.

## Comets

The close approach of a comet is one of the most spectacular sights in the heavens. The comet appears first in the telescope as a small, fuzzy, faintly luminous object. At this point it is far from the earth, but is approaching our planet on the inward leg of a long journey from the edge of the solar system. As it comes closer, solar energy warms the head of the comet and vaporizes gases that were frozen in solid crystals during the many years in which the comet was far from the sun. These gases stream out behind the comet's head. Excited to luminescence by the absorption of solar radiation, the stream of gases forms a spectacular, glowing tail, which becomes clearly visible to the naked eye as the comet nears the sun (See photograph on page 332).

Comets derive their name from the Latin *cometes*, which means "long-haired." Most comets move, in fact, in highly eccentric orbits, sometimes approaching within the orbit of Mercury, and then retreating far beyond the orbit of Pluto on the outward leg of their journeys. The most elongated cometary orbits are estimated to reach one-fifth of the way to the next nearest star. A comet in one of these orbits requires several million years to complete one circuit of the sun, and most comets are estimated to require 10,000 years or more for the round trip. The vast majority of the comets we observe will not be seen again from the earth for thousands of years. The solar system probably contains a reservoir of billions of these long-period comets, of which only a few enter the inner parts of the solar system.

Some comets have shorter periods, ranging down to 3.3 years for Enke's Comet. Most of these short-period comets come close to the orbit of Jupiter, suggesting that Jupiter's gravity has deflected them into new orbits that remain within the solar system. The most famous comet in this group is Halley's Comet, named after Edmund Halley, a friend of

Newton, who studied the records of comets dating back to 1531 and decided that several of these comets were a single body making a repeated appearance in the sky. Halley predicted that this comet would return in 1758, and it reappeared on Christmas night of that year. The last appearance of Halley's Comet was in 1910, and its next appearance is scheduled for 1986.[4]

What is a comet? The most widely held theory suggests that the nucleus of the comet is a swarm of rocky and metallic particles coated with frozen ices of water, ammonia, methane, and carbon dioxide. This collection of dirty, slushy substances resembles the mixture of ice and rocky materials in the original solar nebula out of which the planets formed (page 341). The comet nucleus is a small object, a few miles in diameter, probably with a very loose structure and a low average density. The ice crystals vaporize to form the tail on each sweep around the sun. As the comet moves out toward the edge of the solar system once more, the gases condense and freeze again, and the tail disappears (Figure 13.13).

There is evidence that the earth occasionally passes through the tail of a comet, or through a swarm of particles that have become detached from the tail of a comet that passed by previously. On some evenings, thousands of meteor trails are visible in the sky during the course of the night. These displays are known as meteor showers. The trajectories of the meteorite trails can be calculated by photographing them with special cameras. Most coincide with the orbits of known comets, suggesting that these grains of dust are cometary particles entering the atmosphere. The meteor shower occurring each year around October 21 is produced by particles formerly in the tail of Halley's Comet, while the spectacular shower that occurs about August 12 is identified with the anonymously named comet, 1862 III.

*Figure 13.13 The gradual disappearance of the tail of Halley's Comet as it recedes from the Sun.*

[4] The period of Halley's Comet varies from 74 to 79 years from one orbit to the next as the result of changes produced by Jupiter's gravity.

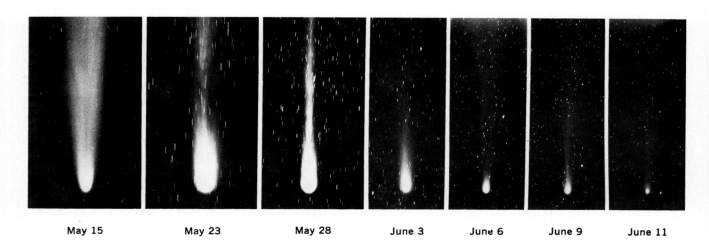

| May 15 | May 23 | May 28 | June 3 | June 6 | June 9 | June 11 |

**Questions**

1. What events that took place earlier in the history of the Galaxy dictated the eventual composition of the earth? Explain.

2. How have planets circling other stars been detected? Which theory of the origin of the solar system do these discoveries support? Why?

3. What is the difference between a star and a planet? Between the sun and Jupiter?

4. What are the principal difficulties in the condensation theory of the origin of the solar system?

5. Would you expect to find a Jupiter-sized planet at the distance of Pluto? Within the orbit of Venus? Explain your answers.

6. A combination of two motions keeps a planet in orbit around the sun. What are the two motions?

7. Using Table 13.1, list the fundamental differences between the inner planets and the outer planets. Explain the differences between the terrestrial and giant planets in terms of their formation and early history.

8. Suppose that a planet circles Barnard's Star at a distance equal to the radius of the earth's orbit. Which body—the earth or Jupiter—would you expect this planet to resemble most closely. Suppose that a planet circled Deneb at a distance equal to the orbital radius of Jupiter. Would this body resemble a giant planet or a terrestrial planet? Explain your answers.

9. On the basis of what you have read in this chapter, what would you expect with respect to the composition of the larger moons of Jupiter and Saturn? What conditions would you expect on the surface of one of these moons?

10. Give a possible explanation for the existence of the asteroid belt.

11. How does the chemical composition and age of meteorites increase our knowledge of the origin of the solar system?

12. Describe the structure of a comet. What do you think would happen if a comet struck the earth?

13. How many years remain in the lifetime of the solar system according to the theory of stellar evolution? How will it end?

# 14    The Earth

The early years of the earth's history are shrouded in mystery. Erosion by wind and running water, and the upheavals that accompany the building of continents and mountain chains—all have combined to erase the record of the earth's past. Although the earth is under our feet, and the stars are far away, writing the history of our planet has turned out to be a far more difficult task than piecing together the life story of the stars, for the skies contain stars of many different ages, and all are available for examination in our telescopes. Through the study of these young, middle-aged, and old stars we have learned the story of the red giants and white dwarfs. But planets of different ages are not available for inspection. We have no direct knowledge of the conditions that might exist on an earth-like planet during its lifetime.

Faced with this problem, students of the earth and its history—geologists—have arrived at ingenious methods for reconstructing the past history of the earth. Some of their investigations are concerned with scattered fragments of rock lying on the surface. The means for studying these rocks range from absurdly simple operations—such as tasting or feeling the

*Irazu volcano in Costa Rica.*

*Figure 14.1 Types of rock: (a) sedimentary; (b) fine-grained igneous; (c) coarse-grained igneous.*

(a)

(b)

(c)

rock—to highly complex, delicate laboratory analyses in which the rocks are taken apart almost atom by atom. To begin with, the geologist brushes the dust off, looks at the clean surface of the rock carefully, hefts it, and then hits it with a hammer. He grinds a smooth face on one of the fragments and inspects it under a low-powered microscope; then he cuts off a thin slice, grinds that down to a transparent slab one-thousandth of an inch thick and sends polarized light through the slab to see what colors are produced. Afterward he bombards another small piece with electrons and x-rays. Finally, he vaporizes the rock and studies it an atom at a time.

Much can be learned about a rock's past by looking at it. Does the rock look like it has been built up layer upon layer? If so, it was probably formed by an accumulation of sediments, perhaps by silt filtering down onto the bottom of a lake or an ocean. Such *sedimentary rocks* (Figure 14.1*a*) are common on the earth, nearly all of whose surface has been covered at one time or another by water. Is the texture of the rock not layered but homogeneous, made up of many fine crystals of different types? If so, it is an *igneous rock*—solidified lava—that came up to the surface of the earth in a molten stream from the deep interior and rapidly cooled there. Is it an igneous rock with large crystals? That means that this rock, again originating in the deep interior, has collected in a pocket under the surface of the earth and cooled and solidified there very slowly, subsequently to be exposed on the surface by a subterranean uplift or by the forces of erosion (Figure 14.1*b* and *c*).

The rocks that lie on the surface of the earth come from depths as great as 60 miles, but no greater. How does the geologist penetrate deeper into the earth to determine the materials of which it is composed and the temperature and pressure at great depths?

The problem is like that of the physician who seeks a knowledge of his patient's interior without taking a cross-sectional cut. Earthquakes provide the geologist with the diagnostic tool equivalent to the x-ray and the electrocardiogram. Vibrations set up by subterranean disturbances travel through the body of the earth at speeds that depend on the properties of the materials through which they travel. By studying the earthquake records—seismograms—from stations located at many points and by comparing the properties and arrival times of the signals from an earthquake, the geologist can deduce a great deal about the earth's interior. In this way he has learned that the center of the earth is largely liquid—a *core* of molten iron 1800 miles in radius. Within the molten core lies an inner solid core, whose atoms are forced into the solid state, in spite of the high temperature at the earth's center, by the enormous pressure of the overlying layers. Surrounding the molten outer core is a *mantle* of dense rock 2200 miles thick. The mantle is capped by a rigid *crust* of lighter rocks with an average thickness of ten miles (Figure 14.3).

The sum of the evidence gives a surprising picture of dynamic change and transformation within the earth's interior and on its surface throughout its history from the moment of its birth. The solid earth, viewed in the span of geologic time, has been the scene of violent activity that belies its seemingly static nature. The pattern of those changes has been assembled painstakingly by geologists through a combination of crude and sophisticated methods, but a key element was missing from earth science until a few short years ago. Prior to the mid-1960's, geologists could see what had happened, but they could not determine *why* it had happened. The missing element, which is connected with the concept of "continental drift," is described in the last section of this chapter.

## THE EARLY HISTORY OF THE EARTH

In Chapter 13 we described how the earth was probably formed. Small pockets of condensed material appeared in the cloud of gas and dust that circled the newborn sun in the first years of the solar system. We do not know how these condensed pockets developed, except that they may have collected under the influence of gravity in a small-scale duplication of the formation of a star. When a condensed pocket of material had grown large enough to exert a strong force of gravity, it constituted the nucleus of a planet. Other bits of material surrounding the planetary nucleus were drawn to its surface by the force of its gravity, and in a relatively short period of time the nucleus developed into a full-sized planet.

As the earth grew to its final size, the force of its gravitational attraction mounted in proportion to the mass of the accumulated material. Toward the end, the force of gravity on the earth's surface was as strong as it is today.

An extraterrestrial fragment of rock—such as a meteorite—drawn down onto the surface of the earth by gravity, crashes into the planet at a speed of about 25,000 miles an hour. The amount of energy liberated by the impact at that speed is greater—pound for pound—than the energy liberated in the explosion of TNT. If the object is substantial in size, an enormous amount of heat is created when it hits the surface. If we can bring our imaginations back in time to the period when the earth was almost fully formed, but there was still a large amount of planetary debris circling around the sun, we can conceive not only of an occasional meteorite hitting the earth, but of a heavy bombardment of rocks of all sizes raining down on the surface.

We can see that the earth must have been heavily scarred by that early bombardment. In fact, calculations indicate that large parts of the earth's outer layers could have melted as a result of the temperature rise caused by the bombardment during the final stages of its birth.

If this were the case, the surface of the earth must have been covered with red-hot lava at one point in its youth, even though it condensed initially from cold grains of rock and ice.

But the interior of the earth would have been little affected by surface bombardment, since heat travels very slowly through layers of rock that are hundreds of miles thick. Calculations indicate that it would take nearly the entire lifetime of the earth for heat to travel a distance of 250 miles through the earth's rocky interior. Two hundred and fifty miles is a small fraction of the earth's radius of 4000 miles. Thus, most of the interior of the earth must have been unaffected by the vigorous bombard-

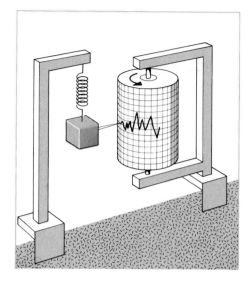

*Figure 14.2   The seismograph: a heavy weight suspended from a spring remains relatively motionless while the ground shakes beneath it during an earthquake. A pen, attached to the nearly stationary weight, traces the vibrations of the earth on a rotating drum.*

*Figure 14.3   The structure of the earth.*

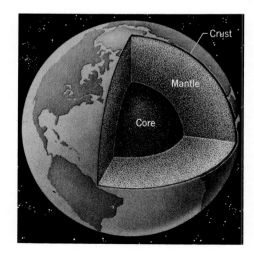

**365**

THE EARTH

ment of the surface that occurred during the final stages of its birth. The earth, when a young planet, may have had a molten surface, but its interior probably was stone-cold.

Yet today the inner 2000 miles of the earth are filled with molten iron, and the temperature at the earth's center is 6000°C or 11,000°F. This temperature is slightly higher than the temperature at the surface of the sun. By comparison, the temperature of a household oven rarely gets above 600°F., and steel furnaces are never hotter than 3000°F.

## RADIOACTIVE HEATING OF THE EARTH

What source of energy could have created the pool of molten iron that now exists in the center of the earth? The answer is believed to be that certain rare elements have raised the temperature of the earth's interior to the point where the iron—originally sprinkled throughout the interior like raisins in a fruitcake—has melted and run to the center.

The rare elements that are the source of the earth's inner heat are uranium, thorium, and potassium. These are the so-called *radioactive* elements. Each of them has the special property—unique among all elements found within the earth—of disintegrating by itself, without any external stimulus, when a sufficient amount of time has passed. In the disintegration, a piece of the nucleus of the radioactive element breaks off and is ejected at high speed, leaving behind a smaller and different nucleus than existed before. The breakup of the nucleus is called radioactive decay.

The fragment of the original nucleus that has broken off, speeding away from the scene of the radioactive decay, crashes through the surrounding atoms of the solid rock in which the radioactive element is located (Figure 14.4). Colliding with these atoms, the nuclear fragment transfers energy to them and heats the rock.

Ernest Rutherford—the discoverer of the atomic nucleus—was the first person who thought about the heating effects of the radioactive elements in the earth. He was also the first person to measure the amount of heat, and the first to realize the implications of this heating for the interior of the earth, and for the understanding of the earth's history. Rutherford measured the temperature of a small amount of radium—a radioactive element derived from uranium—and found that if the radium was carefully insulated from all other sources of heat, it steadily became warmer. The rise in temperature was the result of radioactive decays.

Rutherford calculated the heat that would be released by the radioactive decays in ordinary rock using his measurements of the heat released by the decay of radium atoms. He found that the heating effect was very small as measured by ordinary standards. The radioactive heat released inside a fist-sized piece of rock would take a century to raise the temperature of a thimbleful of water by one degree. Yet Rutherford found that the heat from these decays, accumulating inside the earth for a billion years or more, could raise the temperature of the earth by thousands of degrees.

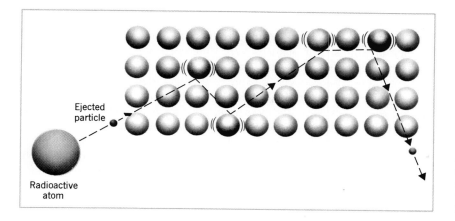

Ejected
particle

Radioactive
atom

*Figure 14.4 The heating effect of radio-
active elements. A particle ejected from a
radioactive atom strikes atoms of the surrounding rock (blue) and transfers energy
to them.*

This was sufficient to melt the solid rock interior of the earth itself.[1]

Calculations indicate that by the end of the first billion years of the earth's existence, enough radioactive heat had been released to melt the iron in the earth's interior. From that point on, the earth had approximately its present structure—a largely molten iron core extending outward from the center, surrounded by a mantle of warm, yielding, but for the most part solid rock.

## Differentiation of the Earth's Interior

After the iron melted within the earth, the release of radioactive heat continued, year after year, and the temperature of the earth continued to rise. Eventually the surrounding mantle of rock melted also. It did not melt earlier, because the temperature required to melt rock is several hundred degrees higher than the temperature required to melt iron.

The melting of the rock mantle may have happened very soon after the iron melted, or perhaps as late as several hundred million years afterward, but in any case it took place sometime during the first billion years of the earth's history. It is not clear whether the whole mantle melted at once; more probably, only a part melted at a time.

Once the rock mantle had melted, or partly melted, the only way in which it could become completely solid again would be by solidifying from the interior of the earth outward toward the surface. You might expect the opposite—that a crust would form on the surface of the molten

---

[1] The heating produced by radioactive decay is supplemented by additional heat released when the iron melts and runs to the center of the earth. As the iron flows toward the center through small channels in the rock, its flow is impeded by frictional resistance, which heats the surrounding rock. The pull of the earth's gravity on the iron is the ultimate source of this heat.

rock first, just as ice forms on the surface of a lake or a pond when the temperature falls. But water is very different from molten rock. When water freezes into ice, the ice weighs less than the water and, therefore, floats on top. That is why ice forms on the surface of a cooling body of water. Solid rock, on the other hand, is denser than liquid rock. If the entire mantle melted, and then a solid crust formed on the surface of the earth, it could not remain there. Being heavy, it must have broken up and sunk into the molten material, melting as it sank. Again and again the crust would have formed, broken up, sank into the liquid rock beneath, and melted. Final solidification could never occur at the surface; it must have started at the "bottom," that is, at the base of the mantle.

It is believed likely that only a part of the mantle was molten at one time. Therefore the picture presented above is oversimplified. The actual history of the earth's mantle probably involved the repeated melting, solidification, and remelting of local regions. The details of the process are complex, but a thorough analysis leads to the same result that one obtains from the simpler picture.

What happened next? As soon as the interior of the earth began to solidify, a complication entered the story. We have been speaking of the rocks in the interior as if they were made of a simple substance like salt or sugar. In actuality, however, the substance we have been calling "rock" is a mixture of many different compounds or minerals. Common rocks generally contain about one dozen separate minerals in varying proportions. For example, consider the rock *granite*. If you look closely at a block of granite on the face of a building, you will see that it contains many small grains of different colors and shapes. Some grains are ivory white in color. These light grains are called *orthoclase feldspar*. Other grains are almost entirely transparent except for a slight cloudiness. They are the mineral *quartz*. The granite is also apt to contain dark specks—nearly black, which are crystals of the mineral *biotite*.

Another common rock is the dark, fine-grained volcanic material called *basalt*. If you have ever seen solidified lava, you probably have seen basalt. Nearly all rocks brought back from the lunar landings are basalts. Basalt usually contains a mixture of three minerals called *pyroxene, olivine,* and *plagioclase feldspar*.

What does the complex mineral structure of rocks have to do with the story of the earth's past? The answer is that each mineral freezes out of a cooling masses of liquid rock at a different temperature. Table 14.1 lists the common rock minerals with their freezing-point temperatures, i.e., temperatures at which a molten mineral becomes solid. The minerals in this table make up 95 percent of the earth's rocks.

As the interior of the earth cooled, beginning at the base of the mantle, the minerals with the highest freezing points were first to freeze. In general, these are also the densest minerals. Consequently, they remained at the bottom of the mantle where they had solidified.

The table indicates that olivine has the highest freezing point temperature. Therefore this mineral must have been the first to crystallize out of the melt. It is also denser than the other minerals in the table. Accord-

| Name | Chemical Formula | Freezing Point, °C,[a] of Single Crystal | Density g/cm³ |
|------|------------------|------------------------------------------|---------------|
| Quartz | $SiO_2$ | 600–900 | 2.7 |
| Biotite | $K(Mg, Fe)_3 (Si_3AlO_{10})(OH, F)_2$ | 900–1000 | 2.7–3.3 |
| Hornblende | $Na, Ca_2 (Mg, Fe^{II}, Al)_5 (Si, Al)_8$ $O_{22}(OH, F)_2$ | 1060–1200 1000–1200 | 3.0–3.5 |
| Feldspar | $KAl(Si_3O_8)$ $NaAl(Si_3O_8)$ $CaAl_2Si_2O_8$ | 1100–1400 | 2.6 |
| Pyroxene | $Ca(Mg, Fe, Al)(Si_2O_6)$ $CaMg(Si_2O_6)$ | 1200–1400 | 3.2–3.6 |
| Olivine | $(Mg, Fe)_2SiO$ $Mg_2SiO_4$ | 1400 | 3.2–4.4 |

Table 14.1
**Table of Rock-Forming Minerals**

[a] Although feldspar has a relatively high melting point, its solid form is less dense than the molten form of other minerals. Therefore, as it formed in the mantle, it floated upward to the crust where it is found today.

ingly, the rocks deep within the mantle must be rich in olivine. In later stages of cooling, pyroxene and hornblende appear. The last minerals to appear, concentrated on or near the surface of the earth, would be biotite and then quartz.

*Degrees of Differentiation.* The process by which the earth was partitioned into core, mantle and crust, and various rock minerals were concentrated at different depths within the planet, is called *differentiation* by geologists (Figure 14.5).

Geologists refer to different degrees of differentiation of a planet. They consider that differentiation has proceeded to an advanced degree within the earth, for example, because of the fact that all or most of the earth's iron has collected at the center to form a nearly pure iron core, while the lightweight minerals have been very strongly concentrated at the surface in the crust.

As we will see in the next chapter, the moon appears to be a partly differentiated planet in which the process of some degree of melting and recrystallization has occurred, but not to the extent that it has occurred on the earth. If a planet has never been melted or partly melted throughout its history, it will probably have the same composition at the surface as at the center. Such a planet is called *undifferentiated*.

*Composition of the Crust.* Oxygen and silicon are the most abundant substances in the earth, and are among the most abundant substances in the Universe. They are the principal constituents of all the rock minerals mentioned above. Quartz, for example, is pure oxygen and silicon

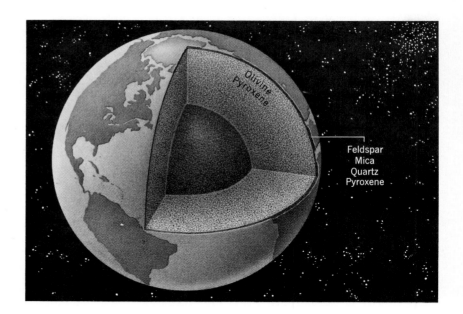

Figure 14.5 Concentrations of minerals in the earth's crust and interior.

in the proportions given by the formula $SiO_2$. Its basic structure is shown in Figure 14.6. A tetrahedron—called the silica tetrahedron—is formed by four oxygen atoms enclosing the relatively small silicon atom at the center. Each oxygen atom is shared by two silicon atoms in neighboring tetrahedrons, hence the chemical formula $SiO_2$ instead of $SiO_4$.

The silica tetrahedron is the basic building block for most of the minerals in the earth's mantle and crust. Quartz consists of pure silica, and other minerals are composed of silica tetrahedra linked by regularly spaced atoms of iron, magnesium, aluminum, sodium, calcium, or potassium. For example, Figure 14.7 shows the structure of olivine.

The smaller the atom that serves as the link between adjacent tetrahedra, the tighter and more compact is the resultant structure. Olivine, for example, consists of silica tetrahedra linked by atoms of iron or magnesium. Being small, the iron atom or magnesium atom fits very easily into the spaces between the other atoms that make up the crystal struc-

Figure 14.6 The silica tetrahedron.

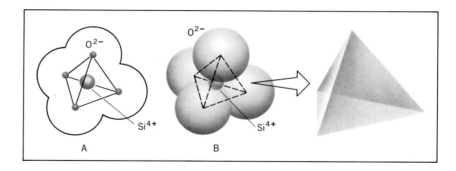

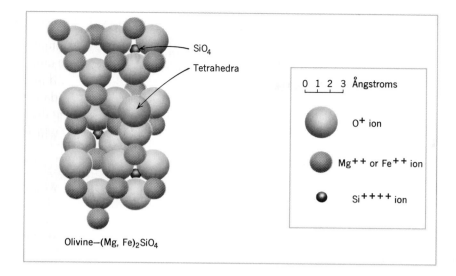

SiO₄

Tetrahedra

0 1 2 3 Ångstroms

$O^+$ ion

$Mg^{++}$ or $Fe^{++}$ ion

$Si^{++++}$ ion

Olivine—$(Mg, Fe)_2SiO_4$

*Figure 14.7   The structure of Olivine.*

ture of olivine. The result is a mineral with a very compact and, therefore, a very dense structure. This is the reason why olivine tends to settle to the bottom of the melt when it crystallizes out of molten rock.

The presence of iron and magnesium atoms in olivine also explains why this mineral is the first to crystallize out of a cooling mass of molten rock, that is, why it has a higher freezing-point temperature than other minerals. Because the iron and magnesium atoms are small, in minerals containing them other atoms come into closer contact. The electrical forces of attraction that tie atoms together to make a crystal are stronger when the atoms are closer together. Suppose now we consider what happens as a molten mass of rock cools down. Collisions occur continually between neighboring atoms in the molten rock, breaking up the bonds that attract one atom to another, and preventing solid crystals from forming. At very high temperatures, these collisions are violent enough to break the bonds of all minerals, including the very strong bonds that tend to tie together the atoms that make up a crystal of olivine.

However, as the temperature drops the violence of the collisions diminishes. When the temperature falls to fourteen hundred degrees, the violence of the collisions is sufficiently diminished so that the atoms that make up a crystal of olivine can lock into place and stay in place in spite of continuing collisions with their neighbors. But at this temperature the collisions are still violent enough to break up the other minerals—with a looser, less compact, and less tightly bound structure—that start to form in the molten mass. These minerals cannot solidify or crystallize out of the melt until the temperature decreases still further.

Thus, as the earth solidifies from the inside out, the small atoms are locked into the structures of the minerals that appear at the base of the mantle, while the large atoms are forced successively upward, layer by layer, because they do not fit into the spaces between other atoms. Finally, at the top there is a residue of the low-density minerals containing larger atoms, which float on top of the denser rocks below. Eventually this layer of low-density minerals solidifies to form the rocks of the crust.

The elements with large atomic ions, having been forced upward into the crust, exist there in great abundance, far in excess of the abundance that they would normally have if they were distributed according to their average proportion as in the original cloud of gas and dust out of which the earth condensed.

Differentiation enriches the crust in all elements having large-sized atomic ions. They include the elements mentioned on page 341, which are intrinsically abundant in the cosmos and, therefore, make up a large fraction of the rock minerals in the crust. They also include elements such as lead and mercury, which are rare in the cosmos, and therefore relatively rare in the crust, but are to be found there in much greater abundance than their average abundance in the earth.

Finally, all the radioactive substances—radium, thorium, potassium, and uranium—have atoms with very large diameters. These radioactive substances are to be found in far greater abundance in the crust than their average concentration in the earth as a whole. Differentiation is estimated to have removed more than half of the radioactive elements from the interior of the earth and to have concentrated them in the relatively thin crustal layer during the course of the earth's history.

If the radioactive substances were still distributed throughout the mantle, their radioactive heat would be sufficient to keep a large part of the mantle in a molten condition. Probably a solid and relatively permanent crust could not exist in that case. In summary, heat released by radioactive substances led to differentiation in the mantle, and the differentiation led to the removal of radioactive substances from the mantle, and hence the removal of the heat source. Differentiation has thus introduced inherent stability into the temperature history of the earth.

## The Floating Crust

The lightweight rocks that form the crust of the earth are the familiar *granites* that are found on the continents, and the somewhat denser *basalts* that make up the rocks underlying the oceans. Like the granites, basalts are also found in substantial amounts on the continents. Granites and basalts together form the light crust that floats on the denser mantle underneath.

The word "floats" is used with care. The rocks of the earth's interior— while solid—are nonetheless warm enough so that they have a kind of plasticity. They yield under pressures, provided that the pressures are applied for a long enough time. They are very much like silly putty, which seems as hard as steel when hit sharply, but yields and flows like a viscous liquid under a steadily applied weight.

As the photographs show (Figure 14.8), a slab of silly putty yields in 30 minutes under the pressure of a weight placed on top. How much time elapses before the solid rocks of the earth's interior yield under the

the weight of a mountain on the surface? The answer is approximately one million years. If a large mass — the size of a mountain — is placed on the surface of the earth, during the course of a very long time the underlying rock of the mantle yields under its weight, and slowly but steadily the mountain sinks into the earth's interior. But several million years must elapse before this occurs. During appreciably shorter periods of time — such as a year, a century, or even ten centuries — very little happens. The atoms of the mantle rock remain locked in fixed positions, each atom bound to its neighbor by an electrical force.

But if solid rock is subjected to a large force, now and then a small layer of atoms somewhere within the rock will slip over an adjoining layer of atoms as a consequence of the pressure on its surface. The sliding motion of one layer of atoms over another occurs only in scattered places in the solid, and only at rare intervals. No change can be seen in the solid rock if you look at it for a short time. But over a sufficiently long period, the accumulation of many tiny displacements adds up to a "flow" of one part of this seemingly rigid, solid body over another part.

The slow movement of a solid body — such as rock — under heavy pressure is called *creep*. The phenomenon of creep is responsible for the fact that the solid rocks of the earth's interior yield and flow under the weight of mountain ranges on the surface.

Mountains are made of crustal rocks — mostly granite — which are lighter than the rocks of the mantle as a consequence of differentiation. The light rocks of the crust are buoyant; they float in the yielding mantle like a block of wood in a tank of water.

Only a small part of a floating block of wood rides above the surface of

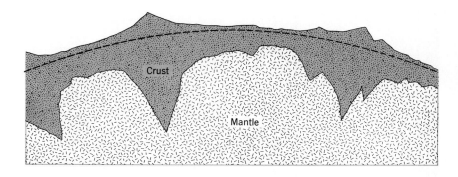

Figure 14.9 *The roots of the continents.*

the water. Similarly, when you see a mountain range—such as the Rockies or the Himalayas—you are only viewing a small part of the complete mass of rock. The remainder is "submerged" below ground. The Himalayas extend upward 4 to 5 miles above sea level and downward about twenty-five miles. Continents also float on the mantle and have deeply submerged roots that extend far into the underlying rock. In fact, the entire crust of the earth floats on the mantle—including the rocks on the floor of the ocean—with the larger part submerged and hidden from view (Figure 14.9).

## CONTINENTAL DRIFT

In recent years evidence has accumulated which makes it indisputable that the continents have not always been located in the positions they now occupy on the globe. Antarctica was once located in a pleasant climate far from the South Pole; North America was once joined to Europe; South America was once joined to Africa; and all these continental masses were far from their present locations.

### The Zone of Weakness

At first thought it seems that these results must be incorrect. How can a continent slide about like a cake of soap in the bathtub? The answer is that a *zone of weakness* exists in the earth's interior at a depth of some 60 miles, well below the deepest continental roots. At this particular depth the mantle rocks—while still below the melting-point temperature—come nearer to melting than at any other place in the interior of the earth. The rocks at a depth of 60 miles are still solid, but only barely so. They are so close to melting that they are as soft as warm butter. The mantle above this deep layer of relatively soft and yielding rock is broken up into a number of slabs, called *plates* by earth scientists. The term conveys a thinness that seems inconsistent with a slab of rock 60 miles thick. However, the plates average thousands of miles in size, or

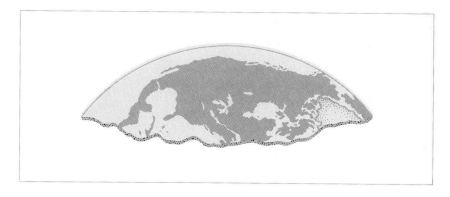

*Figure 14.10   The dimensions of a plate.*

more than twenty times their thickness (Figure 14.10).

If a continent is embedded in a particular plate, the continent moves along with the plate as it slides over the zone of weakness. The movement is slow—usually no more than one inch per year—but over an appreciable period of time in the earth's history (for example, one hundred million years or more) these slow movements of the plates add up to displacements of thousands of miles. They are the explanation of the drifting continents.

## A Map of the Earth's Plates

Between 1965 and 1970, geologists succeeded in mapping the boundaries of most of the great plates into which the earth's surface is divided. The boundaries of the plates are shown in Figure 14.11. Arrows drawn on the plates indicate the direction of the movements. All movements are relative to the Eurasian landmass, which is regarded as fixed in this map. Locations of all earthquakes occurring during the last 10 years are also marked by small circles. The zones of intensive earthquake activity played a major role in defining the plate boundaries. The most striking geological consequences of the map are explained below.

*The African Plate.* In some parts of the globe, two plates appear to be colliding head-on, crumpling the material of the earth's crust in great folds. The map shows that a great plate containing the African continent and part of the Atlantic Ocean is moving northward and plowing into a second plate containing the Eurasian landmass. The Alps are giant wrinkles in the edge of the Eurasian landmass produced by collision between the two plates at the rim of the Mediterranean. The Mediterranean itself is gradually disappearing as the northern movement of the African plate closes the gap between the two continents. The earthquakes that plague Greece and Turkey, of which the most recent occurred in Turkey in 1971, are signs of the subterranean adjustments occurring as one mammoth mass of rock pushes its way into another.

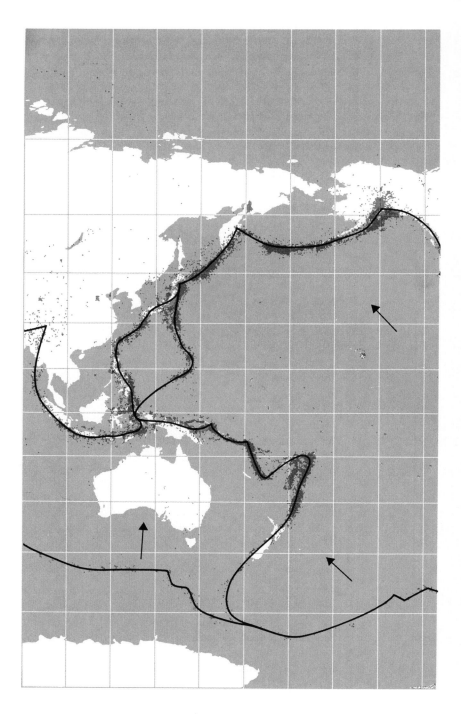

*Figure 14.11 The boundaries of the earth's plates. Arrows indicate plate movements relative to the Eurasian landmass. Shading indicates zones of earthquake occurrence.*

*The India-Australia Plate.* A crustal block in the Indian and Pacific oceans, including the subcontinent of India and the continent of Australia, is currently moving to the north, again at about the rate of one inch a

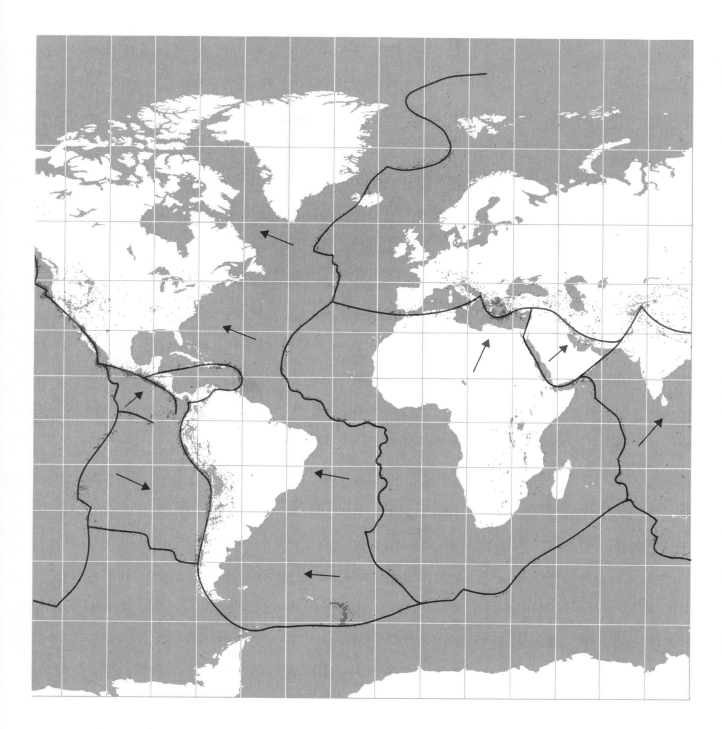

year, crumpling up the lower part of the Eurasian landmass and creating the Himalayan Mountain system. The map indicates a zone of earthquake activity in Asia near the plate boundary, produced again by the subterranean impact of the collision.

*The South American Plate.* The westward movement of the plate containing the South American continent is clearly indicated on the map. The western edge of this plate also marks the western coast of South America. When a plate containing a continental land mass on its leading edge collides with a plate underlying the ocean, the continental plate rides over the ocean plate. Deep-seated earthquakes and intensive volcanic activity occur in Peru and Chile as the adjoining plate underlying the southeastern Pacific Ocean is thrust downward into the interior of the earth, beneath the advancing edge of the South American plate. The edge of the South American plate, forced upward, became the Andes Mountain Range.

*The Pacific Plate.* In some parts of the globe, the edge of one plate slides past another like a ship scraping a pier. A plate underlying the Pacific Ocean basin is moving slowly upward to the northwest, scraping past a plate that includes most of the North American continent. In this case the movement is about 1.5 inches per year. Over the course of a century or so, the eastern rim of the Pacific plate — carrying Los Angeles — will move about 15 feet to the north. San Francisco lies on the other side of the plate boundary. At the present rate, Los Angeles will enter the suburbs of San Francisco in ten million years. The San Andreas Fault marks the boundary of the plate running up the California coastline. The motion of the Pacific plate slowly bends the rocks of the crust where they run across the San Andreas Fault. When the displacement across the fault reaches about 15 feet, the crust breaks in two and the broken ends vibrate, causing earthquake tremors. The crust last snapped in 1906, causing the San Francisco earthquake and fire.

At its northern edges, the Pacific Ocean plate collides with the Eurasian plate near the Aleutian Islands. The collision has forced the Pacific Ocean plate downward into the interior of the earth creating an intensive zone of earthquakes, and volcanic eruptions. The Aleutian island chain is the accumulation of lava produced by the eruptions — called an *island arc*.

## Evidence For Continental Drift

The bulging Brazilian coastline fits nicely into the hollow of the Ivory Coast in West Africa (Figure 14.12). This coincidence provided the first suggestion of drifting continents. Impressed by the agreement between the coastlines, a German meteorologist, Alfred Wegener, put forward the suggestion in 1910 that South America and Africa had once been part of a single land mass. According to Wegener, Africa and South America split apart about 150 million years ago. Since the matching coastlines are about 3000 miles apart, the two continents must have drifted away from each other at an average speed of one inch per year. A speed of one inch per year became accepted as the standard velocity for drifting continents in the Wegener theory.

Wegener's proposal was greeted with scorn by most of the world's

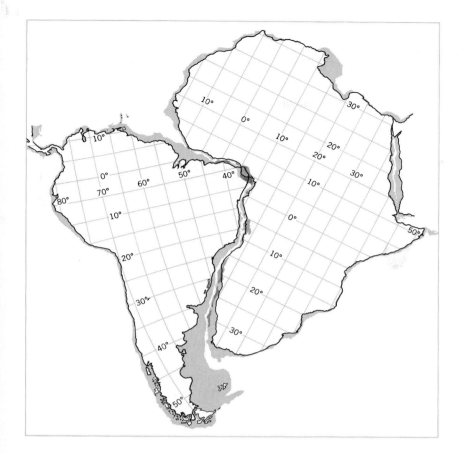

*Figure 14.12  The fit of the African and South American coastlines.*

geologists, although he pointed to other evidence, such as similarities in fossil plants and animals discovered on the two continents. In spite of circumstantial evidence accumulating in its favor, the theory of continental drift continued to be greeted with skepticism.

Interest in continental drift was revived after World War II by a series of geophysical discoveries relating to the floor of the Atlantic Ocean. Each discovery by itself seemed to have little bearing on the Wegener theory until 1965 and 1966, when everything clicked into place and it became clear to geologists everywhere that the theory of continental drift—long ignored or dismissed—was valid.

*The Mid-Atlantic Ridge.* The first important discovery was made by scientists of the Lamont-Doherty geological observatory, who measured the topography of the ocean bottom during a series of cruises of the oceanographic research vessel *Vema.* They found that an underwater chain of mountains ran along the bottom of the Atlantic Ocean from north of Scandinavia to the latitude of Cape Horn at the tip of South America. The subterranean mountain chain was a nearly continuous ridge rather than a series of separate peaks. Because it ran along the middle of the ocean floor, dividing the Atlantic into two nearly equal

Figure 14.13 Map of the Mid-Atlantic Ridge.

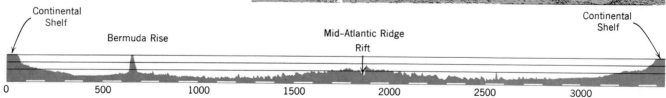

Figure 14.14 Profile across the North Atlantic from Cape Henry to Rio d'Ora. The horizontal scale is in nautical miles and the vertical scale in thousands of fathoms (after Heezen et al., 1959).

parts, it became known as the *Mid-Atlantic Ridge*. A large crack called the *Rift* ran the length of the ridge in its center (Figures 14.13 and 14.14).

During the late 1950's and 1960's, geophysicists carried out intensive studies of submarine earthquakes in the Atlantic Ocean and found that the great majority of these earthquakes were located on or near the Mid-Atlantic Ridge. During the same period, evidence was found of extensive volcanic activity on the Mid-Atlantic Ridge. Finally, measurements of the flow of heat through the floor of the Atlantic Ocean revealed an anomalously large amount of heat emerging from the Central Rift in the ridge.

The flow of heat through the rift suggested a crack in the earth's crust at that point, connecting the surface to the warm interior of the planet. With this thought born, all the facts regarding the Mid-Atlantic Ridge began to fit together. Suppose a large amount of heat appeared in the rift because hot, molten rock was emerging through this crack in the earth's crust and flowing out to either side along the ocean bottom. The material would create fresh rocks continually on the floor of the ocean as it came up from the interior.

The theory could be tested by measuring the ages of the ocean-bottom rocks in the Atlantic. If the ocean floor were continually renewed by fresh lava flowing out of the rift, the ocean-bottom rocks in the vicinity of the ridge should be younger than rocks at a distance from the ridge. The rocks near the continental boundaries should be the oldest rocks in the Atlantic Ocean.

The ages were measured by collecting samples of ocean-bottom sediments and by determining the ages of fossil animals in the deepest layer of sediments lying directly on the ocean floor. The results fully confirmed the theory: the rocks making up the floor of the Atlantic Ocean were very young in the neighborhood of the Mid-Atlantic Ridge, and they became progressively older as the distance from the Mid-Atlantic Ridge increased.

Moreover, no rocks older than 150 million years were found anywhere in the Atlantic between the South American and African continents, indicating that this part of the ocean did not exist prior to that time. Those rocks that were as old as 150 million years were all located very close to the continental shorelines.

No proof could be more convincing. Africa and South America had been a single landmass with no ocean between them until 150 million years ago, when a buried crack—the foreunner of the rift—appeared in the crust, and the landmass broke in two.


## Gondwanaland

Accumulating evidence indicated all the earth's continents—and not only South America and Africa—were once collected into two supercontinents. One, called *Gondwanaland,* was located largely in the Southern Hemisphere. It contained in a single mass the blocks of rock that subsequently broke off to form the separate continents of Africa, India, South America, Australia, and Antarctica. The other, called *Laurasia,* and located largely in the Northern Hemisphere, contained the blocks that broke off subsequently and drifted apart to become the continents of North America, Eurasia, and Greenland.

Painstaking correlation of a mass of items of information has produced

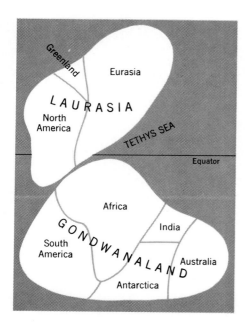

Figure 14.15 *Gondwanaland and Laurasia.*

reconstructions of the supercontinents of Gondwanaland and Laurasia (Figure 14.15). The solid lines in the diagram show the boundaries along which the supercontinents came apart into the separate pieces that form the continents as they are today.

A body of water called the Tethys Sea originally separated the supercontinents. When Gondwanaland and Laurasia broke up, the fragment containing Africa and India moved northward, pushing up into Eurasia and closing up the Tethys Sea. The northward movement of India into Eurasia compressed and buckled the sediments that formerly covered the bottom of the Tethys Sea. The folded layers of ocean-bottom sediment became the Himalayas. Marine fossils were found on the summit of Mount Everest during the first ascent, lifted to the roof of the world from the bottom of the Tethys Sea by the events that followed the breakup of Gondwanaland.

## What Moves the Continents?

With many features of the earth's surface and history explained, one mystery still confronts the geologist: What forces move the continents? As yet, no one knows. One theory proposes that the warm rocks of the mantle move up to the surface of the earth from the deep interior and down again, in steadily circulating currents of solid rock. Similar currents—called convection currents—are set up in any container of fluid when it is heated from below. In the case of the earth, the moving material is a solid rather than a fluid, and the motions are correspondingly slow, with each cycle lasting many millions of years, but the mechanism is the same.

The convection theory proposes that underneath the Mid-Atlantic

Figure 14.16 *The intrusion of magma into the edges of the plates.*

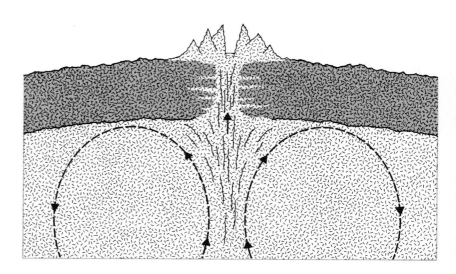

Ridge a current of ascending, warm rock—molten or nearly molten—
rises to the surface, forcing molten rock into the narrow space between
two adjacent plates, and pushing apart these plates—containing Africa
and South America—at the rate of one inch per year (Figure 14.16). The
plates tend to be enlarged steadily by the addition of molten rock forced
into their edges. Since the surface of the planet is constant in area, the
growth in the size of each crustal plate must be balanced by a consump-
tion of crust at its far boundary, where collisions occur with adjacent
plates.

In this way, through the separation of plates along some boundaries,
and collisions between plates along others, the theory explains the volcanism, mountain building, and movement of continents that continuously transform the surface of the earth.

### Questions

1. Describe the formation of the earth.
2. Why is there no evidence on the earth's surface of its early years?
3. What was the major source of the heat that melted the interior of the
   earth? What element melted first? Where did this element concentrate?
4. Why did some minerals begin to crystalize deep in the interior of the
   earth, while others did not do so until the final crust formed? Why is
   uranium, a heavy element, concentrated in the earth's crust instead
   of sinking with iron to the earth's core?
5. Describe the phenomenon of rock creep.
6. What is the zone of weakness? Why does it exist?
7. Explain why earthquakes and volcanoes are confined to long, narrow
   belts.
8. If a two-billion-year-old bedrock were found in the Atlantic Ocean
   floor, what impact would such a find have on the continental drift
   theory? Why?
9. What strikes you as the most significant evidence for continental
   drift? Why?
10. Mountains are generally formed at the edges of plates in collisions
    between adjacent plates. What explanation can you suggest for the
    origin of a mountain range such as the Urals, which are located in
    the middle of the Eurasian landmass far from any existing plate?

# 15    The Moon

When Galileo, the first man to look at the moon through a telescope, turned his primitive instrument on that body in 1609, he saw large, dark areas resembling the earth's oceans, and mountainous light-colored areas that seemed to resemble the continents. This pattern of light and dark regions, visible to the naked eye, makes up the face of the man-in-the-moon (Figure 15.1, p. 386). Galileo thought the dark areas were actually oceans, and called them *maria,* or seas. The light-colored, mountainous regions came to be known as the lunar *highlands*.

Today we know that these similarities to the surface of the earth are illusory. The lunar seas contain no water; no storms rage across the dark plains; no streams flow down from the highlands. And the lunar highlands are not similar in any way to the earth's continents; they do not resemble continental rocks chemically, and the forces that created the lunar mountains were entirely different from the forces that thrust up the great mountain ranges of the earth.

Moreover, a casual inspection of the moon through a telescope reveals that the texture of the moon's surface is completely different from that

*Geologist Jack Schmitt studies a house-sized rock during the Apollo 17 landing.*

of the earth. Photographs of the moon taken through a large telescope show that both the maria and the highlands are pitted by innumerable craters of all sizes. Most of these craters have been produced by the impact of meteorites that have been raining down on the moon's surface for billions of years. Many craters are circled by ramparts ranging up to 10 thousand feet in height. Some of these ramparts must be more than one billion years old, yet photographs taken with a telescope clearly indicate that they have been preserved almost unchanged, with little of the original material worn away (Figure 15.2).

Meteorites have collided with the earth throughout its history, just as they have collided with the moon, and they have produced similar craters; but all traces of the older craters are gone. Only the scars of the most recent collisions, such as the Arizona meteorite crater, formed about 30,000 years ago, are still visible on the earth (Figure 15.3, page 388).

Air and water—the elements that make our planet livable—have worn

*Figure 15.1 (opposite)  Photo of the moon taken from the earth, made by joining two half-moon photos together to provide maximum contrast.*

*Figure 15.2  The ancient surface of the moon.*

*Figure 15.3    The Arizona meteorite crater.*

down the oldest rocks and washed away their remains into the oceans, while the movement of the plates, and associated mountain-building activity and volcanic eruptions, have churned the surface and flooded it repeatedly with fresh lava. These natural forces have entirely removed the materials that lay on the earth's surface when it was first formed. But on the moon there are no oceans and atmosphere to destroy the surface, and there is little or none of the plate movement and mountain-building activity that rapidly change the face of the earth. Over large areas, the materials of the moon's surface are as well preserved as if they had been in cold storage.

## THE LUNAR SURFACE

Pictures of the moon taken by spacecraft provide further proof that the surface of the moon changes very slowly. Many craters that had never been seen before in photographs taken with telescopes on the earth were visible, ranging in size from a few feet up to hundreds of feet (Figure 15.4). These small craters must have existed on the earth as well, but were wiped out almost immediately by the wearing effect of winds and running water. The Arizona Crater probably will last no longer than 10 million years, which is a blink of an eye in the scale of geological time. On the moon the shallow footprints of the Apollo astronauts, six inches deep, will last at least that long (Figure 15.5).

### The Small Rate of Lunar Erosion

Why should the footprints of the astronauts not last forever? The principal force of erosion on the earth is running water; why should the

Figure 15.4  Small craters on the moon. The crater in the center is 500 feet in diameter.

Figure 15.5  Neil Armstrong's footprint.

moon, lacking water and even an atmosphere, suffer any erosion at all?

A part of the explanation is connected with the very thinness of the moon's air, which allows a continuous hail of extremely small meteorites—called *micrometeorites*—to reach the surface of the moon. Micrometeorites bombard both the earth and the moon at all times, but in the case of our planet they are burned up in the outer layers of the atmosphere and never reach the earth's surface.

These tiny grains of rock and metal contribute appreciably to the erosion on the moon because, although very small, they are present in enormous numbers, and collide with the moon at very high speeds. The sizes of typical micrometeorites range from one ten-thousandth to one-thousandth of an inch and they travel at speeds ranging up to 70,000 miles per hour. A micrometeorite moving at 70,000 miles an hour possesses one hundred times as much energy as an equivalent mass of TNT, and it can do one hundred times as much damage. Figure 15.6 shows a pit blasted out of an iron meteorite—picked up on the surface of the moon by the Apollo 11 crew—by a microscopic grain of meteoritic matter one ten-thousandth of an inch in diameter.

The moon's surface is continually fragmented and churned by the impact of these tiny particles, as well as the larger meteorites in the range of sizes that also reach the surface of the earth. Whether the meteorite is small or large, the effect of its impact is to pulverize the surface of the

*Figure 15.6   A lunar micrometeorite crater one-hundredth of an inch in diameter.*

moon, ejecting particles of rock in a spray of fine dust, and, in the case of large meteorites, rock fragments ranging up to 30 or 40 feet in size.

The presence of a substantial layer of rock dust on the moon had long been predicted, and the confirmation of its existence in the first Apollo landing was not a surprise. However, one of the major surprises of the landings did develop when the moon dust was first examined under a microscope. A substantial part of the dust particles consisted of small glass beads, apparently formed when meteorite impacts melted a part of the surface and threw out a spray of molten droplets. Being very small — a few thousandths of an inch in diameter — the droplets cooled rapidly to form glassy spheres, rather than irregular chunks of lavalike material. The photograph of a sample of dust (Figure 15.7) shows several glass beads mixed with irregular fragments of lunar rock.

The presence of the glass beads explains Armstrong's comment, radioed back from Tranquility Base, "These rocks are rather slippery." Although he did not know it, he was skating on a surface of ball bearings.

*Meteorite Erosion.* Each time a meteorite hits the surface, it disturbs the layer of rock dust created by previous impacts, and as the dust particles are shifted about under the succession of impacts, they always tend to slide downhill, filling in all the depressions on the surface. Also, the continuing bombardment steadily wears away the high points of the lunar surface. Thus, the crater edges become steadily more rounded, and the craters themselves are gradually filled in.

The evidence of lunar erosion can be seen clearly in Figure 15.4. This photograph was taken by a lunar satellite orbiting the moon. It shows an area of 100 acres located in the Ocean of Storms near the western limb of the moon. The crater in the middle of the area is 500 feet in diameter. It looks freshly made and probably was formed recently, within the last 200 million years, by the impact of a meteorite about 10 feet in diameter and weighing 500 tons. The freshness of the crater is demonstrated by the sharpness of the edges of the crater wall, and by the fact that the surrounding terrain is still littered with blocks of rock ranging up to 10 or 15 feet in size.

Two craters, indicated by arrows at the right of the fresh crater, demonstrate the effects of lunar erosion by micrometeorites. Their edges are rounded and the craters are partly filled in. These craters are clearly older than the fresh crater. They were formed by a meteorite impact approximately one billion years ago.

A study of many lunar photographs shows that frequently small craters are worn away and filled in, but large craters never are entirely obliterated. The dividing line between the two groups of craters occurs at a crater size of 200 feet and a crater depth of 50 feet. That is, in the 4.6-billion-year history of the moon, meteorite and micrometeorite bombardment has moved the top 50 feet of the moon's crust from place to place, wearing down the high points and filling in the hollows in the moon's surface. Fifty feet of erosion in 4.6 billion years represents a rate of erosion one ten-thousandth of the erosion rate on the earth.

*The Unchanging Surface of the Moon.* **We** referred above to events

*Figure 15.7   Glass beads in moon dust.*

occurring on the moon 200 million years ago as "recent" events. They are recent in the sense that they occurred only a short time ago in comparison to the age of the moon and the age of the solar system. But consider how the surface of the earth has been transformed in the last 200 million years. Two hundred million years ago the continents of the earth were located in entirely different places than they are today: Africa and South America probably were joined in a supercontinent; the Rocky Mountains, the Alps, and the Himalayas did not exist, the Appalachians were just being formed; and the eastern seaboard and most of the southwest region of the United States were at the bottom of a shallow sea. Yet the moon has scarcely changed at all in this period. Fragments of rock, hurled out of the fresh crater in Figure 15.4 when it was formed, still lie on the surface, precisely where they fell 200 million years ago.

## THE MOON PRIOR TO APOLLO

Before Apollo, science was dominated by two views of the moon. According to one, the moon was formed cold and remained cold; according to the other, the moon warmed up gradually, became hot, and remained hot. The first view found the moon to be interesting because it was *unlike* the earth, and might hold a record of the beginning of the solar system. The second view held the moon to be interesting because it was *like* the earth, and would provide an illuminating comparison with our planet.

Which view is correct? The evidence available prior to Apollo was contradictory. There were clear signs that the moon had been the scene of vigorous geological activity, including volcanism and lava-flooding. There were equally clear signs that the moon had been cold, rigid, and geologically lifeless throughout most of its history.

### Evidence for a Geologically Active Moon

Many signs of volcanic activity are evident on the face of the moon. The Hyginus Rille, a huge crack in the surface of the moon about 100 miles long and 2 miles wide, looks like the surface manifestation of a deep-seated rupture within the moon's body (Figure 15.8a). An overhead view reveals a row of craters inside the rille spaced along much of its length (Figure 15.8b). They cannot be the results of meteorite impacts, which would produce a random pattern of craters in the area. The Hyginus Rille craters must be volcanic.

The Marius Hills in the Ocean of Storms are another sign of lunar volcanism (Figure 15.9, page 394). These mounds look like volcanic islands on the earth, with the size and shape of accumulations of lava built up over

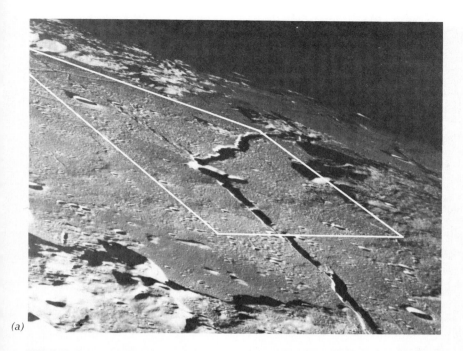

(a)

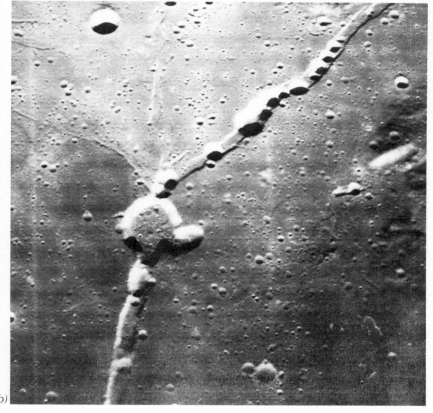

(b)

Figure 15.8  The Hyginus Rille: (a) oblique
view; (b) overhead view.

long periods of time by repeated volcanic eruptions, accompanied by the outpouring of molten rock. The Hawaiian Islands—a volcanic chain built up from the floor of the Pacific Ocean by lava flows—would present a similar appearance to the Marius Hills if the Pacific basin were empty.

The dark, irregular areas on the moon, e.g., the Sea of Tranquility and Ocean of Storms (Figure 15.12, page 397), immediately strike the eye as flows of lava, presumably from a molten or near-molten interior. These regions resemble large lava flows on the earth such as the Columbia River plateau or the Deccan lava field in India.

On the earth, such large beds of lava, covering areas of millions of square miles, and with a maximum thickness of thousands of feet, are formed as a result of repeated flows of lava through multiple cracks or fissures in the crust, each flow spreading out over a portion of the field with a thickness of 10 to 20 feet. The lava field builds up steadily to its final extent and thickness over the course of tens of millions of years. The Apollo 15 exploration along the edge of Hadley Rille—a winding valley in the moon's surface superficially resembling the Grand Canyon— revealed that similar events have occurred on the moon during its history. The Hadley Rille (Figure 15.10) was formed by an unknown process which cut through the moon's crust like a scalpel, exposing

*Figure 15.9 Opposite, The Marius Hills: mounds of lava.*

*Figure 15.10 The Hadley Rille.*

numerous layers that seem, as on the earth, to be composed of successive flows of solidified lava, each some tens of feet in thickness, accumulating to a total depth of a thousand feet (Figure 15.13, page 402).

## Evidence for a Geologically Lifeless Moon

The evidence for lunar volcanism supports scientists who argue that the moon's interior was molten at some time in the past and has been the scene of extensive volcanic eruptions that have covered its surface with floods of lava.

According to this evidence, the moon is a geologically active planet similar to the earth. If that is the case, the principal difference between the two planets is the absence of air and water on the moon, and the abundance of these elements on the earth.

Suppose the earth also lacked air and water; would it look like the moon? Would the most striking feature of its surface be the multitude of meteorite craters that cover the moon?

Undoubtedly more craters would be visible than we can see today, but the earth still would not look like the moon, for many earth craters in certain regions would be distorted from their original circular shape as a result of the continuous crumpling, folding, and fracturing of the surface by the movements of the great crustal plates (Figure 15.11). If the moon were geologically active, it too should have its San Andreas faults breaking circular craters in two. But nowhere on the face of the moon has a photograph ever revealed a broken and displaced crater. The lacework of millions of nearly perfect craters covering the moon's face suggests that it has not undergone much deformation.

*Mascons.* Other evidence, subtler than the evidence provided by the visual appearance of the moon, also suggests that today the moon is relatively cold and geologically inactive. In 1968 scientists at the Jet Propulsion Laboratory noticed that photograph reconnaissance spacecraft were being pulled away from their calculated flight paths in unexpected ways as they circled the moon. These spacecraft—called lunar orbiters—had been placed in orbit around the moon to map its surface in detail in preparation for the lunar landings to follow. The lunar orbiters were being pushed and pulled by excess gravitational forces whenever they passed over one of the circular maria. The extra gravitational force apparently derived from large massive bodies buried in the center of each circular sea. Figure 15.12 shows the location of the buried masses.

The mysterious buried masses came to be known as *mascons*. Some lunar geologists believe that the mascons are the remains of huge meteorites—the size of asteroids or small planets—that crashed into the moon's surface early in its history, melting material and creating each of the circular seas. Other geologists believe that they are large shallow plugs of relatively dense lava that welled up from the moon's interior,

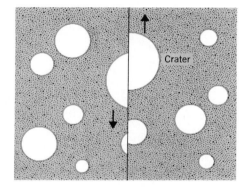

*Figure 15.11   Displacement of craters by a geological fault.*

*Figure 15.12 (opposite)   Locations of principal lunar mascons.*

SEA OF
SHOWERS

SEA OF
SERENITY

SEA
OF
CRISIS

SEA OF
TRANQUILLITY

OCEAN OF
STORMS

SEA OF
FERTILITY

SEA OF
CLOUDS

and the Apollo findings seem to support this view. Whatever the origin of the mascons was, their scientific significance is very clear: if the moon's interior were warm and plastic the mascons could not exist; the moon would yield under their extra weight, and they would sink down into the interior and disappear. The fact that the mascons are present on the moon indicates that the moon's outer layers are strong, unyielding and therefore *cold*.

The excellent preservation of the craters on the moon's surface, the mascons, and the low level of lunar seismic activity described below —all suggest that the outer layers of the moon are cold and rigid, with little of the geological activity that has molded the earth, transformed its surface, and continues to transform the surface today.

## THE APOLLO FINDINGS

Eight hundred thirty six pounds of rocks, hundreds of miles of scientific records, and many thousands of photographs were collected during the Apollo flights. This mountain of facts will be sifted for nuggets of information for years to come. Thus far, lunar scientists have established the following major features of the Apollo results.

### Chemistry of the Moon

The chemical ingredients of the moon's surface have been analyzed with the aid of instruments carried around the moon in orbit during the flights of Apollos 15, 16, and 17, and by studying rock samples in the laboratory. The results show that the entire surface of the moon—including the maria and the highlands—appears to be covered with a thick layer of basalt—a type of rock formed by the cooling of molten lava.

Two kinds of lunar basalt were discovered. One, called mare basalt, came from the maria, and was composed mainly of the same minerals that make up the bulk of the basalts found on the earth. These minerals are pyroxene and plagioclase feldspar. In addition, the mare basalts also contained small amounts of olivine and of a much rarer mineral called ilmenite, which is an oxide of iron and titanium. While olivine is present in about the same concentration as in terrestrial basalts, ilmenite is much more abundant in the lunar basalts than on the earth.

The other type of basalt was collected mainly during the landings on the highlands, and is called highland basalt.[1] The predominant mineral in highland basalt is plagioclase. Some pyroxene was found, but much less than in the mare basalts. Only a trace of ilmenite was detected.

The abundance of ilmenite—a dark mineral—in the mare basalts, and its scarcity in the highland basalts, largely account for the blackness of the lunar seas in comparison to the highlands.

The relative amounts of the various chemical elements also differ greatly in the mare and highland basalts. The most important differences involve aluminum and iron. Aluminum is much more abundant in the highlands than in the maria, but iron is much less abundant.

The relative scarcity of iron and abundance of aluminum in the highland basalts explain the low density of the highland rocks. Their density measured in the laboratory is 2.9 g/cm³, in comparison to 3.3 g/cm³, for the mare rocks.

*Dryness of the Rocks.* In addition to analyzing the basic chemical composition of the rocks, the Apollo investigators also made a very careful search for water. They looked for traces of actual moisture in the rocks, and also for water in the form of molecules locked up in the crystalline structure of the minerals. The laboratory analyses showed the lunar rocks were bone-dry in both respects; they contained neither free moisture, nor minerals with water molecules included in their structure.

Since water is an essential ingredient for life as we know it, the dryness of the moon rocks suggests that life does not exist on the moon at the present time. Moreover, the measurements of the ages of the rocks, discussed below, indicate that they have been in their present state about three billion years. Therefore, we can be confident that no life has existed on the moon for that long interval of time.

The lunar rocks also showed no traces of organic matter, such as would have been left by biological organisms. Laboratory tests failed to detect any fossil organisms or residual molecular building blocks of living matter, such as amino acids and nucleotides, except in amounts so small that it was uncertain whether contamination by a technician's fingerprint might have been responsible for them.

## Evidence for the Early Melting of the Moon

The discovery that the entire surface of the moon—both maria and highlands—is covered with basalt was one of the most important results of the entire Apollo program. On the earth, basalts are created only by the cooling of molten lava. Thus, this single finding from Apollo indicates that the surface of the moon was entirely molten at some point in its past.

What melted the moon's surface? Only two possible answers are known. One is that the moon was melted by an intense meteorite bom-

---

[1] Most highland rocks are actually *breccias*—rocks made up of fragments from pre-existing rocks. These rocks were first broken apart by intense meteorite bombardments, and later packed together again by the force of further impacts.

bardment during a short period at the beginning of its life. This explanation fits in with the currently favored theory on the origin of the solar system, which proposes that all the planets, and their moons, condensed out of particles of gas, dust, and fragments of rock of various sizes. As our moon grew to final size in the last stages of this birth process, gravity pulled the material around it down onto the surface with great force. Each rock generated some heat as it crashed into the surface. The heat would be radiated away slowly to space, but if enough impacts occurred in a short period of time, the total accumulation of heat could be sufficient to melt the moon to a considerable depth.

The alternative explanation is that the moon was melted later by radioactive energy released in the decay of uranium and other radioactive elements scattered throughout the moon's interior. The Apollo measurements showed that the moon rocks contain a substantial amount of radioactivity, suggesting that this could be an explanation for the melting of the moon.

However, radioactive elements release their heat very slowly. In fact, calculations show that it takes about one billion years of steady radioactive heating to bring the interior of a planet like the moon or the earth to the melting point. If the rocks on the surface of the moon were melted when the moon was at least one billion years old, that fact would suggest that radioactive heat was the likely cause of the melting; but if the rocks were melted when the moon was considerably younger than one billion years, it would be necessary to look to another factor — presumably meteorite bombardment — for the explanation.

## The Ages of the Lunar Rocks

These ideas indicate that the times at which the moon rocks were melted could provide the clue to the cause of their melting. With this remark we come to the second critical Apollo result, which is the measurement of the ages of the moon rocks. The ages of the rocks are measured by the technique of radioactive decay, which tells how long they have been in their present crystalline form (see Chapter 14). In other words, it tells how much time has elapsed since they were last melted. Thus, the age measurements provide precisely the information needed to distinguish between the two causes of the moon's melting.

But nature rarely gives up her secrets without a struggle. When the results of the age measurements became available, they yielded the ambiguous answer that *both* causes of melting probably had played roles in the moon's history. The lunar highlands probably were melted by meteorite bombardment early in the moon's life, while the lunar seas were melted later by internal radioactive heat.

This fact did not become clear until the Apollo 16 mission. Apollo 16 was the first flight to land in true highlands terrain. All the highland rocks collected during the Apollo 16 mission turned out to be about 4 billion years old, indicating that these rocks crystallized when the moon had existed for no more than 600 million years. One highlands rock—the oldest rock found on the moon thus far—is 4.2 billion years old. If this rock is representative of the highlands, and is not an exceptionally old rock, it follows that the surface of the moon must have solidified when the moon was only 400 million years old. Four hundred to 600 million years is probably too short a time for the moon to have been melted by the slow process of radioactive heating. Therefore, meteorite bombardment is more likely to have melted the highlands rocks.

The highlands are known to be older than the lunar seas, and are thought to be derived from the moon's original crust, parts of which were later covered over by the darker materials of the lunar seas. Thus, it appears that the original surface of the moon was entirely melted by meteorite bombardment during the moon's birth, or shortly after. This inference from the Apollo findings agrees with the general expectation that an intense meteorite bombardment must accompany the formation of every planetary body such as the moon or the earth.

But the story of the moon's melting does not end there, because other moon rocks—those collected from the lunar seas—have younger ages, ranging from 3.1 to 3.8 billion years. Their average age is roughly three and-a-half billion years, which would indicate crystallization when the moon was roughly one billion years old. This time period would correspond more closely to the time required for radioactive heating of the moon to melt the rocks in its interior, creating floods of molten lava on the surface.

Presumably, the lunar seas are pools of solidified lava that accumulated in the original impact basins of the original highland crust as a result of repeated volcanic eruptions and lava flooding during this later period. Photographs of the walls of the Hadley Rille, taken during the Apollo 15 landing, clearly reveal several layers of the kind that would result from repeated flooding of the moon's surface by lava (Figure 15.13, page 402).

Another important conclusion follows from the age measurements. Apparently, the moon experienced one great episode of sustained volcanism lasting about 700 million years, and ending roughly three billion years ago. The 700-million-year interval of melting and lava flooding on the moon was undoubtedly accompanied by volcanic eruptions, moonquakes, and all the other manifestations of internal heat within a planet, that are so familiar to us through our experience with volcanic activity on the earth. When the volcanism ended, the moon subsided into geological lifelessness, and has remained inactive ever since. Today there seems to be scarcely any volcanism on the moon, apart from a few wisps of gas vented at the surface now and then.

Is it possible that volcanoes have erupted more recently, and may still be erupting, but only in places that have escaped the attention of the astronauts? Another basic Apollo finding—the information obtained from the lunar seismometers—excludes this possibility.

*Figure 15.13  Layers in the surface of the moon exposed in the wall of the Hadley Rille, and photographed from the far side by the Apollo 15 astronauts. The clearly defined layer in the middle consists of four distinct strata, each about 10 feet thick.*

### The Seismometer Results

Five seismometers were placed on the moon during the Apollo program (Figure 15.14). Four are still working and sending back moonquake data daily. According to the seismic data, the energy released by moonquakes per year is only a billionth to a trillionth as much as the energy released in one year by quakes on the earth. The most powerful moonquakes detected by the Apollo seismometers in nearly four years of operation have a rating of 2 on the Richter scale. If you were standing directly over a quake of this size on the earth, it would not produce a perceptible vibration in your feet.

The weakness of the moonquakes detected by the Apollo instruments strongly suggests that there cannot be intense volcanic activity or lava flooding anywhere on the moon at the present time. Volcanoes and lava flows would be accompanied by movements of material within the moon, which would create vibrations detectable by the seismometers left behind on the lunar surface. These vibrations would show up in the records as major moonquakes.

The seismometer signals have provided other details regarding the moon's internal structure. The signals indicated that the quakes occurred at a surprisingly great depth, most originating at depths between 500 and 600 miles, whereas quakes on the earth usually originate within 60 miles of the surface.

The fact that the moonquakes occur at depths of 500 miles or more suggests that little, if any, molten rock or warm and plastic rock exists in the outer 500 miles of the moon. In other words, the moon is capped by a 500-mile thick shell of strong and rigid rock.

Beneath the rigid shell, however, there appears to be a layer of warmer rock, analogous to the earth's zone of weakness. This layer is the source of the moonquakes.

*Layers Within the Moon.* Another important fact to emerge from the seismometer experiment is that the moon appears to have distinct layers, similar to the layers of the earth's interior.

First, the speed of the seismic waves changes sharply at a depth of 15 miles that seems to be produced by a transition from a layer of broken rock — fragmented by repeated meteorite collisions — to solid, intact basalt.

Second, the speeds of the seismic vibrations change sharply again at a depth of 40 miles. This implies that the rocks within the moon change their character 40 miles down. That depth appears to be the boundary between two layers containing different kinds of rock.

The most likely means of producing such a separation into two types of rock is a long-continued process of melting, freezing, and remelting,

*Figure 15.14  Emplacement of a lunar seismometer. The instrument is housed in the drum-shaped container. The side-panels are photocells converting sunlight to electric power.*

leading to chemical differentiation of the moon analogous to the separation of the earth into a crust and mantle. The seismic vibrations suggest that down to 40 miles, moon rocks are similar to basaltic rocks. The signals received from depths below 40 miles suggest a mantle of rock, such as peridotite or dunite. Thus, the seismic results suggest that the moon has a well-differentiated crust and mantle.

Third, the seismic data suggest that the moon has a partly molten core. The discovery of a molten core rests on the fact that one type of seismic wave cannot travel through a liquid. Whenever a moonquake originated on the far side of the moon, the Apollo seismometers located on the near side failed to detect this type of vibration. Since vibrations from quakes on the far side must travel through the center of the moon to reach the Apollo seismometers, it was inferred that the moon's center contains molten rock that blocks the passage of the special vibration, but allows other types of seismic vibrations to pass through. The seismic data do not provide decisive information on this molten core within the moon because of the relatively small number of seismometers involved, and the fact that they are not very widely distributed over the moon's surface. According to rough estimates, the core is 400 miles in radius and contains roughly 5 percent of the moon's mass.

## THE HISTORY OF THE MOON

### The Early Bombardment

Armed with the facts from Apollo, we can now reconstruct the full life story of the moon. Our satellite condensed out of cold matter, but its surface was immediately bombarded by a vast quantity of meteorites. The debris of the newborn solar system—ranging from fragments of rock to meteorites the size of asteroids—rained down on the planet in rapid succession. The rocks in each collision area had no chance to cool off from the heat of one impact before they were warmed up by the next. The temperature rose rapidly as the intense bombardment continued. Eventually, the whole outer layer of the moon melted, and remained molten until the bombardment subsided. As the holocaust ended, the molten outer layers cooled and solidified. The moon became a quiet planet—a sphere of solid rock from its center to its surface.

When the moon was several hundred million years old, the intensity of the bombardment would have fallen off from the earliest years. However, some meteorite collisions still occurred at a reduced rate. These collisions produced the craters visible today in the highlands. Once in a great while a meteorite of exceptional size, perhaps as large as an asteroid, hit the moon. These mammoth chunks of rock, up to 60 miles in diameter, presumably blasted out the basins of the circular maria.

## The 700-Million Year Episode of Volcanism

As the number of meteorite impacts diminished, simultaneously the interior of the moon heated up slowly as a result of the decay of radioactive elements below the surface. When the moon was nearly a billion years old, the radioactive heating partly melted the interior. Occasionally a flood of lava poured across the surface, as molten rock forced its way upward through natural channels provided by fractures in the crust.

The impact of a very large meteorite would produce such fractures in the crust. There must have been numerous fractures under the circular basins that had been the sites of the biggest impacts. This explains how the lunar seas came to be formed; molten rock rose up repeatedly under the surface along these numerous fractures, and flooded the basins, creating the seas as we know them today.

When the lava solidified, it created a plug of dense rock filling the basin of the sea. This disk may have become a *mascon*. The mascon would not sink and disappear, as it must have on the earth, because the crust of the moon was thicker and more rigid than on the earth, and sufficiently strong to support the excess weight.

Volcanic activity and lava flooding—signifying a partly molten interior —continued for about 700 million years, and then subsided, leaving the moon quiet once more. The last major lava flow appears to have occurred 3.1 billion years ago. The next three billion years in the history of the moon have been geologically uneventful.

## The Moon Today

Why did lava flows dry up on the moon three billion years ago? The answer probably is that the moon, being a small planet, lost its heat to space rapidly. As the moon's heat disappeared, and its temperature dropped, the extent of the warm, molten region diminished, and the outer zone of cold, strong rock became thicker. According to the Apollo seismometer data, this cold outer zone today extends from the surface to a depth of 500 miles. Beneath this thick layer, there exists a warm, and perhaps partly molten, zone of yielding rock, analogous to the earth's zone of weakness. Below a depth of 600 miles the moon is probably partly molten.

It is extremely unlikely that molten rock lying at such great depths could force its way upward through the thick layer of solid rock that caps the moon. This casing of cold rock probably explains the absence of volcanic eruptions on the surface of the moon during the last three billion years. Magma from this depth might reach the surface through an intervening layer of 500 miles, but only very infrequently and with great difficulty. A 500-mile outer casing of rigid rock is also too thick to break up into several plates that move against each other, as the earth's plates do, producing earthquakes, volcanic eruptions, and mountain building (Chapter 14). These circumstances account for much of the difference between the geology of the moon and the geology of the earth, and explain why the appearance of the moon today is so different from that of our own planet.

## What the Moon's History Reveals About the Earth

Scientists had always hoped that they would recapture some of the earth's missing past as they unraveled the history of the moon. Now that the Apollo facts are in, what, if anything, can they tell us about our own planet? Consider first the evidence for the early melting of the moon by meteorite bombardment. The meteorites that bombarded the moon at the beginning of its life must also have bombarded the earth. This bombardment would have been even fiercer on the earth because its gravity is six times greater than the moon's. The outer layers of the earth would have been melted by bombardment at about the same time that the first episode of melting occurred on the moon.

After the initial bombardment subsided, the earth, like the moon, would have cooled and been a nearly solid body from the center to the surface. It would then have remained solid, for a period of about 500 million to a billion years, while the radioactive heat within it slowly built up, and its internal temperature rose.

If the abundance of radioactive elements is about the same in the interiors of both the earth and the moon, then, after about a billion years, the earth's interior, as in the case of the moon, would have reached the melting point of rock. At that time, on the earth as well as the moon, molten rock would have started to rise to the surface.

Where did the molten rock first rise? We believe that on the moon it rose in the places where the crust had been punctured or damaged by giant meteorites. The evidence for this conclusion is to be found in the ages of the materials filling the lunar seas. The ages, ranging from 3.1 to 3.8 billion years, indicate that the mammoth craters left by the impact of those meteorites were filled in by lava at just about the point in the moon's history when radioactive melting would have occurred. On the earth, lava must also have filled similar basins left by the largest meteorite collisions, producing terrestrial equivalents of the lunar seas at the same time that they were formed on the moon.

Today, the lunar seas, dating back to that early time when the moon was about one billion years old, are still as well preserved as if the entire sequence of events had happened yesterday. They are preserved because three billion years ago the moon became geologically inactive, and has remained inactive right down to the present. But the terrestrial "seas" of lava were not preserved, and are no longer visible. They have been obliterated by the volcanic activity and mountain building that have

reworked the surface of the earth continuously for nearly four billion
years.


## THE ORIGIN OF THE MOON

The history of the moon has been carried from its first melting by
meteorite bombardment through its second melting by internal heating,
and then through the three billion years of geological inactivity that
followed, down to the structure of the moon as it is today. Left unexplained is the question of the moon's origin: How and where was the
moon formed?

The subject has always been one of the most vigorously debated questions in science. In 1898 the physicist George Darwin, son of Charles
Darwin, proposed that the moon had been wrenched from the earth by
centrifugal forces, when the earth was a young and rapidly spinning
planet. This idea came to be called the fission theory. According to
Darwin, the Pacific Ocean Basin was the scar left by the violent separation that created the moon.

Recently, a variation of the Darwin theory—called the spin-off theory—
has been proposed. This theory holds that the moon was spun off as a
circular ring of gaseous matter from the equator of the young, rapidly
rotating earth, like a stream of mud leaving the rim of a rapidly turning
bicycle wheel.

A third theory is based on twentieth-century ideas relating to the condensation of stars and planets out of clouds of gas. This idea, called the
double-planet theory, suggests that the moon and the earth were formed
simultaneously as a double planet, condensing out of two neighboring
rings or pockets of gas in the solar nebula.

The fourth and most recent theory suggests that the moon was formed
independently of the earth, as a small planet in another part of the solar
system, and was captured subsequently by the earth's gravity as it
hurtled past in a near-encounter. This idea is known as the capture theory.

Very few scientists support Darwin's fission theory today. One reason
is that enormous frictional forces would be created when the moon
started to separate from the earth, and calculations show that the frictional forces would slow down the rotation of the earth to such an extent
that the moon would never get away.

The spin-off theory also suffers from basic difficulties. One relates to
the plane of the moon's orbit. According to the spin-off theory, the material of the moon should have been spun off from the earth's equator,
since the equator is the place on the earth where the centrifugal force
is greatest. The material that was originally on the earth's equator would
continue to circle in the earth's equatorial plane after it has been spun
off. The moon, accumulated out of this material, would move in an orbit
that is also in the earth's equatorial plane. However, the moon's orbit is
actually inclined at a rather large angle to the earth's equatorial plane,

averaging 23.5 degrees.[2] Calculations show that the moon's orbital plane could never have drifted this far from the equatorial plane of the earth, if it had started out in that plane.

The second difficulty with the spin-off theory is that the earth probably was not spinning fast enough to produce the moon in this way. The earth had to turn on its axis once every two and one-half hours, in order to produce sufficient centrifugal force to spin off the moon, but from the rate at which the earth and the moon are rotating today, it can be proved that the earth would have been spinning on its axis once every five hours. That is, it was rotating only half as fast as would have been necessary to spin off the moon.

This and the previous objection also apply to the fission theory. Both objections must be rejected if it happened that another planet flew off at the same time as the moon. It has been suggested that Mars was created in this way, spinning off the equator of the earth along with the moon, in one dramatic episode at the beginning of the solar system. There is no theoretical objection to this hypothesis. However, the observed facts about Mars and the moon seem to work against it. The theory implies that the moon and Mars are made of exactly the same materials: namely, the materials that made up the outer layers of the earth at the time they were formed. This conclusion works well enough for the moon, whose density is about the same as the density of the earth's mantle, but it does not work for Mars, whose density is considerably greater than that of the mantle.

The third major theory of the moon's origin—condensation out of materials near the earth, as a sister planet—also suffers from a basic difficulty. If the moon formed in this way, out of solar nebula materials in the neighborhood of the earth, it should be composed of the same chemical elements as the earth, and in the same proportions. That is, its chemical composition and, therefore, its density, should be the same as that of the earth. But the moon's average density is 3.34, whereas the earth's density, even after allowance is made for the compression of the earth's interior by the weight of the outer layers, is 4.5. The greater density of the earth suggests that it contains different proportions of the chemical elements. This discrepancy seems to exclude the theory that the moon and the earth were formed out of the same materials as a double planet.

The fourth theory—formation of the moon in a different part of the solar system, followed by capture by the earth—has no obvious flaws. The only objection to this theory is that it constitutes an extremely improbable event. For capture to occur, the moon must have moved past the earth at exactly the right distance. If too far away, it would have been whipped around the earth by the force of our gravity, without dropping into orbit; if too close, it would have collided with us. Calculations indicate that the range of conditions that can lead to capture

[2] The angle varies between 17° and 28° in an 18-year cycle.

is very narrow, and the probability of a capture occurring is correspondingly small.

For years prior to the first lunar landing, scientists argued the merits of the four theories of lunar origin with great intensity, but the battle ended in a stalemate, for each theory suffered from at least one major defect. Everyone expected that the lunar landing would promptly settle the debate, for it seemed obvious that as soon as the first samples of lunar rock were analyzed, and we found out what the moon was made of, we would be able to tell where it came from.

These hopes were not realized. The moon's origin is as much a mystery as it was before Apollo. Perhaps the answer is that the moon's formation was, in fact, a very rare accident. Perhaps it was the product of improbable circumstances, so special in their character that we may never be able to separate them from the tangled sequence of events that were involved in the birth-process of the planets.

## Questions

1. Why is the surface of the moon better preserved than the surface of the earth? What forces act to change the surface of the earth? What is the principal force of erosion on the moon?

2. The absence of life on the moon is related to its small mass. Explain the relation.

3. If the earth were as airless and waterless as the moon, in what ways would the surfaces of the two bodies differ in appearance? Explain.

4. What are the lunar "seas"?

5. What are the principal items of evidence for a geologically active moon? For a geologically lifeless moon?

6. What bearing does the presence of a substantial abundance of radioactive elements in the lunar rocks have on the theory of the thermal history of the moon?

7. By what means other than radioactive heating might the moon have been melted?

8. Feldspar is a relatively low-density mineral concentrated in the crust of the earth (see Table 14.1 and associated footnote). One of the principal results of the Apollo 15 mission was the discovery of the "Genesis Rock," composed mainly of feldspar. Explain the significance of this discovery for the moon's history, in the context of the discussion of differentiation in Chapter 14.

9. Discuss the ways in which the history of the moon illuminates the history of the earth. How would you expect the history of a planet to compare with that of the moon if that planet were half the moon's diameter? Twice the moon's diameter? Four times the moon's diameter?

10. Cite evidence for and against the four current theories of the moon's origin.

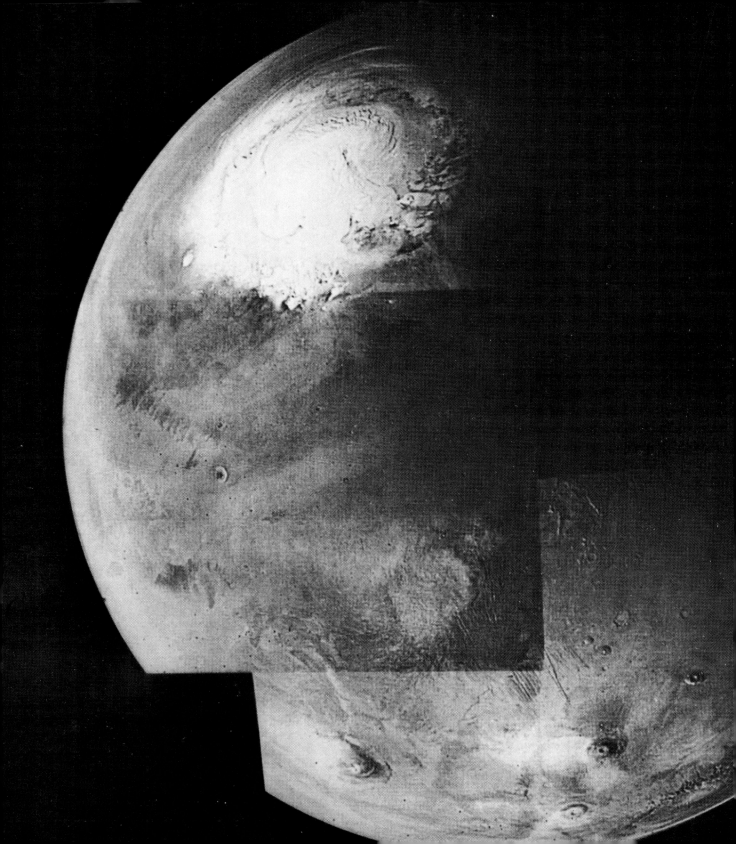

# 16 Venus, Mars and Jupiter

Venus, Mars, and Jupiter have been studied more intensively than any other planet in the solar system except our own. Venus and Mars, lying on either side of the earth, and resembling our planet in some respects, have always been of particular interest because they seemed to hold the promise of bearing life.

Jupiter is a planet of great interest for a different reason. The planetary environment on Jupiter provides an uncongenial atmosphere for oxygen-breathing terrestrial life, which would be suffocated by the noxious gases in its atmosphere and crushed by the force of its gravity, but those very gases are believed to have been the ingredients out of which the molecular building blocks of life were fashioned on the earth in its early years. It is possible that clues to the first circumstances in which chemical evolution commenced on the earth may be found in the materials lying under the ammonia cloud cover of the solar system's most massive planet. Jupiter has the additional interest that its massiveness gives it an intermediate status between planets and stars, affording the opportunity to study the behavior of solar-system materials under extreme conditions of temperature and pressure.

*A composite of pictures of Mars obtained during the 1971 Mariner flight. The north polar cap, the Rift Valley (bottom right), and six large volcanoes appear clearly.*

## VENUS

Venus and the earth are sister planets, closely similar in size and weight, and situated at distances from the sun that are not very different. Conditions on the surface of Venus have always been an enigma because the planet is completely covered by clouds, yet hope has flourished throughout the last 300 years—since Galileo first turned his telescope on Venus in 1610 and discovered that the planet was a round body like the earth and the moon—that beneath the clouds on Venus lie teeming masses of exotic flora and fauna. In 1686 de Fontenelle, in his book *Conversation on the Plurality of Worlds,* described the characteristics he expected to find in the people on Venus:

> "I can tell from here . . . what the inhabitants of Venus are like; they resemble the Moors of Granada; a small black people, burned by the sun, full of wit and fire. . . ."

### Planetary Properties

The density of Venus is very close to the density of the earth—indicating that it is probably composed of a similar mixture of iron and rocky materials. Assuming that Venus has the same concentration of radioactive elements in its interior, it can also be assumed that the iron within the planet melted early in its history and collected at the center to form a molten core.

It is also safe to assume that a mantle and a crust are located outside the core and that at some depth below the crust in the interior of the planet there is a zone of molten or nearly molten rock. The external manifestations of the internal melting have undoubtedly been a substantial degree of movement of surface plates during the history of the planet, analogous to the movement of the great plates of material that have determined the position of the earth's continents. Volcanism, earthquakes, and mountain-building activity must accompany the movement of these plates, as on the earth.

Evidence for mountain ranges on Venus has been found in the analysis of radar signals reflected off the surface of the planet. These signals are sent in the direction of Venus and irregularities in the radar echo are analyzed to determine the large-scale topography of the planet.

The planet circles the sun inside the earth's orbit, completing one circuit in 225 earth-days. As Venus revolves around the sun, it goes through phases similar to those of the moon (Figure 16.1). The phases of Venus are visible in Figure 16.1 (inset) taken with the 36-inch telescope at Lowell Observatory. At the "full Venus" (1) the planet is on the opposite side of the solar system from the earth, and its face is fully illuminated by the sun. At the "half Venus" (2), the planet has moved halfway around its orbit toward the earth. The "new Venus" (3) is on the same side of the sun as the earth; because it is directly between the earth and the sun, it can

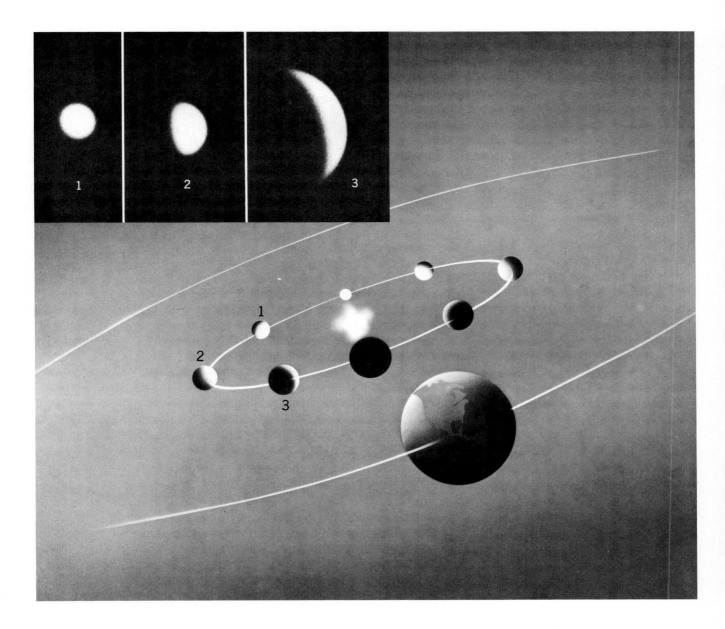

hardly be seen, although at this point in its orbit, it is at minimum distance and assumes its largest apparent size.

A heavy cover of clouds veils Venus at all times. Breaks may occur in
the clouds but would not be visible in our telescopes because of the blurring effect of the earth's atmosphere, which obscures all features less than
50 miles across. The detailed appearance of Venus will remain a mystery
until the planet is photographed from close range by spacecraft.

The rate of rotation of the planet about its own axis, and the orientation
of the axis of rotation in space, can also be deduced from the radar

*Figure 16.1   The phases of Venus.*

**413**

echoes, because the reflection from the rotating surface of the planet produces an observable Doppler shift in the reflected radar signal. According to these measurements, Venus rotates on its axis once in 243 days. Its direction of rotation on its axis is opposite to the direction of its movement around the sun. That is, looking down on the north pole of Venus from "above," the planet will be seen to revolve about the sun in a counterclockwise direction—as is the case for all planets in this solar system—while rotating on its axis in a clockwise direction. This type of motion is called *retrograde* rotation. Venus is the only planet in the solar system whose rotation is unmistakably retrograde.

Why is the rotation of Venus so peculiar? When the planet was born, it probably rotated on its axis at about the same speed as the earth. However, being closer to the sun, it felt the effect of the sun's gravity more strongly. The pull of solar gravity must have slowed down the rate of rotation until the planet was rotating on its axis at very closely the same rate at which it revolved around the sun. This means that Venus would present the same face to the sun at all times. The moon, which is controlled by the earth's gravity, rotates on its own axis in the same period of time in which it repeats one revolution around our planet, and presents one face to the earth at all times for this same reason.

This would explain why the length of the Venus day is so much longer than the length of our day. The explanation for the retrograde rotation is probably connected with the fact that the rotation rate observed for Venus is precisely such as to assure that the planet will present the same face *to the earth* every time it comes closest to the earth in its orbit. Apparently, once the sun had slowed the rotation rate of Venus down to 200 days or so, the weaker pull of the earth was then effective enough to swing the same face of Venus around to the earth whenever the two planets were at their minimum distance during their orbits around the sun.

## The Surface of Venus

The planetary properties of Venus suggests that it should provide an agreeable climate for living organisms. The planet is 67 million miles from the sun, while the distance of the earth is 93 million miles. Because Venus is closer it receives 1.9 times, or approximately twice, the intensity of sunlight falling on the earth. This is true because the intensity of sunlight falls off in proportion to one over the square of the distance from a planet to the sun. Therefore, the intensity of the sunlight received by Venus in comparison to the sunlight received by the earth is equal to:

$$(93/67)^2 = 1.9$$

The heavy cloud cover on Venus keeps out 80 percent of this excess of solar energy, but the predicted average temperature on the planet is, nonetheless, slightly higher than the average temperature on the earth, and very comfortable by terrestrial standards.

But 12 years ago radio astronomers obtained evidence which suggested that the climate on Venus might be far from balmy. In fact, the measurements indicated that Venus is a very hot planet. The results obtained by the radio astronomers depended on the fact that a heated object radiates energy at all wavelengths, although the most intense radiation occurs at one particular wavelength that is determined by its temperature. Figures 5.5 to 5.8 in Chapter 5 show radiation curves for several different temperatures. An object that is only hot to the touch, but not visibly glowing, such as a heated iron, radiates most of its energy in the infrared region; but, like every heated object, it also radiates energy at all other wavelengths from the shortest to the longest. For example, the relative amount of energy radiated by a hot iron in the short ultraviolet wavelengths or at very long radio wavelengths, is very small compared to the amount emitted at the peak intensity in the infrared; but nevertheless, the radiation at these wavelengths is present.

A planet, heated by the rays of the sun to a temperature of a few hundred degrees, is like the hot iron. Most of the energy radiated from its surface is in wavelengths in the far infrared region, around 100,000Å, or 10 microns.[1]

The long-wave radiation includes waves ranging from a fraction of an inch in length up to many miles. These waves fall into the parts of the electromagnetic spectrum that are called the *radar* or the *microwave* region, and the *radio* region. This radiation is of great interest to astronomers because of the fact that after it has been emitted from the surface of the planet, it passes through the atmosphere of the planet relatively unhindered, and escapes freely to space. Furthermore, that part of it which reaches the earth also passes through the atmosphere of the earth relatively unhindered and thus can be detected by radio telescopes on the earth's surface (see Chapter 3).

Especially when the surface of the planet is covered by clouds, as is the case for Venus, the long-wave radiation, which can penetrate through these clouds, is the only means available to the earthbound astronomer of obtaining information about the surface conditions on the planet.

Motivated by this fact, radio astronomers at the Naval Research Laboratory pointed their antenna in the direction of Venus in 1955 and tried to pick up long-wave radiation coming from the direction of that planet. They were immediately successful in detecting the signal from Venus. In fact, it was considerably stronger than their calculations had led them to expect.

The calculations were based on the assumption that the temperature on Venus was about the same as the temperature on the earth. The fact that the signal from Venus was more intense than expected indicated that the temperature on Venus must be considerably higher than the temperature

---

[1] A micron, which is a convenient unit for radiation in the far infrared, is one-millionth of a meter. The visible band of wavelengths extends from 0.4 to 0.7 microns.

on the surface of the earth. Fitting their data to a radiation curve, the Naval Research Laboratory astronomers deduced that the temperature on Venus was a sizzling 800°F.

This temperature is hot enough to melt lead. It is also high enough to break apart all the delicate molecules that make up the essential ingredients of a living cell. It is certain that no organisms remotely resembling any form of life on earth can exist on the surface of Venus.

Yet hope lingered on for the discovery of a green world on Venus. Some astronomers argued that the intense radiation might come from the atmosphere of Venus and not from its surface. Others suggested that life might be supported at the north and south poles, which should be cooler. Another suggestion was that organisms might have developed on Venus with a very specialized form, consisting of gas-filled bladders, something like beach balls, whose buoyancy would cause them to float high in the atmosphere, where the temperature should be considerably cooler than it is on the ground.

## Spacecraft Exploration of Venus

Between 1967 and 1971, a series of Russian and American spacecraft reached the planet and carried out measurements that removed the last trace of doubt regarding the sizzling hot temperatures of its surface. The

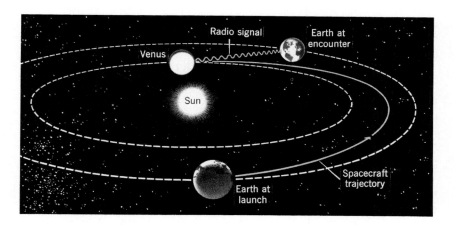

*Figure 16.2 Spacecraft trajectories to Venus. The spacecraft, starting out from the earth's orbit and moving more slowly than Venus, meets Venus after completing half an orbit around the sun.*

spacecraft were the USSR Venera series and the US Mariner series. Both spacecraft traversed curved paths of approximately 200 million miles, and arrived in the vicinity of the planet four months after launch (Figure 16.2).

The histories of the spacecraft diverged as they drew near Venus (Figure 16.3). Drawn toward the planet by its gravitational pull, the US spacecraft crossed over parts of the sunlit and dark sides of the planet and continued

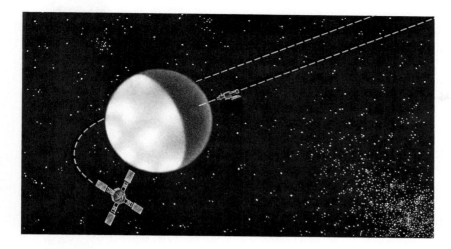

Figure 16.3 Trajectories during the rendezvous with Venus.

on around the sun. As it passed behind Venus, its radio signals probed the atmosphere to a height of 20 miles above the surface. Venera, the USSR spacecraft, headed directly for the dark side of the planet and crashed near the equator. As it entered the atmosphere, a daughter probe separated from the main USSR spacecraft and parachuted down toward the surface[2] radioing information on atmospheric conditions during its descent.

*The American Experiment.* The Mariner spacecraft contained scientific instruments in eight compartments making up the body of the spacecraft (Figure 16.4). A dish-shaped antenna beamed the main radio signal to the earth. The dish swivels about in space and can be pointed at the earth by radio command. The four paddles, which give the spacecraft its windmill appearance, are solar panels covered with photoelectric cells, which generate 550 watts of electricity from sunlight.

The trajectory of the Mariner spacecraft was planned so that it would miss Venus by a distance of several thousand miles, swinging behind the planet and then continuing on its way around the sun. As Mariner passed behind the planet, it was hidden from our view for about 15 minutes. During this period, radio contact between the spacecraft and the earth was broken. The important segments of the flight were the intervals of time just prior to the passage of the craft behind the planet, and just after it reemerged on the other side. During these brief moments—lasting about 60 seconds—the radio signals from Mariner passed through the Venus atmosphere on their way to the earth. The atmosphere slowed down the waves and bent or refracted them, just as the earth's atmosphere bends rays of light (Figure 16.5). We have already mentioned that this effect, acting on starlight as it passes through the atmosphere of the earth, causes the twinkling of the stars in the sky (see Chapter 3).

[2] The landing craft ceased functioning before it reached the surface in the first two attempts. The third flight, in 1970, was fully successful. The craft transmitted data for 23 minutes after coming to rest.

*Figure 16.4 The Mariner spacecraft undergoing prelaunch tests at the Jet Propulsion Laboratory.*

The slowing down of the radio waves from the Mariner spacecraft, as these waves passed through the atmosphere of Venus en route to the earth, was easily detected by Jet Propulsion Laboratory scientists who were monitoring the Mariner signals. Working backward they calculated the density of the atmosphere that would be required to produce the observed effect on the Mariner signals. From the changes in the density with height above the surface they could also determine the temperature.

*The Soviet Experiment.* The Soviet flight to Venus marked the first attempt to land on the surface of another planet. The spacecraft was con-

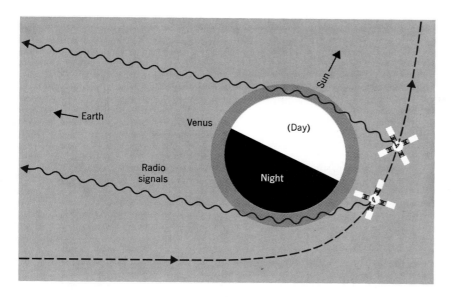

Figure 16.5 The trajectory of Mariner as the spacecraft passed behind Venus. The radio waves are bent by the Venus atmosphere.

structed on the piggyback principle. A large mother ship designed for the interplanetary cruise carried a smaller craft designed for the actual landing. The mother ship ejected the landing craft as it approached Venus, and then plunged into the atmosphere and burned. The landing craft,

Figure 16.6 The descent of the Soviet landing craft to the surface of Venus.

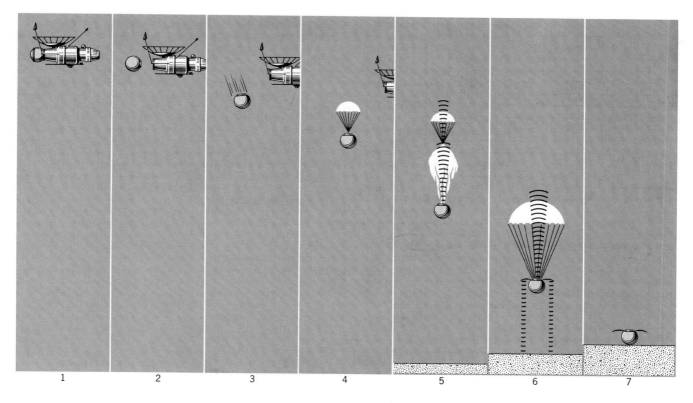

1    2    3    4    5    6    7

equipped with parachutes, descended gently toward the surface (Figure 16.6).

Figure 16.7 shows the landing craft lying in a field in the USSR following a parachute test. It is approximately hemispherical in shape and 24 inches in diameter. Instruments and batteries are located within the craft so that, on landing and rolling along the surface of Venus, it always comes to rest with its radio antenna pointed "upward" toward the earth.

*Figure 16.7   The Soviet landing craft after a parachute test.*

The landing craft contained electrical thermometers for the measurement of temperature in the atmosphere and on the surface of Venus, pressure gauges and density gauges for measuring pressure and density, and several instruments for determining the gases present in the atmosphere. Some of these devices were metal containers in which samples of the Venus atmosphere were collected. Each container had a chemical chosen to react with a particular gas, including oxygen, carbon dioxide, nitrogen, and water vapor. Simple, automatic tests were built into the spacecraft to observe the results of these chemical reactions.

**The Venus Atmosphere**

What did the American and Russian flights reveal? *First,* they showed that the temperature on the surface of Venus is approximately 800°F. This measurement removed the last doubts regarding the temperature deduced by the radio astronomers. Venus is, indeed, hot enough to melt lead, and there is no reasonable chance of finding life on its surface.

*Second,* the atmospheric pressure at the surface turned out to be nearly one ton per square inch—one hundred times the pressure on the surface of the earth.

*Third,* the Russian experiments showed that carbon dioxide made up

90 to 95 percent of the atmosphere, compared to 0.03 percent on the earth. They also indicated the presence of a small amount of water — sufficient, if condensed to liquid form, to cover the surface of the planet to a depth of one foot. The composition of the remainder of the atmosphere remained uncertain, although the tentative Soviet results suggested some oxygen.

What would existence be like in this atmosphere? Human lungs probably would not be able to take in and expel gases at the pressure and density of the ''air,'' because a pressure of 100 atmospheres corresponds to the pressure in the ocean at a depth of 3000 feet. The greatest depth to which unprotected divers have descended is 800 feet.

*Figure 16.8   Sunset on Venus.*

Assuming a human being could survive the pressure, the denseness of the atmosphere would create an eerie scene for the observer on the surface of the planet. As the light from the sun passes through the successively denser layers of the atmosphere on its way to the ground, it is bent or refracted into a curved path. This effect occurs in the earth's atmosphere, but is very small. In the dense atmosphere of Venus, the refraction curves rays of light through an angle of 180°. This means that if the sun is setting on Venus, and you face in the opposite direction, you will still see the sunset, but the image of the sun will be spread out into a band of light stretching around the horizon (Figure 16.8). The horizon itself disappears, and the world looks to you as though you are standing at the bottom of a gigantic fishbowl. You can theoretically see the back of your head on Venus without mirrors.

## The Greenhouse Effect

Why is Venus so hot? What determines the temperature on the surface of this planet, or any planet? A planet's temperature is determined primarily by its distance from the sun, which controls the intensity of the solar heat falling on the planet's surface. Most of this heat reaches the planet in the form of visible light. A part of the light is reflected back to space by clouds and scattered in the atmosphere; the remainder passes through the atmosphere to the surface, which is warmed by the absorption of this solar energy. The warm surface radiates heat back to space in the form of infrared radiation. Over a long period of time a planet must radiate as much heat back to space as it receives in the form of visible sunlight.

At first the surface of the planet will rise steadily in temperature as it receives the solar energy, but as it becomes hotter it radiates more heat to space, until eventually the outgoing heat just balances the inflow of sunlight. For any planet, the temperature can be calculated at which the radiated heat equals the influx of sunlight. In the case of the earth, we find in this way that the average ground temperature should be −20°F.

But the actual average temperature of the earth is 60°F, and not −20°F. The increase in temperature is produced by our atmosphere, which acts as an insulating blanket, trapping a part of the radiation from the ground and returning it to the planet, where it adds to the heat provided by the absorption of sunlight. In this way the average temperature of the earth is raised from the theoretical value of −20°F to the level of 60°F, which is actually observed.

The outgoing heat is trapped by the trace amounts of water vapor, carbon dioxide, and ozone in the atmosphere. Although these constituents together make up less than one percent of the earth's atmosphere, they absorb 90 percent of the heat radiated into the atmosphere from the surface.

The heat insulation provided by the atmosphere is called the "greenhouse effect." A greenhouse has a glass cover that is transparent to the sun's visible radiation (just as the earth's atmosphere is transparent), but blocks the heat radiated from the plants within. Thus the heat is trapped within the greenhouse and warms the interior, just as water vapor, carbon dioxide, and ozone retain heat in the earth's atmosphere and warm the surface of our planet.

The greenhouse effect provides the answer to the mystery of Venus's high temperature. Carbon dioxide, of which only a trace exists in the earth's atmosphere, is present in the atmosphere of Venus in such massive amounts that it makes that atmosphere almost impervious to infrared radiation. Nearly 100 percent of the heat radiated from the surface of Venus is trapped by this insulating blanket of carbon dioxide, causing the temperature of the surface to soar far above the value it would have if the atmosphere of Venus were similar to ours.

## The Dryness of Venus

It is also a great puzzle to students of the planets that the equivalent of only one foot of liquid water was formed on Venus. According to current views on the origin of the solar system, Venus and the earth condensed out of the same materials, containing similar amounts of water. During the condensation, water was trapped in the interior of the earth. Later the trapped water rose to the surface and escaped through cracks in the crust to fill the oceans. The water in the earth's oceans, if spread uniformly over the surface of the planet, would form a layer about 8000 feet deep, and a layer of water of approximately the same thickness should also exist on Venus. Because of the high temperature on the surface of Venus, the water would be present not in liquid form as an ocean, but in the form of water vapor in the atmosphere. The Soviet measurement showed that most of this water is missing.

Thus, by terrestrial standards, Venus is a dry, hot planet and not a very suitable environment for the development of life.

## MARS

Mars has generated more speculation regarding extraterrestrial life than any other planet in the solar system. Several characteristics of the planet have contributed to the growth of these speculations. Its surface, free of clouds, reveals changes during the Martian year that resemble the march of the seasons on the earth. In each hemisphere a polar cap grows larger in the fall and winter and diminishes in the spring and summer. Dark regions appear each spring that are suggestive of the seasonal growth of vegetation. Toward the end of the nineteenth century, some observers reported a planetary network of canals presumably engineered by intelligent life.

More accurate observations of Mars were made from the earth in recent years, and as the new evidence accumulated, the prospects for Martian life diminished. However, they have by no means dwindled to zero. The detection of a small amount of water vapor in the atmosphere, confirmed in 1969 by spacecraft observations, indicates that Mars may harbor life, although its climate is far less congenial than that of the earth. Pictures of greatly improved quality, acquired by the Mariner 9 spacecraft in 1971 and 1972, suggest that Mars may have had more surface water at an earlier time than it has today. This discovery has increased the level of optimism among observers of the planet regarding the prospects for finding life or the remains of life on Mars.

## Planetary Properties

Mars circles the sun at a distance of 142 million miles, one and one-half times more distant than the earth. The average temperature on the planet is 235°K, or approximately −40°F. Noontime temperatures on the equator rise to about 70°F. Its radius is 2100 miles—about half the radius of the earth. The density of the planet is less than the density of the earth, suggesting that it is made of different substances. However, the rocks in the interior of Mars are less heavily compressed than the rocks in the interior of the earth, because of the smaller mass of Mars. If the earth's density is corrected for this effect, the densities of the two planets turn out to be approximately the same. Accordingly, Mars probably is composed of a mixture of rocky materials and iron similar to the materials that make up the body of the earth.

Mars completes one circuit around the sun in 687 days or 1.9 earth-years. It rotates on its axis in 24 hours and 37 minutes. Its axis of rotation is inclined at an angle of 24° to the plane of its orbit. By a coincidence, the length of the Mars day and the inclination of the Mars axis to the plane of the planet's orbit are nearly the same as the corresponding quantities for the earth.

## The Surface of Mars

The surface of Mars, unlike the surface of Venus, is almost entirely free of clouds. Dust storms and haze occasionally are conspicuous, but most of the time the face of the planet is open to photographic surveillance.

In spite of the thinness of the Martian air, it is impossible to obtain good photographs of Mars from telescopes on the earth because of the blurring effect of the earth's atmosphere on the rays of life reaching us from it. Features on the surface of Mars cannot be seen from the earth,

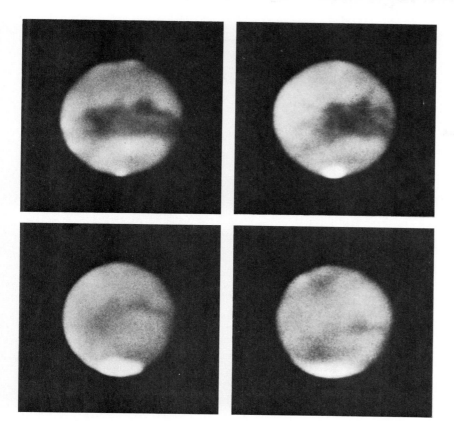

*Figure 16.9 The change of seasons on Mars.*

unless they are at least 50 miles in diameter. It is impossible to tell from the earth whether Mars has continents, basins that might once have held oceans of water, or other features that might indicate geological activity or the presence of life today or at an earlier time.

*The Change of Seasons on Mars.* In spite of their poor resolution, photographs of Mars taken from the earth have revealed several interesting properties. The photographs in Figure 16.9 were taken at intervals during the Martian year, stretching from summer to winter in the southern hemisphere. The most conspicuous feature of these photographs is the polar cap, resembling the cover of ice and snow at the poles of the earth. The polar cap grows in size from 200 miles in the upper photograph to a maximum of 2000 miles in the lower photograph. At one time the polar caps were believed to be composed entirely of ice. However, temperature measurements made over the polar caps with the aid of instruments carried on the Mariner spacecraft (see below) have revealed that the caps are composed of a thin layer of frozen carbon dioxide (dry ice), probably covering a less extensive but thicker cap of water ice possibly mixed with dry ice.

*Canals on Mars.* In 1877, the Italian astronomer, Giovanni Schiaparelli, reported seeing a network of canals on Mars. His report was subsequently confirmed by other astronomers, and Percival Lowell described

**425**

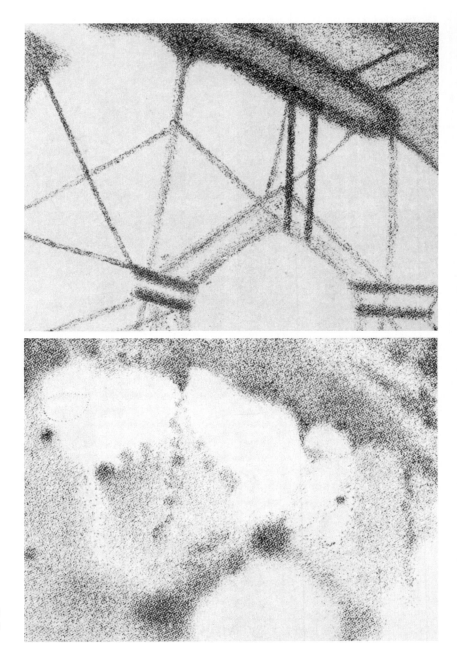

Figure 16.10 Martian canals (above) drawn by Schiaparelli; (below) the same region drawn by Antoniadi.

the canals as "a vision of a thread stretched across orange seas." However, not all astronomers saw them. A drawing of Mars made by Schiaparelli on the basis of telescope sightings shows many canals clearly marked (Figure 16.10a). Figure 16.10b showing the same region on the planet as observed by E. M. Antoniadi has similar markings, but no hint of canals.

The canals, although sighted visually through telescopes, never appeared on photographs from the earth. Recent close-ups of Mars from passing spacecraft also show no signs of them. It is now generally believed that they were imagined by astronomers straining at the limits of visibility.

*The Mariner Photographs of Mars.* Clearer pictures of Mars were taken by Mariner spacecraft in 1969, as they swept past the planet at a distance of a few thousand miles. The clearest photographs revealed features as small as 100 feet across. The photographs showed that the Martian surface was pitted with meteorite craters resembling meteorite craters on the moon. Figure 16.11 shows hundreds of craters in an area approximately 400 by 1500 miles in size. The largest crater in the mosaic is 160 miles in diameter.

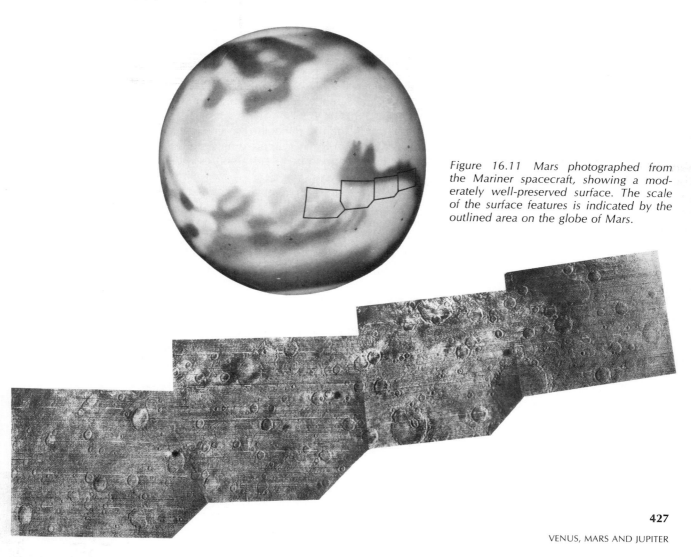

Figure 16.11 Mars photographed from the Mariner spacecraft, showing a moderately well-preserved surface. The scale of the surface features is indicated by the outlined area on the globe of Mars.

Lunar craters that are similar to the ones visible on Mars are surrounded by circular ramparts ranging up to 15,000 feet in height and contain central peaks that rise thousands of feet from the crater floor. The ramparts and central peaks are missing from many of the Mars craters, indicating stronger forces of erosion than exist on the moon. The Mariner pictures of the 1971 dust storm revealed that the abrasive action of dust-laden winds is probably one of the major sources of erosion on Mars.

### The Atmosphere of Mars

During the Mariner flights to Mars in 1969 and 1971 the spacecraft passed behind the planet at one point, so that it was hidden from the view of the earth for a brief period. As the craft disappeared behind Mars, and again as it reappeared on the other side a short time afterward, its radio signals passed through all levels of the Martian atmosphere on their way to the earth. Scientists analyzed the effect of the atmosphere on the signals and deduced the atmospheric density, as in the case of the Mariner flights to Venus.

The results indicated that the Martian atmosphere is extremely thin. The density at ground level is approximately one two-hundredth of the density of the earth's atmosphere, and the pressure at ground level is one-twentieth of a pound per square inch — approximately the same as the pressure in the earth's atmosphere at a height of 120,000 feet.

Other Mariner instruments measured the spectrum of Martian gases in the infrared region and determined the composition of the gases in the atmosphere. The primary constituent turned out to be carbon dioxide, just as in the atmosphere of Venus. An ultraviolet spectrometer designed to search for nitrogen — the principal constituent in the earth's atmosphere — failed to reveal the presence of this gas. However, a modest amount of nitrogen — up to 5 percent — could have been present and still escaped detection because of the limited sensitivity of the instrument.

Water vapor was also detected in the atmosphere. The water was present in trace amounts, equivalent to a film one-thousandth of an inch thick if condensed to liquid form.

### Volcanoes on Mars

Mars stands midway between the moon and the earth as a planetary body. It is approximately twice as large as the moon, and one-half the size of the earth; its mass is approximately 10 times the moon's mass, and one-tenth the mass of the earth. These facts lead to a prediction regarding volcanism on Mars. We know from the study of the earth that volcanism results from the release of radioactive heat within the body of an earthlike planet. The source of the heat is the radioactive decay of the elements uranium, thorium, and potassium, which exist in the interior of the earth in small concentrations of a few parts per million. The Apollo findings indicate that these elements exist in very roughly the same concentrations in the moon, and presumably they also exist in the interior of Mars.

The heat that the radioactive elements create must have gradually increased the temperature in the interior of the moon and Mars, as well as the earth, during their early years. Volcanism may have commenced on the earth, the moon and Mars at about the same time, when the three planets were about one billion years old, as the product of this chain of circumstances.

Although the volcanism probably began at about the same time on each planet, it did not last for the same length of time on each, because of the differences in their sizes. The reason is that the heat generated in the interior of a small planet has to travel only a short distance to reach the surface. Heat is lost from the surface of a small planet at a relatively rapid rate because of this fact. If the planet is very small, its temperature will never reach the melting point of rock, and volcanism will never commence.

If the planet is somewhat larger, its internal temperature may reach the melting point of rock, but will not remain there for more than a short time. This appears to have been the case for the moon, whose episode of volcanism lasted only 700 million years.

If the planet is very large, its internal heat must travel through a relatively thick layer to reach the surface. This thick layer acts as an insulating blanket, bottling up the radioactive heat, and causing the temperature in the interior to remain at a high level for a relatively long period of time. This is the case for the earth, which has remained volcanically active throughout most of its lifetime, and is still active.

Mars, intermediate in size between the moon and the earth, must have retained its radioactive heat longer than the moon, but not as long as the earth. Therefore, Mars must have been volcanically active for longer than 700 million years, and may have been active in relatively recent geological eras. This remark implies that volcanism may have persisted on Mars for as long as several billion years.

Water vapor and other gases released by internal heating are the sources of probable water on the earth's surface, and gaseous elements in its atmosphere. If the Martian episode of volcanism lasted for a sufficiently long time, water may have accumulated on its surface to a considerable depth. In fact, if volcanic activity persisted on Mars for two or three billion years, and the relative amount of water trapped in the interior of Mars is roughly the same as the amount of water trapped within the earth, the surface of Mars may once have been covered by water to an average depth of hundreds of feet or more. A substantial atmosphere also may have existed during this long period of volcanic activity.

## Evidence for Geological Activity on Mars

The suggestion of volcanism on Mars was confirmed in 1971 and 1972, when television cameras, mounted on a satellite placed in orbit around Mars by the Mariner 9 spacecraft, obtained excellent pictures of the entire planet. The televised pictures were taken at altitudes varying between 775 and 10,500 miles. When the first pictures were obtained, the surface of the planet was obscured by a violent dust storm that persisted for nearly two months. Eventually the dust settled, and the television cameras revealed details of the surface of Mars that had never before been seen by man.

The most conspicuous feature was the huge volcanic mountain, Nix Olympica, shown in Figure 16.12. The crater at the summit of this mountain is about 40 miles in diameter. The entire mountain is 300 miles across at its base, and rises at least 70,000 feet above the surrounding floor.

The roughly conical shape of the mountain and the presence of a crater at its summit clearly establish it as a volcano. The summit crater, called a caldera, is a familiar feature of terrestrial volcanoes. It is formed by the collapse of surrounding material into the molten pit of lava at the top of the volcano. The volcanic nature of the mountain is also indicated by high-resolution pictures that show individual tongues of frozen lava emanating from the summit crater and descending the flanks of the mountain (Figure 16.13a and b, page 432).

In every respect the Mars mountain resembles the mounds of congealed lava that form on the earth, when many successive outpourings of molten rock occur at a single spot over a period of millions of years. If the Pacific Ocean basin could be emptied, the Hawaiian Islands would be revealed as similar mounds of lava, rising out of the floor of the ocean basin in the same way that the Mars mountain rises out of the surrounding terrain.

About a dozen large volcanic mountains have been discovered on Mars. Six can be seen clearly in the photograph on p. 410.

*The Great Rift Valley.* On the earth, volcanism is associated with the movement of large slabs of rock, generally thousands of miles in size, that slide over underlying layers of warm and yielding material. These slabs of rock are the earth's plates (Chapter 14). When the edge of one plate slides past another, cracks such as the San Andreas fault appear in the surface of the planet. When two plates collide head-on, their edges crumple, and huge masses of rock are thrust upward to form mountain ranges such as the Himalayas. When two plates move apart, a shallow cleft, called a rift valley, appears between them. The Great Rift Valley in Africa, which runs through the Jordan Valley, the Dead Sea and the Red Sea, and continues into East Africa, is an example.

Are there also signs of the movement of plates on Mars, to accompany the clear evidence of Martian volcanism? Thus far, no evidence has been found of features like the San Andreas fault or the Himalayan Mountains, but a feature clearly resembling the great African Rift Valley has been discovered. The Mars Rift Valley is 3000 miles long, roughly

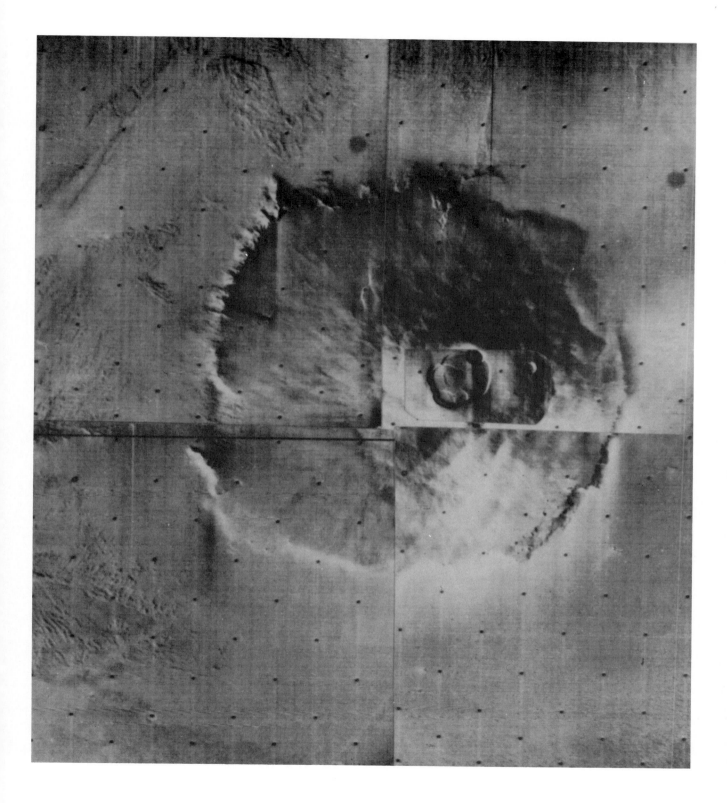

Figure 16.13 (a) The crater at the summit of the Mars volcano shown in Figure 16.12. The area in the white rectangle is 27 by 34 miles. (b) A high-resolution picture of the area outlined in white in Figure 16.13a, showing numerous individual flows of lava cascading down the slopes of the volcano.

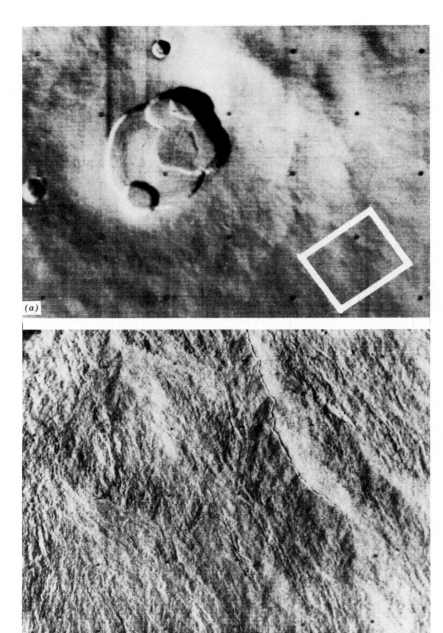

(a)

(b)

75 miles across and 15,000 feet deep at its lowest point. It is visible in the composite photograph on p. 410 as a jagged crack at the lower right, near the edge of the planet. The Mars Rift Valley resembles the African Rift Valley in dimensions, and is generally regarded as a geologically similar phenomenon. An overhead view of a 300-mile segment of the Mars Rift Valley is shown in Figure 16.14.

## Evidence for an Early Abundance of Water

Figure 16.14 also shows a clear pattern of branching channels that cut through the rims of the Mars Rift Valley. They resemble the tributaries of rivers or water-drainage systems on the earth. Figure 16.15 shows a similar pattern of branching tributaries cutting into the rim of the Grand Canyon (visible at the bottom), that were eroded by running water during the last few million years. It is difficult to imagine a process that could have produced the channels in Figures 16.14 and 16.15, other than the flow of large volumes of water.

*Figure 16.14   The Mars Rift Valley.*

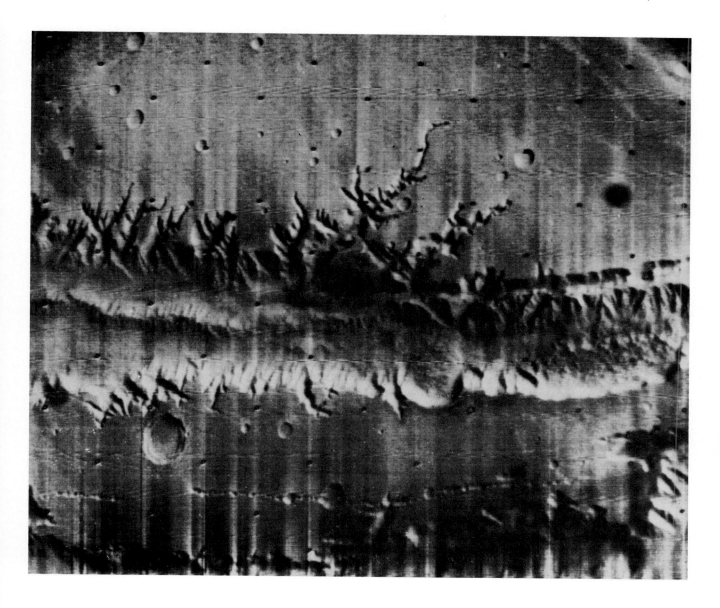

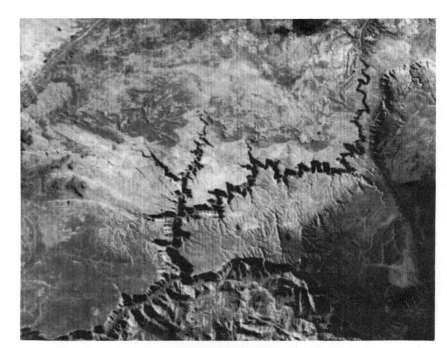

Figure 16.15  Branching tributaries cutting through a rim of the Grand Canyon (bottom).

Figure 16.16  A meandering channel on Mars, with tributaries feeding into the main channel at several points. The channel is about 250 miles long and three miles wide.

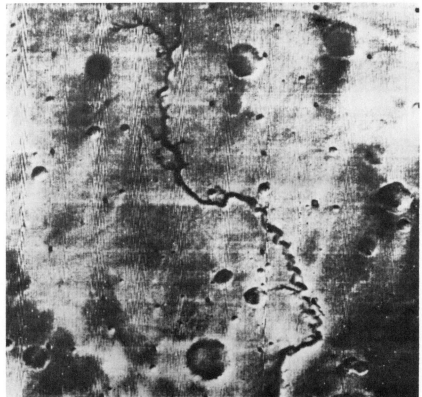

Other indications of an abundance of water appear in the Mariner pictures. Figure 16.16 shows a winding channel resembling a river bed, with tributaries feeding in at the upper end, and a distinct pattern of meanders in one section. From the angle at which the tributaries enter the main channel, it can be concluded that the direction of the flow of the water or other fluid was from the top to the bottom of the photograph.

Figure 16.17 shows the strongest evidence for the presence of an abundance of water on Mars at one time. This mosaic of pictures reveals a braided pattern of nearly parallel channels, such as are formed on the earth when the silt carried along by a river is deposited on the river bottom in large amounts. The silt eventually blocks the flow in the existing channel, and forces the river to create a new channel nearby. The channel segment shown in Figure 16.17 is 50 miles long and several miles wide.

The Mariner 9 pictures provide circumstantial evidence for the presence of large amounts of fluid on Mars at one time, but no direct proof that the fluid was water. However, geologists think that water is the most likely possibility. If water was once present on Mars in large amounts, it has long since disappeared, because observations made both from the earth and from the Mariner spacecraft indicate that only a trace of water exists on the surface of the planet at the present time. This small amount could not account for these riverlike features, which can only have been produced by the rapid flow of a very large volume of fluid.

Combining our knowledge of the present scarcity of water on Mars with the circumstantial evidence for its abundance in the past, we are

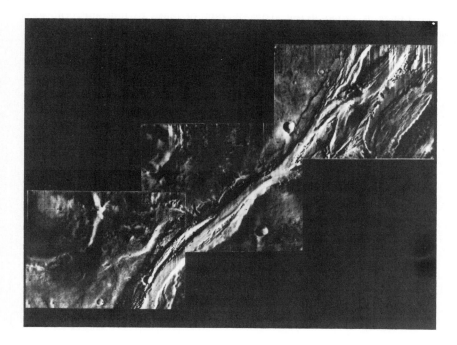

Figure 16.17 A braided pattern of channels on Mars, resembling riverbeds on the earth.

led to a tentative picture of the history of the planet. At one time, Mars may have had a substantial amount of water, which accumulated during a relatively long period of volcanism, possibly lasting billions of years. During that period its surface may have been traversed by many streams and rivers, but when the volcanism subsided, the source of the water disappeared. Individual molecules of water, leaking steadily into space, gradually diminished the amount of available moisture, and the streams and rivers dried up. Finally the planet reached the condition in which we find it today, with only a trace of moisture in the atmosphere, and no liquid water visible on the surface.

## Prospects for Life on Mars

Will life be found on Mars? Possibly. Mars is dry, cold, and less favorable than the earth for the support of life, but not implacably hostile. Experiments have shown that some terrestrial plants could exist in the Martian climate, although they would not flourish there. Cosmic-ray and ultraviolet radiation bombard Mars with an intensity that would be lethal to terrestrial life, but biologists have suggested ways in which Martian organisms could have evolved natural shields against these deadly rays.

*The Relationship Between Water and Life.* Water is the key element in a discussion of the prospects for Martian life. Water is an essential ingredient for the development of the kinds of life with which we are familiar, because it provides a fluid medium in which the complex molecules of the cell can move freely. This movement leads to frequent collisions between neighboring molecules and, as a consequence of these collisions, to chemical reactions that make up the ongoing process of life. The basic building blocks of living matter—immersed in the shallow Martian seas—would have collided ceaselessly; now and then the collisions would have linked them into the large molecules—proteins, DNA, and RNA—which are the essence of the living organism. The linking of small molecules to form large ones would have marked the first step along the path from nonlife to life.

The relationship between water and life lends a special interest to the Mariner evidence for large amounts of water on Mars at an earlier time. If life exists on Mars today, it probably can be traced back to a golden age on the planet, when emission of gases from volcanoes maintained a substantial average depth of water on the surface, as well as a denser atmosphere, and the climate rivaled the climate of the earth. The transition to the drier climate of today may have occurred very slowly, over a period of millions of years and a very large number of generations. In this case, Martian life could have adapted progressively to the gradual onset of severe conditions. During this long period of slowly increasing aridity, the weakest individuals in each generation would be eliminated, and the hardiest would remain, propagating their qualities of strength to their descendants. There seems no reason to doubt that varied and interesting forms could exist on Mars today as a result of this long-continued process of natural selection, if the planet once had an abundance of water.

Martian organisms, highly specialized for survival on a nearly water-free and airless planet, doubtless would present an unusual appearance; their forms, internal arrangements, and methods of reproduction might seem bizarre; the fundamental differences between the plants and animals, as we know them, might be blurred. Nonetheless, this extraterrestrial life would have much to teach us about the nature of life on the earth, for the basic chemistry of Martian life — product of an independent line of evolution, and adapted to very different conditions — conceivably would not be identical with the chemistry of terrestrial life. The comparison between two living structures, parallel but distinct, might yield insights into the metabolism of all living organisms, including man, that could not be acquired in decades of laboratory research on earth.

## Tests for Life on Mars

An important series of tests for Martian life is planned for 1976. In that year, a large spacecraft called the Viking will be launched, carrying an unmanned biochemical laboratory that will descend to the surface of Mars in a soft landing. The laboratory will be operated by remote control from the earth, and will be equipped with an array of instruments designed to search for signs of life on an alien planet. The instruments will include a device for the detection of some of the molecular building blocks of living matter, which will react either to living organisms or to the remains of life. Experiments will also be performed on samples of Martian soil after the soil has been moistened with water. If Martian organisms have evolved a way of life in which they hibernate during dry spells, awaiting a rare change in climatic conditions, these experiments may detect them.

In one of the experiments, a scoop mounted on the end of a long boom will be scraped across the landing site and then withdrawn into the body of the craft, presumably with particles of Martian soil adhering to it. The particles will be deposited in culture dishes containing ingredients that make excellent food for terrestrial bacteria. This food is made up of atoms of carbon, nitrogen, oxygen, and other elements. If microorganisms are present on Mars and their chemistry resembles that of bacteria on the earth, they will consume the food and multiply. The heart of the experiment lies in the fact that the carbon in the food is *radioactive* carbon, and not the nonradioactive carbon normally found in nature. The radioactive carbon, if ingested by the Martian organisms, would be incorporated into the chemicals of their cells in the reactions that constitute the life processes of the organisms. In these reactions, some of the radioactive carbon would be combined with oxygen to make radioactive carbon dioxide. The radioactive carbon dioxide would be exhaled by the bacteria.

Adjoining the chamber with the culture dishes would be a second chamber containing an instrument sensitive to radioactive substances. The second chamber is separated from the first by a filter through which no particle larger than a molecule of gas can penetrate. The food particles containing the original radioactive carbon cannot pass between the two chambers, but carbon dioxide gas can do so. The instrument will send a signal to the earth if it detects radioactivity. The receipt of this signal would indicate that a Martian microorganism has been eating the food containing radioactive carbon.

## JUPITER

The general properties of the giant planets were discussed in Chapter 13. Jupiter is the largest of the giant planets, and also the closest and most easily observed (Color Plate 16). It has been the subject of major theoretical studies which have led to a more detailed picture of its interior and atmosphere than is available for any other planet in this group. Figure 16.18 shows the internal structure yielded by these studies. The pressure at the center of the planet is estimated to be 10 million pounds per square inch. The temperature at the center of the planet may be as high as 100,000 degrees Fahrenheit. The high temperature is a result of the condensation of this massive body under the force of its own gravity. If Jupiter were more massive, its temperature could have risen to a level sufficient to ignite nuclear reactions, converting it into a small star.

It might be expected that the interior of Jupiter would be in a gaseous state, in view of the high temperature at its center. However, calculations on the properties of hydrogen subjected to a pressure of 10 million pounds per square inch and a temperature of one million degrees indicate, surprisingly, that the extreme pressure locks the atoms into a regular lattice of solid hydrogen with the properties of a metal. This solid core of metallic hydrogen is believed to extend out to a distance of 36,-000 miles from the planet's center.

A thick envelope of highly compressed, gaseous hydrogen and helium overlies the solid hydrogen core, extending upward with gradually diminishing density. The gaseous envelope is topped by two thick decks of clouds. The lower deck consists of water droplets and ice crystals. The upper layer of clouds consists of crystals of frozen ammonia compounds at a temperature of −300°F. These clouds of frozen ammonia present the visible face of the planet to the observer.

Conditions beneath the clouds are concealed from our view, but below the level of the cloud tops the temperature must rise, just as it does on the earth or any other planet with a fairly dense atmosphere. The rise in temperature is the result of the greenhouse effect, in which the atmosphere acts as an insulating blanket, sealing in the planet's heat and increasing its

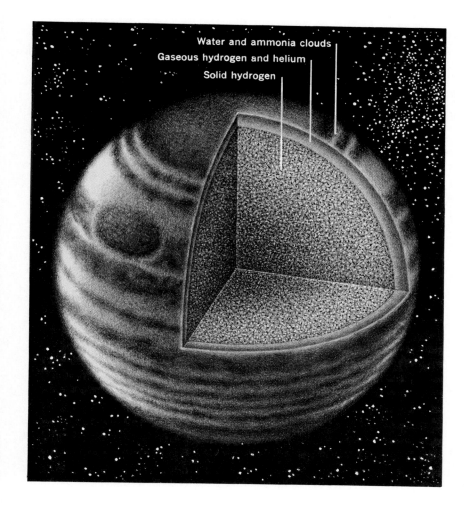

Water and ammonia clouds
Gaseous hydrogen and helium
Solid hydrogen

Figure 16.18   The structure of Jupiter.

temperature (see page 422). The temperature on the surface of the earth is increased from −20°F—the value which it would have if the earth were an airless body of rock—to a global average of 60°F as a consequence of this effect.

We suspect that Jupiter does not have a well-defined surface; probably, it possesses instead an atmosphere that grows steadily denser with increasing penetration into the interior of the planet. With or without a surface, Jupiter probably has a region, at some depth below its clouds, in which the temperature passes through a comfortable range for the development and continued support of life.

## Prospects for Life on Jupiter

Is it possible that life exists on Jupiter? The planet contains an abundance of hydrogen as well as the compounds of hydrogen with relatively common elements such as carbon, nitrogen, and oxygen. These compounds — among which ammonia and methane have been observed, and water vapor probably exists also — were present in abundance in the primitive atmosphere of the earth and are believed to have played a critical role in the events that led to the development of life on our planet. Their importance in evolution on the earth has ended, and they have long since escaped, but their continued presence on Jupiter leads us to wonder whether the initial steps along the path to life have not also occurred on that planet.

At first this seems unlikely, because Jupiter is nearly 500 million miles from the sun and five times as distant as the earth, and receives very little solar heat. But the greenhouse effect introduces a ray of hope into the situation. In the region below the clouds in which the temperature is comfortable for life as we know it, the very gases exist out of which life is believed to have evolved on the earth in its early years. Perhaps a kind of life has developed on Jupiter as well. It could not be the oxygen-breathing life with which we are familiar; and we assume that it would be quite different. The answer will not be known until — perhaps, at the turn of the century — spacecraft equipped with instruments for the detection of alien forms of life make their first trip to Jupiter.

## Questions

1. Why are Venus and the earth thought to have had roughly the same starting conditions at their origin? Why is Venus assumed to have approximately the same internal geological history as the earth?

2. Why does Venus rotate so slowly on its axis? What is the probable explanation for its retrograde rotation?

3. Briefly describe the methods by which the Venus atmosphere has been explored.

4. What are the temperature, pressure, and atmospheric composition at the surface of Venus? How has the abundance of $CO_2$ in the atmosphere produced these conditions?

5. In what respects is the surface of Mars different from the moon's surface? The earth's surface?

6. On the basis of the Apollo findings and our knowledge of the earth's history, would you expect to find evidence of geological activity on Mars? How long would you expect the activity to have persisted? What do the Mariner photographs indicate in this connection? Cite the evidence in support of your answers.

7. Cite the evidence for the presence of substantial amounts of liquid water on Mars during its history.

8. Has the probability of life on Mars increased or decreased since the the Mariner 9 findings? Explain.

9. Describe an experiment designed to search for life on Mars. Can you devise one or more additional experiments designed to search for life on an alien planet. Your experiments should be different from those described in the text.

10. Should one expect to find an earthlike core at the center of Jupiter? Why or why not?

11. Draw a diagram roughly to scale showing cross-sectional views of the internal structures of the earth and Jupiter. Draw additional diagrams showing your guesses regarding the internal structures of the moon, Mars, and Venus. Explain the reasoning on which your guesses are based.

12. Why is the seemingly hostile planet Jupiter considered a possible candidate for the presence of life?

# 17 The Earth and Venus: A Study in Contrasts

Why did two planets, probably formed out of similar materials and situated at comparable distances from the sun, evolve along different paths? Why is the surface of Venus baked by a searing heat, while the earth luxuriates in a climate friendly to all known forms of life? These are among the most important and interesting questions in astronomy from the viewpoint of the nonscientist, because they bear directly on the probability of finding life on other earthlike planets in other solar systems. The answers depend on an understanding of the factors that govern the evolution of a planet's atmosphere.

## THE PRIMITIVE ATMOSPHERE OF A PLANET

When a planet is newly formed, the ingredients of its atmosphere consist of the gaseous elements and compounds that were abundant in the solar nebula out of which it condensed. The foremost among them is hydrogen—the most abundant element in the Cosmos. Next is helium,

*An artist's concept of the Venus Inferno.*

and then oxygen and nitrogen, the third and fifth most abundant elements in the Cosmos and presumably in the solar nebula. Carbon, fourth in abundance after oxygen, would not be present in the atmosphere as an element, because it does not exist in gaseous form at the temperatures that probably prevailed on the surfaces of the new planets. However, carbon combines with hydrogen readily to form the gas methane, or $CH_4$. This gas must have been a major constituent of the primitive atmospheres of the planets. Oxygen and nitrogen also combine with hydrogen readily to form water vapor and ammonia, respectively. These gases also must have been abundant in the primitive atmosphere of every planet. Free oxygen reacts readily with iron and silicon in the crust of the planet to form solid compounds, and would not exist in the atmosphere. However, free nitrogen would be present.

Neon is the next most abundant element after oxygen that would exist in the form of a gas at atmospheric temperatures. It would therefore be an important constituent of primitive planetary atmospheres. Beyond neon, we expect to find a small amount of primitive argon and a still smaller amount of xenon. Only trace amounts of other elements would be expected, either because of their intrinsic rarity in the cosmic abundances of the elements, or because of a strong tendency to enter into chemical combinations that remove the element from the atmosphere.

These, then, are the gases that must have made up the atmospheres of the earth and Venus when they were young planets: hydrogen, helium, methane, ammonia, nitrogen, neon, a small amount of argon, and trace amounts of other substances.

**The Escape of the Primitive Atmospheres**

One of the most surprising things about the present atmospheres of the earth and Venus is that they bear little resemblance to the elements and compounds listed above. These substances must have formed a gaseous layer around each planet when it first condensed out of the solar nebula, but today they are either missing—with the exception of nitrogen and water vapor—or greatly diminished in abundance in comparison with theoretical estimates. At the same time, other gases not included in the list are abundant, including, in the case of Venus, the gas carbon dioxide, which constitutes the major constituent of its atmosphere.

Even more surprising is the fact that the earth and Venus—two closely similar planets in size and mass as well as distance from the sun, and undoubtedly possessing closely similar atmospheres at the time of their formation—should today have completely different atmospheres. Contrast the two planets: the atmosphere of Venus is 90 to 95 percent carbon dioxide with the remainder unknown. The atmosphere of the earth is composed mainly of nitrogen and oxygen, with only 0.03 percent of carbon dioxide.

What is the explanation for these gross differences? The scarcity of hydrogen and helium in the atmospheres of the two planets can be explained easily, because these gases are too light to be held by the gravitational forces at the surfaces of the earth and Venus. Calculations show that they must have leaked away promptly at the beginning of each planet's life. The remainder of the puzzle is not as easy to solve. The first clue lies in the fact that the noble gases, neon, argon, and xenon, are present in vastly smaller amounts than expected. Neon should be even more abundant than nitrogen in the earth's atmosphere, but only one ten-billionth as much is present, and only one-millionth of the expected concentrations of argon and xenon. The importance of these gases lies in the fact that they are chemically inert and, therefore, their absence cannot be explained on the grounds that they have reacted with other elements and have been removed from the atmosphere as solid compounds.

The only possible explanation for the scarcity of the noble gases is that they escaped to space early in the history of the earth. When they disappeared into space, the other major constituents of the primitive atmospheres must have disappeared with them, for each of these constituents is far lighter than xenon. The forces that broke the grip of the earth's gravity on the atoms of xenon must also have broken its grip on the atoms of all other substances in the primitive atmosphere.

Perhaps the cause of their disappearance was a shortlived increase in the outpouring of energy from the sun, which evaporated the atmospheres of all planets circling in relatively close orbits. Or, perhaps, in the final stages of their accumulation, the earth and Venus were raised to high temperatures during the bombardment by planetesimals and pieces of rocky debris of all sizes, and the primitive gases evaporated.

Whatever the forces were that stripped away the earth's primitive atmosphere, they must also have removed the primitive atmosphere of Venus. At some point early in their lives, the two planets probably circled the sun as airless bodies of rock.


## THE PRESENT ATMOSPHERES OF THE EARTH AND VENUS

Today each planet is blanketed by a substantial atmosphere. Where did these atmospheres come from? At the beginning, the two bodies were airless. Their atmospheres could not have condensed later on out of the space surrounding each planet, because after the initial formation of the planets the solar system was as devoid of interplanetary gases as it is today. The only remaining possibility is that the gases of the present atmospheres were exhaled from the earth's interior at some point during its history. There is evidence in support of this theory. Large amounts of gas are exhaled from the interior of the earth in volcanic eruptions. These gases consist primarily of water vapor, nitrogen, and carbon dioxide. The amount of water vapor coming up to the surface of the earth could suffice, if volcanoes have been as frequent throughout the earth's history as they are today, to fill the earth's oceans without any other source being provided. Furthermore, the amount of nitrogen entering the atmosphere of the earth in the form of volcanic gas, could suffice—again if it has been accumulating throughout the earth's history at the same rate—to account for all the nitrogen in the atmosphere today.

These ideas support the theory that the oceans and the atmosphere of the earth have accumulated either sporadically or steadily in the life of the planet, as a result of the exhalation of gases from the interior of the planet through cracks and fissures in the crust.

One major discrepancy remains: carbon dioxide makes up a substantial fraction, about 10 percent, of the gases emitted in volcanic eruptions. If this gas has been accumulating steadily throughout the history of the earth as hypothesized for the nitrogen and water vapor emitted in volcanic eruptions, carbon dioxide should be the principal gas in the atmosphere, with a concentration nearly 100 times greater than the concentration of nitrogen. But the observed abundance of carbon dioxide in the atmosphere is three-hundredths of one percent.

**Accumulation of $CO_2$ in the Venus Atmosphere**

This difficulty brings us to the matter of the difference between the earth and Venus. The two planets undoubtedly condensed out of the same materials in the solar nebula, and similar amounts of carbon dioxide must have been trapped in the interior of each planet as it formed. The planets have comparable masses, and during the first billion years of their lives they probably were heated to the same internal temperatures by the decay of radioactive elements. Consequently, they must have had the same level of volcanic activity, and carbon dioxide, along with other volcanic gases, must have escaped through the crust of each planet at about the same rate during its history. It should now be accumulated in the same concentration in the atmospheres of both planets.

According to calculations based on today's rate of emission of volcanic gases through the earth's crust, the amount of carbon dioxide accumulated in each atmosphere should be half a ton per square inch. The atmosphere of Venus is observed to have just this amount of carbon dioxide, but the earth has ten thousand times less. Why did the histories of the two planets diverge?

Carbon dioxide is an extremely effective agent in trapping a planet's heat and in preventing it from escaping into space, while continuing to admit solar heat in the form of visible light. This "greenhouse effect" produced by the carbon dioxide in the Venus atmosphere explains the 800° temperatures on the surface of that planet (see Chapter 16). The temperature of the earth would also be unbearable if the planet were covered by a blanket of carbon dioxide as dense as that on Venus. When we understand why the earth has a scarcity of carbon dioxide relative to Venus, we will also understand why the earth offers a pleasant climate for life, while Venus is an inferno.

## Removal of $CO_2$ from the Earth's Atmosphere

The solution to the carbon dioxide problem is connected with the fact that the rocks of the earth's crust have a capacity for absorbing carbon dioxide. The absorption takes place in a chemical reaction in which atmospheric carbon dioxide combines with rocks to form carbonates, somewhat as oxygen in the atmosphere combines with iron to form rust. Geochemists have long been familiar with chemical reactions taking place at the surface of the earth which demonstrate this process. An example of such reactions is,

$$CO_2 + CaSiO_3 \rightarrow SiO_2 + CaCO_3$$

in which carbon dioxide combines with calcium silicate, a variety of rock, to form quartz plus calcium carbonate. Sand is an example of nearly pure quartz, and marble and limestone are examples of calcium carbonate.

This and similar reactions, going on slowly but steadily at the surface of the earth over millions of years, presumably removed carbon dioxide from the atmosphere and converted it into solid compounds nearly as fast as it entered the atmosphere through volcanic eruptions.

A partial confirmation of this theory is found in the fact that the amount of carbon locked up in the sedimentary rocks in the earth's crust in the form of carbonates is approximately equal to the amount of carbon contained on Venus in the form of atmospheric carbon dioxide.

*The Role of Marine Animals.* Perhaps the removal of carbon dioxide by the rocks of the crust was the reason why the temperature on the earth stayed at a comfortable level long enough for life to take hold here. But as soon as animal life appeared in abundance, its own chemistry could have provided a means for keeping down the level of carbon dioxide in the earth's atmosphere. Marine animals absorb carbon dioxide from seawater and convert it within their bodies to the solid substances called carbonates.[1] For every molecule of carbon dioxide removed from the water by this process, one molecule eventually enters the ocean from the atmosphere. Thus, the absorption of carbon dioxide from seawater by marine animals is equivalent to its removal from the atmosphere.

When the earth was young, life was either absent or scarce, and

[1] Seashells, for example, are nearly pure calcium carbonate.

carbon dioxide could not have been removed in this way. But the gas still would have been absorbed from the atmosphere of the young earth by the chemical changes involving rocks rather than living organisms.

### The Runaway Greenhouse Effect

Venus may be lifeless but its crust must contain the same silicates that removed carbon dioxide from the earth's atmosphere. Why were the silicates less effective on Venus? Why did Venus not follow the same path as the earth when both were young planets?

The answer is connected with a special property of the reactions that cause carbon dioxide to combine with rocks in the form of carbonates. The effectiveness of these reactions in removing carbon dioxide from the atmosphere depends very strongly on the temperature of the rocks. The higher the temperature of the rocks, the less effective they are in removing carbon dioxide. When the earth was a relatively young planet—old enough to have cooled from its birth process, but still too young to have accumulated its insulating blanket of atmospheric gases—its average temperature was about $-20°F$. At this low temperature the rocks of the earth's crust would absorb nearly all carbon dioxide from the atmosphere.

When Venus, on the other hand, was a young planet without an atmosphere, its average temperature was higher—about $140°F$—because of its closeness to the sun. At this higher temperature, the rocks on its surface absorbed far less carbon dioxide than was absorbed by rocks on the earth in the same period. According to the measurements on the temperature-dependence of the reactions, the amount of carbon dioxide remaining in the atmosphere of Venus must have been 10 to 100 times greater than the amount in the earth's atmosphere.

This blanket of carbon dioxide, although still thin at that early time in comparison with the heavy $CO_2$ atmosphere on Venus today, would still have been great enough to produce an appreciable greenhouse effect. The greenhouse effect, raising the temperature of the surface of Venus still higher, would further diminish the effectiveness of the Venus rocks in absorbing the $CO_2$ that continued to pour into the atmosphere through volcanoes. As more $CO_2$ accumulated in the atmosphere, the greenhouse effect became stronger, the temperature of the rocks rose higher, their effectiveness in absorbing $CO_2$ was still further diminished, the greenhouse effect was further enhanced, and so on. In this way, a runaway greenhouse effect may have started, leading to the massive carbon dioxide atmosphere and ovenlike conditions that characterize Venus today.

### Presence of Oxygen on the Earth

These ideas explain the difference in carbon dioxide on the earth and Venus. They do not account for the presence of free oxygen, the second most abundant constituent of our modern atmosphere. This gas should not exist in the atmosphere in appreciable concentrations despite its high abundance in the Cosmos, because its strong tendency to enter into chemical reactions removes it promptly from the atmosphere and deposits it on the surface in the form of solid oxides. Again, the explanation lies in the presence of life on the earth. The main source of atmospheric oxygen is generally believed to be plant life. Plants—especially marine algae—absorb carbon dioxide from the atmosphere, convert the carbon into compounds that enter the structures of their bodies, and return the residual oxygen atoms to the atmosphere. These biological activities continually renew the supply of free oxygen in the atmosphere, which would otherwise be depleted rapidly—in 2000 years or so—by chemical reactions with the materials of the earth's crust.

### Scarcity of Water on Venus

Water, like carbon dioxide, was one of the gases that must have been trapped in the interiors of both the earth and Venus when they first condensed out of the solar nebula. The release of trapped water vapor through cracks and fissures in the crust of the earth has led to the accumulation of oceans covering three quarters of the surface of the planet to a mean global thickness of 8000 feet.

A comparable amount of water should exist on the surface of Venus. Because of the planet's high temperature, this water should be in the form of a vapor in the atmosphere, rather than a liquid filling the natural basins in the topography of the crust. But it is known from the results of the measurements performed by the Russian spacecraft that the equivalent of an ocean of water is not present on Venus. In place of the 8000 feet of liquid water that cover the face of the earth, the atmosphere of Venus has no more than the equivalent of a foot of liquid water.

The surprising fact is not that Venus has so little water, but that the earth has so much. Water is lost very readily from the atmosphere of a planet with the mass of the earth or Venus, because solar ultraviolet radiation breaks up the water molecule into its basic hydrogen and oxygen atoms, and the hydrogen atoms, being extremely light, quickly evaporate to space. The oxygen atoms remain behind, trapped by the earth's gravity, and enter into chemical combinations with elements in the crust of the planet.

As a result of this process—called *photodissociation*—water should be removed from the atmosphere of the earth almost as rapidly as it reaches the surface. But the earth has retained its water in spite of the photodissociation effect. Two factors appear to be responsible for this. *First,* the temperature of the air drops with increasing height above the surface of the earth, reaching a low point of about $-80°F$ at a height of approximately 30,000 feet. As the temperature falls, the water vapor condenses into droplets of liquid water or tiny crystals of ice, forming clouds. The water droplets or ice crystals are immune to destruction by solar ultraviolet radiation, which can only destroy individual molecules of water. Only one-tenth of one percent of the water in the atmosphere remains in the form of a vapor at high altitudes, as a result of the so-called "cold trap" at 30,000 feet. *Second,* solar ultraviolet radiation coming from the sun is absorbed rapidly as it enters the atmosphere from above, partly by photodissociation and partly by reactions in which ultraviolet photons eject electrons from atoms of air. Oxygen atoms, which are abundant at great heights in the earth's atmosphere, are particularly effective in absorbing the ultraviolet wavelengths that break up water molecules. By the time the solar radiation has penetrated to 30,000 feet, its ultraviolet component has been almost entirely removed.

Thus, in the atmosphere below 30,000 feet, where an abundance of water vapor exists, the ultraviolet intensity is negligible because it has been screened out by oxygen; and in the upper atmosphere, where the ultraviolet intensity is still strong, the amount of water vapor is negligible because of the cold trap. This combination of circumstances has protected the water on our planet. Consequently, water has accumulated steadily on the surface of the earth throughout its history, gradually filling up the empty basins of the earth's crust to form the oceans.

*Disappearance of Water from Venus.* What are the corresponding circumstances on Venus? First, the temperatures in the atmosphere of Venus are higher than at the comparable levels in the earth's atmosphere. The consequence of this fact is that 200 times more water can remain in the atmosphere in the form of vapor than in the case of the earth. Second, Venus lacks the shielding screen of oxygen molecules that make up 20 percent of the terrestrial atmosphere and protect the water vapor at lower levels from photodissociation. The amount of solar ultraviolet radiation penetrating to the lower levels of the Venus atmosphere is 1000 times greater than in the case of the earth.

The product of the two factors — 200 times more water vapor, and 1000 times more solar ultraviolet radiation — corresponds to an increase by a factor of 200,000 in the rate of removal of water molecules from the Venus atmosphere by photodissociation, in comparison with the rate of removal of water from the earth by the same process. This factor is adequate to account for the scarcity of water on Venus and its abundance on the earth.

## Conditions for Life

How far from the sun must a planet be to maintain a congenial climate for life? We do not know. The exploration of Venus has told us only that that planet was too close, and that was its undoing. If it had been only a few million miles farther away, the temperature of its surface might have climbed slowly enough to permit life to gain a toehold, and once life began, it would have held the abundance of carbon dioxide in check, and would have prevented the temperature from climbing out of bounds subsequently. But Venus, having lost its chance to harbor life when the solar system was first formed, could never again recapture the opportunity.

## A WORD ABOUT MARS

This comparative study of the histories of the atmospheres of the earth and Venus demonstrates that an earthlike planet must not be too close to its sun if conditions are to remain favorable for the evolution of terrestrial forms of life. In the case of our solar system, the critical threshold lies somewhere between the orbits of Venus and the earth.

How far on the other side of the earth's orbit can a planet be, and yet maintain a comfortable climate for the support of life? The answer depends sensitively on the size of the planet. It may be quite far from its sun, and the warming effect of the sun's rays correspondingly small, but if the planet is large enough to build up a high level of internal temperatures through radioactive heating, volcanic eruptions will be frequent, with copious emissions of gases from the interior of the planet, and its atmosphere can be expected to accumulate to a high density. Thus, a significant greenhouse effect will develop, trapping the heat of the sun and building up a comfortable temperature on the planet's surface.

It is quite possible that a planet the size of the earth could reach a suitable temperature for the evolution and support of life at distances from the sun equal to that of Mars or greater.

If, however, the planet is modest in size—the size of Mars, for example—its internal temperatures will not rise to a sufficiently high level to produce substantial volcanic activity with copious emission of gases. Moreover, the relatively weak gravitational field of the planet will render it less effective in preventing the lighter gases from escaping to space, further diminishing the density of the atmospheric blanket. For these two reasons, the greenhouse effect will be small, and the temperature on the planet's surface will be low. The scarcity of water—as a further consequence of the diminished emission of gases from the interior of the planet—will also tend to act against the onset of chemical evolution.

Mars is an example of a planet described by the second set of circumstances. Its distance from the sun is greater than that of the earth, and the mean intensity of solar heat is correspondingly diminished. Consequently, the mean temperature on the surface of the planet is 40° below zero Fahrenheit. More serious for the prospects of Martian evolution is, however, the small size of the planet, which has undoubtedly resulted in a fairly low level of internal heating, and a diminished output of volcanic gases. It is mainly as a consequence of this factor that the Martian atmosphere is so thin, and the greenhouse effect relatively modest. If Mars were in its present distance from the sun, but were as large as the earth, it could present a very agreeable environment for living organisms.

**Questions**

1. Explain why the primitive atmospheres of the earth and Venus are believed to have disappeared.
2. What is believed to be the source of the gases now present in the atmosphere of the earth and Venus? What is the source of the water in the earth's oceans?
3. How was $CO_2$ removed from the earth's atmosphere?
4. Why did $CO_2$ accumulate in the Venus atmosphere?
5. What is the origin of the free oxygen in the earth's atmosphere?
6. What happened to the water on Venus? Why was water retained on the earth?
7. Explain the runaway greenhouse effect.
8. Mariner IX observations in 1971 indicated a region of high heat flow in a limited area on the surface of Mars. Assuming that the "warm spot" is permanent, and not a transient phenomenon, would this area be a particularly good place in which to begin the search for life on Mars? Explain.
9. List and explain the principal factors that determine the conditions on a planet that are favorable for life as we know it. According to your answer, do these factors define a limited "zone of life," within which conditions are favorable for the support of life? Again, according to your ideas, how would the luminosity of the star affect the position of the "zone of life"? Where would you expect the "zone of life" to be for planets circling a Main-Sequence blue giant? A Main-Sequence red dwarf?

THE PLANETS

# 18    Life in the Cosmos

Discoveries in astronomy and biochemistry during the last decades have linked the Universe of the stars to the world of life in a chain of cause and effect that extends over billions of years, commencing with the formation of stars in our Galaxy and ending with the appearance of man on the earth. At the beginning of this history there existed only atoms of the primeval element, hydrogen, which swirled through outer space in vast clouds. These clouds were the raw stuff out of which stars and planets were made. Occasionally the atoms of a cloud were drawn together by the attractive forces of gravity; with the passage of time the cloud contracted to a small, dense globe of gas; heated by self-compression, it rose in temperature until, at a level of some millions of degrees, its center burst into nuclear flame. Out of such events, stars were born.

Within the newborn star a series of nuclear reactions set in, in which all the other elements of the Universe were manufactured out of the basic ingredient, hydrogen. Eventually these nuclear reactions died out, and the star's life came to an end. Deprived of its resources of nuclear energy, it collapsed under its own weight, and in the aftermath of the collapse an

*Fossil bacteria several billion years old.*

explosion occurred, spraying out to space all the materials that had been created within the star during its lifetime.

In the course of time, new stars, some with planets around them, condensed out of these materials. The sun and the earth were formed in this way, four and a half billion years ago, out of materials that were manufactured in the bodies of other stars earlier in the life of the Galaxy, and then dispersed to space when those stars exploded.

When the earth was formed, it must have been barren, but within one billion years or so life appeared on its surface. How can we explain this fact? If we confine the inquiry within the boundaries of science, discoveries made during the last few decades suggest a tentative scientific explanation for the presence of life on the earth.

## THE ORIGIN OF LIFE

It now appears possible that the first living creatures on the earth evolved spontaneously out of chemicals that filled the atmosphere and oceans of the planet in the early years of its existence. The following facts suggest this hypothesis.

*First,* biologists have shown that all living organisms on the face of the earth depend on two kinds of molecules—amino acids and nucleotides—which are the *basic building blocks of life*—just as the physicists have shown that all matter in the Universe is constructed out of three building blocks—the neutron, proton, and electron.

*Second,* chemists have manufactured these molecular building blocks of life in the laboratory out of the gases that are believed to have filled the atmosphere of the earth when it was a young planet.

*Third,* an object has been discovered that links the nuclei, atoms, and molecules of the physical universe to the complex organisms of the living world. This object, called the *virus,* lies on the borderline between inanimate matter and life. Its existence gives credibility to the notion that life evolved out of nonliving chemicals.

### The Building Blocks of Life

The basic building blocks of living matter are more complicated than the building blocks of the physical world. Twenty different kinds of amino acids play a critical role in living creatures, and five different kinds of nucleotides. Furthermore, each amino acid or nucleotide is, itself, a rather complex molecule made up of approximately 30 atoms of hydrogen, nitrogen, oxygen, and carbon, bound together by electrical forces of attraction. An example of the structure of a typical amino acid is shown in Figure 18.1.

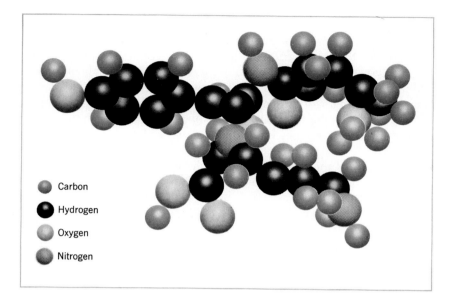

Carbon

Hydrogen

Oxygen

Nitrogen

*Figure 18.1    Three amino acids linked into a short segment of a typical protein.*

The amino acid and the nucleotide have very different functions in the chemistry of life. Within the cell the amino acids are linked together into very large molecules called *proteins*. One class of proteins, called structural proteins, makes up the structural elements of the living organism — the walls of the cell, hair, muscles, and bone. The structural proteins are like the steel framework and walls of a building. The other type of protein is called the enzyme. Many kinds of enzymes exist; each kind controls one of the many chemical reactions that are necessary to sustain the life of the organism.

All proteins, in all forms of life, plant and animal, are constructed out of the same basic set of 20 amino acids. One protein differs from another only in the way in which their constituent amino acids are linked together. However, these differences are all-important. The distinction between a man and a mouse, both in appearance and personality, depends entirely on the differences between the proteins contained in the cells of their bodies.

Proteins are assembled within living organisms by the second set of building blocks — the nucleotides. Nucleotides are joined together within the cell to form very long chains, called *nucleic acids*. The most important type of nucleic acid is called deoxyribonucleic acid, or DNA for short. DNA is the largest molecule known, containing, in advanced organisms such as man, as many as 10 billion separate atoms. The size of the DNA molecule is understandable when we consider the complexity and importance of its functions in the living cell. The DNA molecule is the most important molecule in every living organism, even more important than the protein, because it determines *which* proteins will be assembled; the DNA molecule has the master plan for the organism.

How does DNA control the assembly of proteins in the cell? The general features of the process began to emerge during the decade of the 1950's, although many of the details are still not clearly understood. It appears that the 20 amino acids and five nucleotides float freely in the fluid of the cell. The DNA molecules that direct the assembly of proteins are located in the center of the cell. In the first step, free nucleotides are attracted to a segment of one of the DNA molecules at the center. They line up alongside the DNA segment to form a replica of it. In the second step, the replica detaches itself from the master DNA chain, and drifts off into the cell; it is a messenger that carries instructions from the DNA into the body of the cell for the assembly of one particular kind of protein. In the third step, another molecule enters the picture. This molecule serves as a connecting link which brings the amino acids in the fluid of the cell to the appropriate places alongside the messenger. There are 20 kinds of connecting links, one for each kind of amino acid. Each of the connecting links attracts one and only one of the 20 amino acids. When, in the course of chance collisions, the right kind of amino acid comes into contact with the end of the connecting link designed for that particular amino acid, it is held fast there. At the other end of the connecting link is another set of molecules, making up a surface of nooks and crannies so constructed that it can only fit into the appropriate place along the length of the messenger. When the connecting link takes its place along the messenger, it adds the amino acid to the chain of amino acids that has already been built up. When a chain of amino acids has been assembled along the full length of the messenger, the assembly of amino acids into a protein is complete. The assembled chain then detaches itself from the messenger and drifts off into the fluid of the cell.

By this rather complicated process, the essential proteins are built up within an animal in accordance with the order of the nucleotides in its DNA molecules. The segments of the DNA molecule are "read" like the words of a book. Each DNA segment, controlling the assembly of one protein, is a word; each nucleotide within a segment is a letter; the order of the letters provides the meaning of the word, that is, the protein to be assembled. The full set of DNA molecules contained within a cell is the library of genetic information for the organism. The DNA molecules in the cells of a human being direct the assembly of the amino acids in the human body into human proteins; the different DNA molecules in the cells of a mouse direct the assembly of its amino acids into mouse proteins.

## The Mechanism of Inheritance DNA

How is the plan for the assembly of the right kind of proteins passed from one generation to the next? How do progeny inherit their characteristics from their parents. The answer lies in a most extraordinary property of the DNA molecule—the ability to make a copy of itself. The mechanism by which DNA copies itself was discovered in 1953 by an

Figure 18.2 James Watson and Francis Crick.

Anglo-American team, James D. Watson of Harvard University and Francis Crick of Cambridge University (Figure 18.2). This discovery is one of the most important single scientific events of the twentieth century to date. We have described the DNA molecule as a chain of nucleotides; but Watson and Crick found it is not a *single* chain of nucleotides; it consists, rather, of *two* chains, joined at regular intervals by molecules between them like rungs of a ladder (Figure 18.3). In each rung of the ladder there is a weak spot, which is easily broken. During the early history of a cell the two strands remain connected, but when the cell has attained its full growth, and the division into two daughter cells is about to commence, the weak connections running down the middle of the ladder break, and the double strand separates into two single strands. Each of the single strands then collects unto itself new nucleotides out of the pool of nucleotides floating in the cell, and assembles them into a new double-stranded molecule. There now exist two identical DNA molecules, where formerly there was one. The two DNA molecules separate, and move to opposite corners of the cell; the cell then divides into two daughter cells, each containing one complete set of DNA molecules. Thus, each of the daughter cells contains a copy of the volumes of genetic information that has been in the parent cell. This is the way in which the shape and the character of a plant or an animal are transmitted from generation to generation.

In summary, the DNA molecule controls the assembly of proteins, and the proteins determine the nature of the organism. Each living organism has its own special set of DNA molecules; no two organisms have the same set unless they are identical twins. However, *the basic*

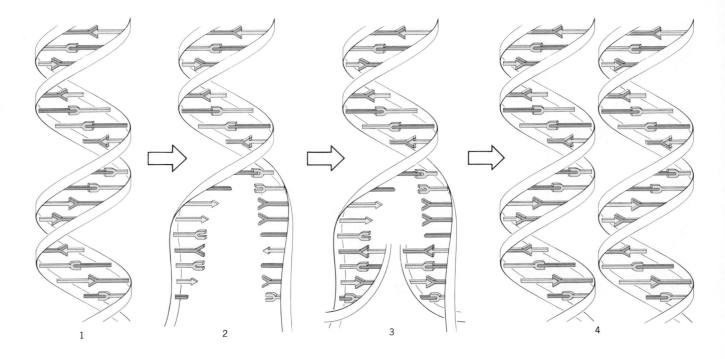

1 2 3 4

Figure 18.3 The mechanism of reproduction: DNA resembles a twisted ladder. (1) Each rung of the ladder is a linked pair of nucleotides represented here by the four symbols ⟹ ⟹ ⟹ ⟹. When a cell is about to divide, the ladder untwists, each rung breaks at the midpoint (2) and new nucleotides are collected from the surrounding molecules in the cell. (3) The result is two DNA molecules, replicas of the original DNA (4).

nucleotides and amino acids are the same in every living creature on the face of the earth, whether bacterium, mollusc, or man.

### Formation of the Building Blocks Out of Atmospheric Gases

With this fundamental property of living creatures in mind, one can appreciate the importance of a critical experiment performed in 1952 by Stanley Miller, at that time a graduate student working on his Ph.D. thesis under Harold Urey. At the suggestion of Urey, Miller mixed together the gases—ammonia, methane, water vapor, and hydrogen—which were abundant in the parent cloud of the earth, and which were probably abundant in the atmosphere of the young earth. He circulated the mixture through an electric discharge. At the end of a week, Miller found that the water contained several types of amino acids. Figure 18.4 is a schematic diagram of Miller's apparatus.

Subsequently, nucleotides were created in the laboratory under similar conditions. In some experiments, amino acids and nucleotides have been manufactured out of a variety of gas mixtures, using various sources of energy—bombardment by alpha particles, irradiation with ultraviolet light, and simple heating of the ingredients. The results of all these experiments, taken together, demonstrate that the molecular building blocks of life could have been created in any one of many different ways during the early history of the earth.

Amino acids and nucleotides might have been formed on the earth out of these same ingredients 4 billion years ago, by the discharge of lightning in primitive thunderstorms, or by the action of solar ultraviolet rays from the sun. We can guess what happened thereafter. Gradually the critical molecules drained out of the atmosphere into the oceans, building up a nutrient broth of continuously increasing strength. Over a long period of time the concentration of amino acids and nucleotides built up, until eventually a chance combination of building blocks produced still more complex molecules—primitive proteins and nucleic acids. With the further passage of time, cells developed; many-celled organisms appeared; and living organisms were started on the long road down to the complexity of the creatures that exist today.

## The Virus

This theory of the origin of life emerges from the union of astronomy, biology, and chemistry—three solidly established branches of modern science. Yet it is hard to believe that existing forms of life, in all their variety and sophistication, can be traced back to simple chemicals. Is

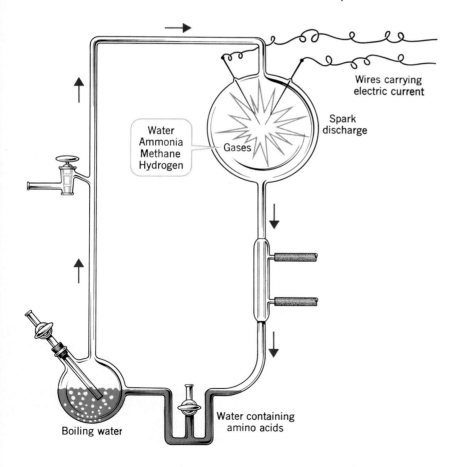

*Figure 18.4 The apparatus in which Miller first produced amino acids from gases of the primitive atmosphere.*

Wires carrying electric current

Spark discharge

Water
Ammonia
Methane
Hydrogen

Gases

Boiling water

Water containing amino acids

there any direct evidence for the development of life out of nonliving molecules?

The answer is: Yes, there is an entity, very common in the world today, that possesses, at the same time, the attributes of a nonliving molecule and the attributes of a living organism. This entity is the virus—the smallest and simplest object that can be said to be alive.

The existence of viruses first came to light at the end of the nineteenth century, in the course of a series of experiments designed to reveal the cause of a disease affecting tobacco plants. It was found that the juice pressed from the leaves of infected plants could transmit the infection to other plants. Apparently, the infection was transmitted by bacteria contained in the fluid. But when pressed through a fine filter, which screened out all visible bacteria, the fluid still retained its power of infection. In 1898 a Dutch botanist, Beijerinck, suggested that the disease was not caused by a germ, but by a poisonous chemical. Beijerinck called the chemical a "virus," which is the Latin name for poison.

Further research revealed that viruses are the cause of many diseases, and medical interest in viruses intensified during the first decades of the twentieth century. Gradually, the suspicion developed that the virus was no ordinary chemical. A variety of experiments suggested that the virus, although too small to be seen under the microscope, possessed the basic attribute of living organisms—the ability to reproduce itself.

Still, the evidence for the living virus was indirect; no one had yet seen one in the act of reproduction. But in the years after World War II a new instrument was perfected, which provided the biologists with a powerful tool for the study of small organisms. This instrument was the electron microscope. Ordinary microscopes, in which the object under study is illuminated by rays of light, are limited to a magnifying power of approximately 2000. The smallest bacteria, which are a hundred-thousandth of an inch in size, can just barely be seen in these microscopes. But the electron microscope, which directs a beam of electrons at the object instead of a beam of light, can produce magnifications as high as several hundred thousand diameters.

Under the electron microscope the virus finally became visible, and all the important details of its structure were revealed. It was found that viruses come in many shapes—round, cylindrical, polyhedral, and with tails (Figure 18.5). They also come in many sizes. The largest is as large as a small bacterium; the smallest, which is a millionth of an inch in diameter, is smaller than many nonliving molecules. Viruses bridge the gap in size between the inanimate and the animate worlds.

Yet, these tiny particles are indisputably alive. Chemical studies show that they contain DNA—the molecular blueprint of life and the means by which every living creature reproduces itself. They also contain a substantial amount of protein, in the form of a protective coat wrapped around the precious, delicate strands of DNA. But they contain very little else. In particular, they have none of the sugar and fat molecules that provide energy for the chemical reactions in other living creatures.

*Figure 18.5  Particles of the influenza virus magnified approximately 100,000 times by an electron microscope.*

They also lack the free nucleotides and amino acids out of which all other organisms make proteins and assemble copies of themselves.

How can viruses live if they lack an energy source and the raw materials needed for growth and reproduction?

The answer is clearly revealed by the electron microscope. A virus, by itself, is not alive. If a collection of virus particles is carefully hydrated the viruses will stick together in a symmetric pattern to form a crystal as geometric — and as lifeless — as a crystal of salt or a diamond (Figure 18.6); left undisturbed, the crystal remains inert for years. But, dissolved again in water, and placed in contact with living cells, the molecules of the crystal spring to life; they fasten to the walls of the cell, dissolve a small opening in the cell wall, and, through the opening, inject their DNA into the cell. Once inside the cell, the virus DNA seizes control, displacing the original DNA of the cell and establishing itself as the master of all further chemical activity. The virus gathers the free nucleotides floating in the cell fluid and assembles them, not into copies of the DNA of the invaded cell, but into copies of itself. In a final step, the virus secretes an additional enzyme that dissolves the cell walls. An army of virus particles

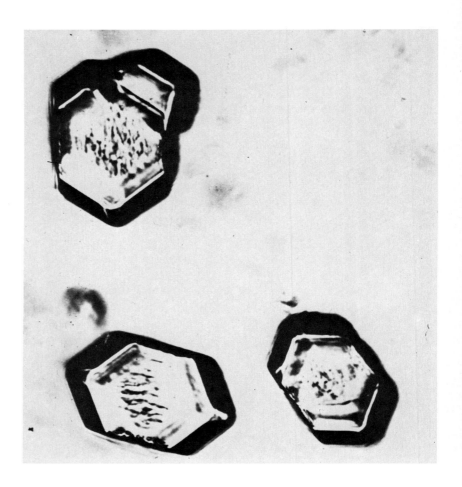

*Figure 18.6   Virus crystals.*

marches forth, each seeking new cells to conquer, leaving behind the empty husk of the invaded cell. (Figure 18.7) The operation has been executed by an organism that is, in the smallest viruses, only 200 atoms in diameter. In simplicity as well as size, the virus is the link between life and nonlife.

## THE CHANCE OF LIFE ELSEWHERE

The mind of the scientist has seized on these items of evidence and has fashioned out of them a picture of the origin of life on the earth. No living form existed on our planet in its infancy; the atmosphere was filled with a noxious mixture of ammonia, methane, water, and hydrogen; peals of thunder rumbled across the sky; flashes of lightning occasionally illuminated the surface, but no eye perceived them; minute amounts of amino acids and nucleotides were formed in each flash, and gradually these critical molecules accumulated in the earth's oceans; collisions occurred between them now and then, linking small molecules into

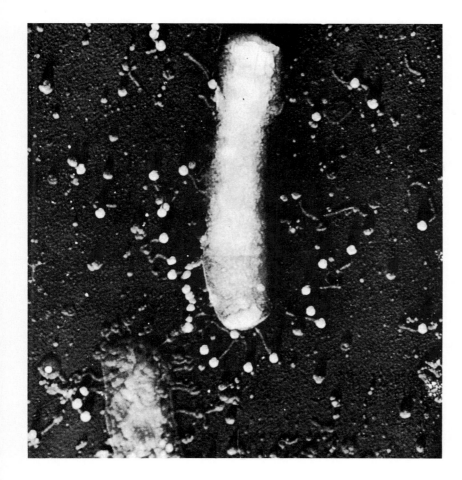

*Figure 18.7   Virus attack on bacterium. At lower left are the remains of a bacterium attacked one hour earlier.*

larger ones. During the course of a billion years the concentration of complex molecules increased; eventually a complete DNA chain appeared. Thus, the threshold was crossed from inorganic matter to the living organism.

According to this story, life can appear spontaneously in any favorably planetary environment, and evolve into complex beings, *provided vast amounts of time are available.*

How much time is needed? Studies of the fossil record suggest that life appeared sometime during the first two billion years of the earth's history. Apparently one or two billion years or so is the length of time required.

Our knowledge of the life cycle of a star indicates that the necessary period of several billion years will be available for the chemical evolution of life on any planets that circle around a star similar to our sun. Stars larger than the sun burn out too quickly to provide the needed time. Stars smaller than the sun are suitable, provided that they have planets close enough to them to raise their temperatures to a comfortable range. All stars the size of the sun, surrounded by one or more planets that are approximately the same distance from them as the earth is from its star, should certainly provide very favorable circumstances for the development of living organisms.

There are 100 billion stars belonging to the cluster we call our Galaxy. Ten billion other galaxies, each with 100 billion stars—and probably a like number of planets—are within the range of the largest telescopes. Perhaps only a small fraction of them are earthlike planets, but that would mean millions of earthlike planets in our Galaxy alone.

Can we maintain our belief in the uniqueness of life on the planet earth in the face of these numbers? We can on the astronomical evidence alone, because all planets except ours could be dead bodies of rock; but the biological discoveries described in this chapter suggest that this is not the case. *First,* all life on the earth depends on a few basic molecules, and these molecules have been created out of simple atoms in the laboratory; *second,* the atoms that compose the basic molecules of life are the same as the atoms that exist on every other star and planet in the Universe; *third,* there is reason to believe that the same laws of physics apply in every corner of the cosmos. If a chain of physical and chemical reactions led to the appearance of life on the earth, a similar chain could also have occurred on other planets.

This reasoning leads to the following conclusion: if sufficient time is available, a kind of life resembling ours can develop on many planets in our Galaxy, and on planets in other galaxies as well. What is the probability of this happening?

Neither theoretical calculations nor laboratory experiments can yield an answer to that question. Nature required several hundred million years of ceaseless experimentation to discover the chemical pathways to life on the earth, and the scientist's ingenuity may never be equal to the task of imitating her in any finite span of time. We will probably never learn in the laboratory the probability of the spontaneous generation of life on an earthlike planet.

We are more likely to find the answer on Mars. Mars is one of the only two earthlike planets in the solar system that could conceivably support life. We know that on one of them—the earth—life has, in fact, developed. That may be an improbable accident, but it seems unlikely that two such accidents would occur in one solar system. If life is discovered on Mars, we will be forced to conclude that on any earthlike planet the development of life out of nonlife is not a rare accident, but a relatively probable event. We will know that millions of inhabited planets surround us. No scientific discovery more significant for mankind can be imagined.

## THE HISTORY OF LIFE ON THE EARTH

Throughout the present discussion of stars, galaxies, and planets, our viewpoint has been that of the physical scientist, seeking to understand the essence of the world around him in terms of a few simple principles—the laws of physics and chemistry—which are the distillation of man's knowledge of the physical world acquired during thousands of years of observation. These laws have carried us from the origin of the Universe through the formation of stars and planets to the birth of the sun and the earth, and—with speculation and uncertainty—across the threshold of life on the earth.

Now we come to the explanation of the subsequent course of events in the history of life, leading from the first simple organisms to man. Here, for the first time, the principles of physics are no longer helpful. The stars and planets have yielded the secrets of their history to the physicist; the molecular foundations of living organisms are beginning to be understood; but the *complete* organism—even of the simplest and most primitive kind—is incalculably more complicated than any star, or planet, or giant molecule. New insights are needed for the understanding of its structure and evolution. A new law must be found.

## The Theory of Natural Selection

The new law was discovered by Charles Darwin more than a century ago (Figure 18.8). Darwin showed that evolution is controlled by a "force" that acts on plants and animals slowly, over the course of many generations, to produce changes in their forms. This force has no mathematical description; it is not to be found in any textbook of physics, listed alongside the basic forces that control the world of nonliving matter; but, nonetheless, it guides the course of evolution from simple cells to advanced creatures—on this planet and on all planets on which life has arisen—as firmly and as surely as gravity controls the motions of the stars and the planets.

Darwin began with an almost self-evident set of observations on the nature of life: All living things reproduce themselves; reproduction is the essence of life; *but the process of reproduction is never perfect.* The offspring in each generation are not perfect copies of their parents; brothers and sisters differ from one another; no two individuals in the world are exactly alike.

Usually the variations are small; brothers and sisters resemble one another, all human beings look more or less alike, and all elephants look more or less like other elephants.

Yet, Darwin asserted, these small variations are critically important; for, in the struggle for existence, the creature that is distinguished from its brethren by a special trait, giving it an advantage in the competition for food, or in the struggle against the rigors of the climate, or in the fight against the natural enemies of its species—that creature is the one most likely to survive, to reach maturity, and to *reproduce its kind.* The offspring of the favored individual will inherit the characteristic that has given it its advantage. Some among them will possess the desirable

Figure 18.8    Charles Darwin in 1881.

characteristic to a greater degree than others. These individuals, doubly favored, have a still greater chance to survive to maturity and to produce offspring of their own.

Thus, through successive generations, the advantageous trait appears with ever-increasing strength in the descendants of the individual who first possessed it.

Not only does the strength of the trait increase with the passage of time, but the number of individual animals possessing it also increases. For these individuals have slightly larger families than the average for the species; in each generation they leave behind a greater number of offspring than their less-favored neighbors; their descendants multiply more rapidly than the rest of the population, and in the course of many generations, their progeny replace the progeny of the animals that lack the desirable trait.

In *The Origin of Species,* Darwin gave this process the name by which it is known today:

"This principle of preservation or the survival of the fittest, I have called *Natural Selection.*"

Through the action of Natural Selection, a trait, which first appeared as an accidental variation in a single individual will, with the passage of sufficient time, become a pronounced characteristic of the entire species. So the deer became fleet of foot, for the deer that ran fastest in each generation escaped their predators and lived to produce the greatest number of progeny for the next generation. So did man become more intelligent, for superior intelligence was of premium value: the intelligent and resourceful hunter was the one most likely to secure food. Thus developed the brain of man; thus, too, in response to other pressures and opportunities in their environments, developed the trunk of the elephant and the neck of the giraffe.

Of course, the incorporation of one new trait does not create an entirely new animal. But if we count all the births that occur to a single species over the face of the globe in one year, an enormous number of variations will appear in this multitude of young creatures. On all these variations the same process of selection works steadily, preserving for future generations the new traits that give strength to the species, and eliminating those that lend weakness. The changes are imperceptible from one generation to the next, but over the course of many generations the accumulation of these many favorable variations, each slight in itself, completely transforms the animal. According to Darwin,

"Natural selection is daily and hourly scrutinising, throughout the world, the slightest variations, rejecting those that are bad, preserving and adding up all that are good; silently and insensibly working at the improvement of each organic being in relation to its . . . conditions of life,"

Natural selection is the force that molds the forms of living creatures.

Under the continuing action of the law of natural selection the qualities and shapes of animals change with time; old species disappear in response to changing conditions, and new ones arise. Few of the species of animals that roamed the face of the earth 100 million years ago still exist today, and few of those existing today will exist 100 million years hence. From *The Origin of Species:*

> "...Not one living species will transmit its unaltered likeness to a distant futurity."

But natural selection works its effects slowly. Its influence is not felt in one individual or in his immediate descendants. Ten thousand generations may elapse before a change becomes noticeable; in man that amounts to a quarter of a million years.

Darwin knew that this concept of infinitesimally slow change was bewildering when first encountered. He wrote in the *Origin:*

> "The mind cannot grasp the full meaning of the term of even a million years; it cannot add up and perceive the full effects of many slight variations accumulated during an almost infinite number of generations...We see nothing of these slow changes in progress, until the hand of time has marked the lapse of ages, and then...we see only that the forms of life are now different from what they formerly were."

A vast amount of time was needed for natural selection to produce the highest forms of life out of these simple beginnings. Yet, ever since Rutherford measured the age of the earth, we have known that enough time is available. Our planet has existed for billions of years; that is the secret strength of Darwin's theory.

## The Origin of Intelligence on the Earth

Certain highlights in the history of life — revealed by the fossil record — supply clues to the origin of intelligence on the earth. Innumerable skeletons and fossil remains mark the path by which life climbed upward from its crude beginnings. The first steps along the path are not known, for those early forms were very fragile and no trace of them remains. The earliest signs of life to appear in the record, already far advanced beyond the molecule lying on the threshold between life and nonlife, are the deposits of simple one-celled plants called algae, and the shells of rod-shaped organisms resembling bacteria. These are found in rocks solidified 3 billion years ago, when the earth was approximately one billion years old.

Little else appears in the fossil record during the first three billion years. One of the mysteries in the study of life is the fact that suddenly, in rocks 600 million years old, the record explodes in a profusion of living forms. A great variety of animals appears in the record at that time. Perhaps the forms of life were nearly as numerous and populous just prior to this critical date, but left no trace of their existence because they lacked the body armor that is most easily preserved.

Somewhat more than 400 million years ago an event occurred that is of great consequence for the development of man. There appeared, for the first time, a new kind of creature—one with an internal skeleton and a backbone. This animal—the vertebrate—evolved out of a wormlike ancestor resembling the modern lancelet, a small, translucent creature, lacking fins and jaws, but possessing gills and, most important, a primitive version of the backbone.

Among the descendants of the first vertebrates were the fishes. Some of the early fishes contained crude lungs for gulping air at the surface of the water, as well as gills. These lungs were lost or converted to other uses in most instances, but in some forms of fish, perhaps those living in small bodies of water such as ponds and tidepools, the lungs came into frequent use. Whenever a drought developed and the water level in the ponds dropped, the fish with the best lung capacity survived where others perished. They lived to produce progeny that inherited their superior capacity for breathing air. In this way, an efficient lung evolved gradually among the fish that inhabited shallow bodies of water.

Some of the air-breathing fish were doubly favored in possessing strong fins that enabled them to waddle over the land from one pond to another in search of water. By a slow accumulation of favorable mutations, the muscle and bone of the fin gradually changed into a form suitable for walking on land. In this way, the fin evolved into the leg. The metamorphosis took place over a period of perhaps 50 million years, and a like number of generations. The result was a four legged, air-breathing animal, known as the amphibian.

The amphibians were born in water, lived most of their adult lives near water, and almost always returned to the water to lay their eggs. For 50 million years they flourished on shores and river banks. Some became large, aggressive carnivores as much as 10 feet in length, fearing no other animals of their time. The amphibians attained the peak of their size 250 million years ago, and thereafter they declined. Today their common descendants are the diminutive frog, toad, and salamander.

In the course of time, some of the ancient amphibians, again by the chance occurrence of a succession of favorable mutations, developed the ability to lay their eggs on land. These eggs were encased in a firm, leathery shell, which retained moisture and provided the embryo with its own private pool of fluid. Other mutations led to a tough, leathery hide, which preserved the water in their bodies without the need for continual immersion. Such creatures were completely emancipated from the water. They were the first reptiles.

The reptiles marked a very successful step in evolution, for they had access to rich resources of food previously denied to the fishes and the amphibians. The reptiles flourished and developed into a great variety of forms, including the ancestors of every land animal with a backbone now on this earth. They reached their evolutionary zenith in the dinosaurs. These animals ruled the earth for 100 million years. They displayed an extraordinary vigor, evolving into such extreme forms as the giant vegetarian swamp-dweller, *Brontosaurus,* 70 feet long and weighing 30 tons; and the meat-eating *Tyrannosaurus rex,* 40 feet high, with a 4-foot skull filled with daggerlike teeth—unquestionably the fiercest land-living predator the world had ever seen.

Two hundred million years ago, somewhat before the appearance of the first dinosaurs, another branch of the reptile class veered off on an entirely different course. This particular group may have lived in places on the edge of the temperate zone where the weather was relatively severe. Through the action of natural selection on chance variations, the new branch of the reptiles acquired a set of traits that fitted them uniquely for survival in a rigorous climate. They developed the rudimentary characteristics of a warm-blooded animal. The naked scaly skin of the reptile was replaced in these animals by insulating coats of hair and fur that kept them warm in low temperatures, while sweat glands under the skin, controlled by an internal thermostat, cooled the body by evaporation when the temperature rose too high.

In spite of this and other advantages, the mammals remained subordinate to the dinosaurs for more than 100 million years—small, furry animals, inconspicuous, keeping out of sight of the rapacious reptiles by living in the trees or in the grasses.

But 70 million years ago, the dinosaurs died out. The reasons for their disappearance are still obscure. It is likely that their downfall was the consequence of a worldwide change in climate, which they were ill-equipped to survive. Dinosaurs, like all reptiles, were cold-blooded animals; that is, they lacked the internal heat controls that could maintain the temperature of the body at a constant level regardless of the rigors of the climate. We know that the period during which they disappeared was marked by repeated upheavals of the earth's crust in which many new mountain ranges were formed. The Rocky Mountains were among the ranges created in these upheavals. Most probably, the upward thrust of huge masses of rock disrupted the flow of currents of air around globe; perhaps the climate of the temperate zone was changed in this manner from one of uniform warmth and humidity, agreeable to a coldblooded animal, to a climate marked by major changes of temperature from season to season.

As the population of dinosaurs dwindled, the mammals came down from the trees and up from their burrows in the ground, and they inherited the earth. Quickly they spread out across all the continents. Within 20 million years, the basic mammalian stock evolved into the forebears of most of the mammals with which we are familiar today—bats, elephants, horses, whales, and many others.

But one group of mammals remained in the trees. These mammals—the primates—were singled out, by the circumstances of their tree-dwelling existence, to be the ancestors of man. They were small, insect-eating animals, the size of a squirrel, and similar in appearance to the modern tree shrew of Borneo. Man owes his remarkable brain to the fact that these animals required two physical attributes for survival in their arboreal habitat: first, they needed hands and an opposable thumb for securing a tight hold on branches; and second, they needed sharp binocular vision to judge the distances to nearby branches. In the competition for survival among primitive tree-dwelling mammals, 100 million years ago, those who possessed these characteristics in the highest degree were favored. They were the individuals most likely to survive and to produce offspring. Through successive generations the desirable traits of a well-developed hand and keen vision, passed on from parents to offspring, were steadily refined and strengthened. Fifty million years ago they existed in advanced form in the animals from which the modern tree shrew, lemur, and tarsier are descended. They became even better developed in some of the immediate descendants of these animals, under the continued pressure of the struggle for survival in the trees. Gradually, the evolutionary trends established by the requirements of life in the trees transformed some of these early primates into animals resembling the monkey.

Animals with hands also had the potential capacity to exercise rudimentary manual skills; when this potential was combined with the development of the associated brain centers, such animals had, almost by accident, the ability to use tools. For those who had this ability, great value became attached to the mental capacity for the remembrance of the usage of tools in the past, and for the planning of their use in the future; thus, by the action of natural selection on a succession of chance mutations, those centers of the brain developed and expanded in which past experiences were stored and future actions were contemplated. These mental qualities proved to be of great value in meeting the general problems of survival. As a result, the brain evolved and expanded under the continued pressure of the struggle for existence. It doubled in size in 10 million years, and nearly doubled again in the next two million. These circumstances established the line of ascent leading to intelligent life.

## COSMIC EVOLUTION

In Chapter 1 of this book we suggested that the study of astronomy can illuminate basic problems of interest to every person, regardless of his professional interest in science. What am I? How did I get here? What is my relationship to the rest of the Universe? These questions acquire a new meaning when viewed in the 10-billion year perspective of stellar lifetimes. The human life span seems so short in comparison with astronomical time scales that the student of these questions cannot help but feel a new humility as he contemplates the 4½ billion years that have elapsed in the history of our solar system and the 6 billion years that lie ahead before the sun becomes a red giant.

The evidence provided by astrophysics and cosmology suggests that we owe our physical existence to events that took place billions of years ago in stars that lived and died long before our solar system was formed. According to the results of astrophysical investigations, the atoms in our bodies—and the atoms that make up the body of the earth—are drawn from the interiors of countless stars that have ended their own lives in supernova explosions earlier in the history of the Galaxy. The substance of each of those stars in turn was condensed out of an interstellar medium that had been enriched by innumerable supernova explosions that occurred still earlier in the history of the Universe. Step by step, astronomy traces the origin of our material substance back through time and into the parent cloud of hydrogen. The philosophical implication in this union of astronomical, geological, and biological concepts is that man has appeared on the scene as the product of an unbroken sequence of events, extending over more than 10 billion years, in which the Universe expands and cools, stars are born and die, the sun and earth are formed, and life arises on the earth.

*"We are brothers of the boulders, cousins of the clouds."* Harlow Shapley

## Questions

1. Describe the chain of events leading from the beginning of the Universe to the threshold of life on the earth. Include the major points of relevance in cosmology and stellar evolution. Limit your answer to not more than six critically important steps in this chain of events. Briefly describe the conditions that existed at the time of each critical event.

2. What role do proteins play in living organisms? What is the structure of a protein?

3. What role does DNA play in living organisms? What is its structure?

4. How have the building blocks of life been produced in the laboratory? What gases were used? Why?

5. Give your own definition of a living organism. For each property that you consider essential to the property of "life" try to think of a non-living object which shares this property with living organisms. Is it possible to isolate one characteristic that is unique to living organisms? Explain your answer.

6. Is a virus in crystal form "alive"?

7. Briefly describe the scientific picture of the origin of life on earth.

8. Do you think that there is life elsewhere in the Universe? Why?

9. Summarize in five or ten sentences the reasoning that led Darwin to his theory of evolution by natural selection.

10. Describe how environmental pressures produced a number of critical evolutionary advances in the chain of events leading from the

threshold of life to man. Pick one of these important advances, and invent a different set of circumstances in the environment that might have deflected terrestrial life into a different line of evolution. Apply the Darwinian reasoning on natural selection to predict the alternative path along which life would have evolved on the earth in these alternate circumstances. Will your answer lead to the evolution of intelligent life? Why?

11. Do you think that there are many intelligent societies in the Universe? What is the likelihood that contact will be made with these societies either by radio communication or by space travel? Cite the scientific bases for your answer.

# Appendix A
# Methods of Measurement

In numerous places in the text, references have been made to stellar distances and masses without an indication of the way in which these important properties are measured. The determination of the distances and masses of astronomical objects presents a formidable problem that has taxed the ingenuity of astronomers for centuries. The solutions are a triumph of observational astronomy, and provide the empirical foundation for the description of stellar evolution in Chapters 5 and 7.

## MEASUREMENT OF ASTRONOMICAL DISTANCES

The determination of stellar and galactic distances depends on several distinct methods, beginning with the method of trigonometric parallax and the moving-cluster method for nearby stars, and extending outward to greater distances by a succession of other methods.

Each method is calibrated by applying it to objects whose distances have already been measured by a previous method. Ultimately, the entire system of distance determinations rests on a set of known distances

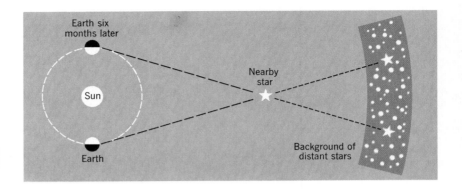

Figure A.1 *Shift in the apparent position of a nearby star against the background of distant stars.*

to nearby stars, which have been measured with the aid of the trigonometric parallax and moving-cluster methods. This scheme of distance determination may be compared to an inverted pyramid, with the parallax method and the moving-cluster method at the bottom of the pyramid, bearing the weight of the entire structure.

### The Method of Trigonometric Parallax

The trigonometric parallax method is often called simply the method of parallax. It consists in observing the position of a nearby star against the background formed by the distant stars in the sky, and then measuring the position again six months later, when the earth is on the other side of its orbit. Because the star being studied is close to the solar system, its position seems to shift against the background of the distant stars when the position of the earth changes (Figure A.1).

The observer can demonstrate the parallax effect by holding up a pencil one foot away and closing first one eye and then the other as he views the pencil against a background of more distant objects, such as the edge of the blackboard or a tree across the street. This is the parallax effect. If the sighting is repeated with the pencil held farther away from him at arm's length, the shift in apparent position is not as great as before. In other words, the amount of parallax depends on the distance to the pencil.

Figure A.2 illustrates the change in parallax when the pencil is held close (*a*) and at arm's length (*b*).

The shift in apparent position of a star, as observed from the earth on opposite sides of its orbit, is called *stellar parallax.* The stellar parallax is defined as one-half of the apparent shift in angular position observed when the earth moves across its orbit (Figure A.3). It is measured in seconds of arc.

The largest stellar parallax is 0.763", determined for Proxima Centauri (the companion of Alpha Centauri), the closest star to the sun.

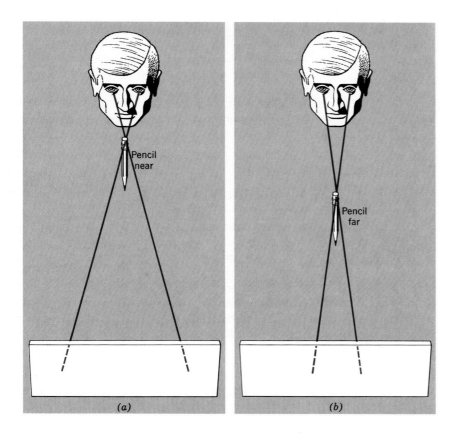

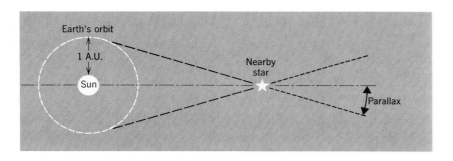

Figure A.2 Illustration of the parallax effect.

The best parallax measurements are uncertain by ±0.005". Because of this uncertainty, the method is limited to a useful range of roughly 100 light-years. A star at that distance has a parallax of 0.03", and a probable error of 15 percent or 15 light-years in the distance derived from its parallax.

Because the observation of parallax is the basis of all distance-measuring techniques, astronomers often use the term "parallax" as a synonym for distance to an astronomical object, regardless of how that distance has been measured.

Figure A.3 Stellar parallax.

Another unit of stellar distance, frequently used in place of the light-year, is derived from the parallax method. This unit, called the parsec (abbreviated pc) is the distance at which a star has a parallax of one second of arc. One parsec = 3.26 light-years. The distance of Proxima Centauri, for example, is 1.3 pc.

For very large distances, units of the kiloparsec ($10^3$ pc, written kpc) and the megaparsec ($10^6$ pc, written Mpc) are employed.

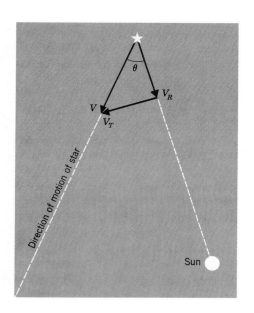

*Figure A.4 The arrow labelled V is the velocity of the star relative to the sun. The arrows $V_R$ and $V_T$ are the components of v along (radial) and across (tangential to) the line of sight, respectively.*

## Beyond the Parallax: The Moving-Cluster Method

As noted, a star more distant than about 100 light-years has a trigonometric parallax too small to be accurately measured. However, if the star is a member of a cluster, another method of direct distance determination can be employed to determine the mean distance to the stars in the cluster. This technique, called the moving cluster method, is more complicated than the parallax method and involves a series of steps. The first step is to determine the direction in which the star is moving through space relative to the sun. In Figure A.4, the direction of motion of the star is shown by a dashed line. The method for determining the direction of the star's motion through space is explained below. The second step is to measure the Doppler shift in the spectrum of the star. From the Doppler shift and the formula on p. 35, the speed with which the star is approaching or receding from the sun can be calculated. This means the speed along the line of sight to the observer. In Figure A.4, the speed along the line of sight is indicated by the arrow labelled $V_R$ (for radial velocity).

It is important to note that $V_R$ is not the velocity with which the star is travelling through space relative to the sun. It is only the *component* of that velocity *along the line of sight to the observer.*

The third step is to calculate the velocity of the star *across* the line of sight. In Figure A.4, this is labelled $V_T$ (for tangential velocity). If $\theta$ is the angle between the line of sight to the observer and the direction of the star's motion, $V_T$ is obtained from the formula (Figure A.4).

$$V_T = V_R \tan \theta$$

The fourth step is to measure the shift in the apparent position of the star against the background of the fixed stars during the course of an extended period of time. Usually a number of years is needed for an accurate measurement of the shift in a star's position. This measurement yields the *rate* at which the position of the star is changing. The rate of change of a star's position is called its *proper motion,* usually denoted by $\mu$.

Finally, the distance of $d$ to the star follows from the relation

$$d = \frac{V_T}{\mu}$$

The most difficult observational part of the procedure occurs in the first step. How do we determine the direction in which the star is moving through space relative to the sun? The answer can be clarified with the aid of an example.

If you have driven along a straight road at night with telephone poles on either side, you may have noticed how the pairs of poles on either side of the road ahead of you appear to move apart as the car approached them. The apparent motions of the telephone poles are such that they appear to diverge from a point in the distance. This point represents the direction on the horizon toward which the car is traveling.

A second example is the flight of a flock of birds coming toward the observer. As they approach the observer and fly overhead, their flight paths also appear to diverge from a point in the distance, because of the same perspective effect. The line drawn from the observer to that point is the direction of travel of the flock.

To apply these ideas to the direction of motion of stars, consider a cluster of stars, such as the galactic cluster discussed in Chapter 8. All the stars in the cluster move with about the same speed relative to the sun, as a part of the general motion of the entire cluster through space. In this respect, they resemble the flock of birds. When the proper motions for the members of the cluster are plotted, they will be seen to converge toward or diverge from a single point, for the same reason that the flight paths of the birds appear to originate in a single point.

The chart on page 480 (Figure A.5), showing the proper motions of stars in the Hyades Cluster, provides an example of the convergence to a point.

The proper motions converge in this example because the relative motion of the Hyades is away from the sun. If the relative motion is toward the sun, the proper motions diverge from a point in the sky.

Why is a cluster of stars necessary for this method? In principle, one star is sufficient if its direction of motion relative to the sun is known. In general, however, the direction is not known. The proper motions for the entire cluster are needed to give the direction of relative motion, as explained.

The moving-cluster method breaks down at great distances because the proper motions become too small to be measured accurately in a reasonable interval of years. The effective limit on accurate distance determinations using this method is about 500 light-years.

Actually, the moving-cluster method is used primarily for two particular stellar groups—the Hyades cluster and the Scorpio-Centaurus association, at distances of 120 light-years and 500 light-years, respectively. These are the only groups of stars that are close enough to permit an accurate measurement of the proper motions of their members and, at the same time, contain a sufficiently large number of stars so that the directions of motion can be located with high precision.

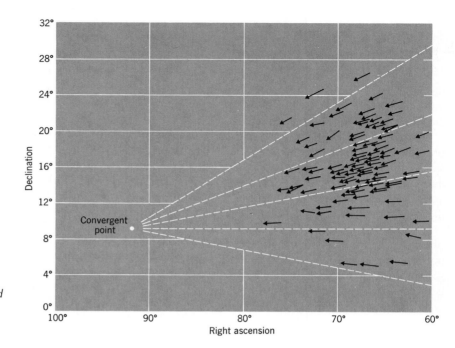

*Figure A.5 The moving-cluster method applied to the Hyades Cluster.*

## The Method of Spectroscopic Parallax

The method of spectroscopic parallax depends on using the spectrum of a star to locate its place on the H-R diagram. When the star has been located on the H-R diagram, its absolute luminosity or magnitude can be read off the diagram. At the same time, the apparent magnitude of the star is measured by photometry. Knowing the star's absolute magnitude and apparent magnitude, the distance to it can be calculated from the formula on page 483. Solving for distance:

$$d \text{ (parsecs)} = 10^{\frac{m - M + 5}{5}}$$

where $m$ and $M$ are the apparent and absolute magnitudes, respectively.

The critical step in the spectroscopic parallax method is the location of the star on the H-R diagram. Once that is done, the absolute magnitude is known, and the distance follows. The key question is: How is a star's spectrum used to locate its place on the H-R diagram?

If the star is on the Main Sequence, the measurement of its spectral type or temperature is sufficient for this purpose, since the observed Main-Sequence "line" (although really a band of finite width) provides a fairly well-defined relationship between spectral type and absolute magnitude.

If the star is not on the Main Sequence, the same remarks apply, but the precision of the method is not as great as for Main-Sequence stars.

Suppose, for example, that the star is known to be a giant. The giants also lie in a definite region of the H-R diagram—a broad band with luminosities ranging from $10^2 L_\odot$ to $10^4 L_\odot$. If a star is classified as a giant and its spectral type is also known, we have a rough knowledge of its position on the diagram and its absolute luminosity.

But how do we know whether or not a star is on the Main Sequence? The star's spectrum provides the answer. The absorption lines in the spectrum of a giant are much narrower than the same lines in the spectrum of a Main-Sequence star, because giants are distended stars with tenuous atmospheres, in which collisions are relatively infrequent. Therefore, collisional broadening (see p. 110) is much less pronounced in the spectral lines of giants than it is in Main-Sequence spectra. In Figure 4.34, which compares the spectra of a giant and a Main-Sequence star of the same spectral type, the difference in the widths of the lines is clearly evident to the eye.

The classification of stars according to the widths of their spectral lines has been refined into a system of luminosity classes. All stars can be assigned to one of these classes on the basis of spectral line widths and related properties of their spectra. The luminosity class is denoted by a Roman numeral between I and VI. Stars in luminosity class I have the highest absolute luminosities, in the range from $10^4$ to $10^6$ solar luminosities. These are the supergiants. The most luminous supergiants, such as Deneb and Rigel ($L \sim 2 \times 10^5 L_\odot$), are called Ia, and the less luminous supergiants, like Antares ($L \sim 4 \times 10^4 L_\odot$), are designated Ib. Bright giants like Epsilon Canis Majoris are placed in luminosity class II, which includes luminosities between $10^3$ and $10^4$. Normal giants such as Arcturus ($L \sim 100 L_\odot$) are in class III. Stars lying between the giants and the Main Sequence are placed in class IV, with luminosities ranging from $10 L_\odot$ to $10^2 L_\odot$. Main-Sequence stars comprise luminosity class V. Stars less luminous than Main-Sequence stars—the subdwarfs—sometimes are placed in a luminosity class VI. The white dwarfs—the least luminous stars in a given spectral class—are given a special classification wd or D.

Figure A.6 shows luminosity classes superimposed on an H-R diagram of the form used in Chapters 5 and 7. Normally a star is labeled by its spectral type and luminosity class in that order, for example, A2 I for Deneb or G2 V for the sun. This specification in terms of the spectral type and luminosity class is a two-dimensional system of classifying stellar spectra, known as the Morgan-Keenan or MK system. Classification of a star's spectrum in the MK system permits the position of the star to be located approximately on the H-R diagram, fixing its absolute luminosity, and, therefore, its absolute magnitude. The distance follows from the formula on page 476. The distance to a star determined in this way is called the star's *spectroscopic parallax*.

Distances determined by the method of spectroscopic parallax are accurate to about 15 percent. This fractional error is roughly independent of the distance of the star. In the method of trigonometric parallax, on the other hand, the error has a fixed absolute value, and, expressed as a fraction, is much smaller than 15 percent for nearby stars, but larger for distant stars. The two methods yield the same error for stars at a distance of about 100 light-years. If a star is closer than 100 light-years, the method of trigonometric parallax yields more accurate results, while for stars more distant than 100 light-years, the method of spectroscopic parallax is superior.

*Figure A.6   Location of luminosity classes on the H-R diagram.*

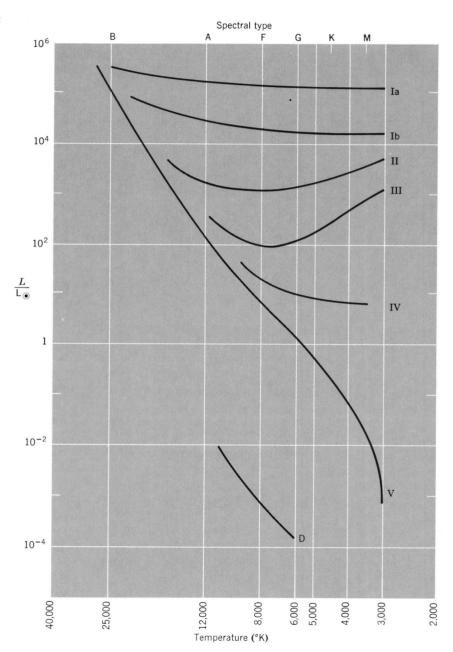

*Main-Sequence Fitting.* In principle, the method of spectroscopic parallax is applicable to single stars. In practice, the method achieves its greatest accuracy when applied to a cluster of stars, rather than to a single star, because it is nearly always possible to identify the Main-Sequence stars in a cluster unambiguously. As noted, Main-Sequence stars are the most suitable candidates for accurate distance determinations with this method. The application of the method of spectroscopic parallax to clusters is known as Main-Sequence fitting.

As an illustration of the use of the method of Main-Sequence fitting, consider first one of the relatively young clusters, such as the Praesaepe or Pleiades clusters, which are discussed in Chapter 8. In a young cluster, many of the stars are still on the Main Sequence; few have yet evolved off the Main Sequence into the red giant or red supergiant regions.

The first step in the method is to plot the directly observed H-R diagram for the stars in the cluster, giving their *apparent* magnitudes versus spectral types. The stars will be strung out along a line on the diagram, which is clearly the Main-Sequence line.

The next step is to compare this H-R diagram, plotted in terms of apparent magnitude, with a calibrated H-R diagram plotted in terms of *absolute* magnitude. This diagram is made up by using stars whose distances and, therefore, absolute luminosities have been determined previously by the trigonometric parallax or the moving cluster method. This is the first instance of the procedure described in the introductory paragraph, whereby each successive method of distance measurement is based on a previous one.

Since all the stars in the cluster are at approximately the same distance, the difference between the two H-R diagrams, one in terms of absolute magnitudes and the other in terms of apparent magnitudes, will only be a shift in the vertical scale. This can be seen from the formula,

$$M = m + 5 - 5 \log d$$

in which $M$ and $m$ are the absolute and apparent magnitudes, respectively, and $d$ is the distance in parsecs (Chapter 5).

The determination of the distance is equivalent to laying one diagram over the other and sliding it vertically until the two Main Sequences fit. The distance is calculated from the difference in the two magnitude scales, using the formula above.

An example of Main-Sequence fitting is shown in Figure A.7.[1]

All the Main-Sequence lines for the clusters in Figure A.7 are anchored to the Main-Sequence line for the Hyades Cluster. The absolute magnitudes for the Hyades Cluster are known with great accuracy because this cluster happens to be particularly suitable for the application of the moving-cluster method of distance determination. Therefore it is used as the anchor line for the other clusters. The solid line in the diagram gives

[1] Adapted from *Basic Astronomical Data,* edited by Strand, Volume III, p. 407.

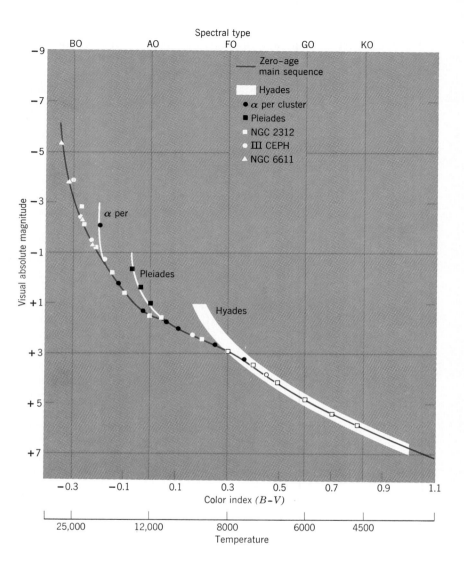

*Figure A.7   Main-Sequence fitting.*

a calibrated "Zero-Age Main-Sequence" line determined by overlapping the main sequence lines for the six clusters listed in the diagram.

If the Hyades' distances are known so well, why use the other clusters in addition? The reason is that while the Hyades Cluster calibrates the lower Main Sequence, it does not provide all the needed information for the upper Main Sequence, because its more massive stars have evolved away from that region. The upper Main-Sequence line is established by fitting the five other overlapping clusters to the Hyades Main-Sequence line. These clusters, which are younger than the Hyades, extend the "Zero Age Main-Sequence" line successively upward from the K, G, and F stars into the region of O and B stars.

The precision of the Main-Sequence fitting method is typically ±15%. As in the case of the method of spectroscopic parallax, Main-Sequence fitting is more accurate than the trigonometric parallax method for distances greater than 100 light-years.

It will be seen below that most of the methods for measuring extragalactic distances rest on the spectroscopic parallax or Main-Sequence fitting methods. These, in turn, rest on the moving-cluster method applied to one particular cluster—the Hyades. Thus, the moving-cluster method carries a greater weight than any other single method in the scheme of cosmic distance determinations. The other independent methods—trigonometric parallax and statistical parallax (described below)—also play important roles as primary yardsticks, but neither is quite as important as the moving-cluster method.

## Globular Clusters as Standard Candles

The methods discussed thus far are not usable for measuring distances to objects outside the galaxy.[2] Globular clusters provide the first method of determining extragalactic distances.

Globular clusters are plentiful in our galaxy, approximately 150 having been observed thus far. They are also plentiful in some other galaxies. If we assume that the extragalactic globular clusters have the same absolute magnitudes as those in our galaxy, and also assume that the absolute magnitudes of the brightest globular clusters in our galaxy are known, the distances to these extragalactic clusters can be determined by comparing their apparent and absolute magnitudes.

The first step in translating this idea into practice is the determination of the distances to globular clusters in our galaxy. The method of Main-Sequence fitting is one way of measuring these distances. However, a complication arises through the fact that globular clusters are composed of stars that were formed early in the history of the Galaxy, when relatively little conversion of hydrogen and helium to heavier elements had occurred. When the absolute magnitudes of these stars are plotted against spectral type, they form a Main Sequence that is somewhat different from the Main-Sequence line for clusters of stars formed more recently, such as the Hyades Cluster. The stars formed early in the history of the Galaxy are called population II stars, and stars formed more recently are described as belonging to population I (see Chapter 8). The calibrated Main-Sequence line described in the previous section is based on population I stars, and therefore cannot be used for globular clusters. However, a separate Main-Sequence line for population II stars can be constructed by observing the stars of this type that happen to be near the sun. These stars are identified as belonging to population II either

[2] Except the Magellanic clouds, whose brightest stars fall within the range of the method of spectroscopic parallax.

by their large velocities or by direct spectroscopic evidence for a low abundance of heavier elements. Some of the population II stars are close enough to permit a determination of distances by trigonometric parallax and thus can be used to calibrate the Main-Sequence line for population II stars. The calibration is not very accurate because only a few population II stars exist in the neighborhood of the sun.

When the calibrated Main-Sequence line for population II stars has been constructed, distances and therefore absolute magnitudes can be determined for globular clusters in our galaxy. Assuming the same range of absolute magnitudes for globular clusters in other galaxies, the distance *d* to these galaxies can be calculated from the formula on page 480.

Globular clusters provide a method for measuring distances up to 30 million light-years. Beyond that distance they are too faint to be useful.

### RR Lyrae Stars as Distance Indicators (The Method of Statistical Parallax)

One method for distance determination that is independent of Main-Sequence fitting involves the use of the RR Lyrae variable stars as standard candles. As will be seen, the method based on the RR Lyrae stars also leads to a determination of the distances to the globular clusters in our galaxy and, thus, to an independent calibration of these globular clusters as standard candles for determining distances to other galaxies.

The RR Lyrae method depends on a comparison between the true velocity of a star through space, called its peculiar motion, and its rate of change of angular position, or proper motion. It is clear that the peculiar motion, proper motion and distance of a star are related. If two stars are moving through space with the same speed, but one star is much farther away than the other, the more distant star will change its apparent position more slowly than the nearer one; that is, its proper motion will be less. If both the peculiar motion and the proper motion of a star can be measured, its distance can be determined through this relationship, and its absolute magnitude can be deduced again from the formula on page 480. The determination of the peculiar motion involves (1) measuring the radial velocity by the star's Doppler shift and (2) assuming that in a statistical average over space velocities, the radial component is one-third of the total space velocity. Because of this dependence on a statistical average over space velocities, the method is called the statistical parallax.

When the absolute magnitudes of a number of RR Lyrae stars are measured in this way, they turn out to be close to 0.6. Apparently, all the members of this particular class of variables have approximately the same absolute magnitude. This fact permits the RR Lyrae stars to be used as standard candles.

The RR Lyrae stars are not bright enough to be visible in external galaxies other than the Magellanic clouds and a few small galaxies in the

Local Group (p. 15) and therefore cannot be used as standard candles for measuring extragalactic distances in general. However, they can be used extensively within the Milky Way Galaxy. In particular, since RR Lyrae stars exist in the globular clusters of the Milky Way Galaxy, they can be used to determine the distances to the globular clusters, and therefore the absolute magnitudes of these clusters. As discussed above, a knowledge of the absolute magnitudes of the globular clusters provides an important means for measuring extragalactic distances. Thus, the RR Lyrae stars help indirectly to expand the range of distance measurements beyond the boundary of the Galaxy.

## Cepheid Variables as Distance Indicators

A relationship between period of pulsation and absolute luminosity is one of the characteristic properties of the Cepheid Variables. This relationship permits the absolute magnitude of a Cepheid Variable, and therefore its distance, to be determined from its observed pulsation period.

Since the Cepheids are considerably brighter than RR Lyrae stars, with absolute luminosities exceeding those of the RR Lyrae stars by more than a factor of a thousand in the case of the brightest Cepheids, they are visible in many external galaxies. Thus, the Cepheids, unlike the RR Lyrae stars, provide a direct method for crossing the void between the Milky Way Galaxy and neighboring galaxies, in the scheme of distance measurements.

Two factors complicate the use of Cepheids as distance indicators. First, two distinct types of Cepheids exist with a separate period-luminosity curve for each. One type is called the Classical Cepheids because they resemble the prototype Cepheid, Delta Cephei. The classical Cepheids are mostly yellow supergiants of population I and are relatively rare. The other type, called the Type-Two Cepheids, are found in globular clusters in the galactic halo and in the center of the Galaxy, and are also relatively rare. The Type-Two Cepheids are population II stars.

The separate period-luminosity lines for the two groups of Cepheids are shown in Figure A.8. Either type of Cepheid can be used for extragalactic distance determinations, but it is important to know which type is being used in each case.

When the Cepheids were first used for distance determinations, astronomers were not aware of the existence of two separate period-luminosity curves and misapplied the Type-Two curve to Classical Cepheids in their basic calibration of extragalactic distances. The error, which amounted to 1.5 on the absolute magnitude scale, was discovered by Baade in 1952. The resultant correction doubled the size of the Universe.

The second complication is a dependence of the period of pulsation on the spectral type, or temperature, of the star. As a result of this dependence, the line representing the period-luminosity relationship on

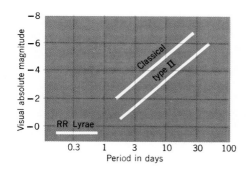

*Figure A.8 Period-luminosity law for Cepheid and RR Lyrae variables.*

**487**

the graph should be broadened to a finite band, with a vertical extent of about one magnitude.

Distance determinations based on Cepheids have both advantages and disadvantages, relative to the use of globular clusters as standard candles, for extragalactic distance measurements. An important advantage is their greater accuracy; a disadvantage is the lower luminosity of the Cepheids, which limits the range of the method to smaller distances than the globular-cluster method. Whereas the globular-cluster method extends to about 30 million light-years, the range of the Cepheids for this purpose is only about 10 million light-years.

### Bright Blue Stars as Standard Candles[3]

A third technique for estimating extragalactic distances depends on the assumption that the brightest, hottest blue stars in any one galaxy such as our own have the same luminosities as the brightest stars in any other galaxy.

The brightest blue stars are the blue supergiants. The assumption of a known luminosity for all blue supergiants is a reasonable one, since these are believed to be the most massive stars that can be formed in a galaxy, and the upper limit of masses in star formation is probably nearly the same in all galaxies.

If the distances and, therefore, the absolute magnitudes can be determined for a number of blue supergiants in galaxies of known distance, the blue supergiants can serve as standard candles for determination of distances to other galaxies.

Since the absolute luminosities of the blue supergiants range up to $10^6 L_\odot$, or approximately the same as the luminosities of the brightest globular clusters, the range of effectiveness in distance determinations is about the same for the two methods.

The absolute magnitudes of the blue supergiants are determined by applying the Main-Sequence fitting method to a few relative young clusters in our galaxy. The clusters must be young because the blue supergiants are massive stars that only live about 3 million years, and will have completely evolved in any cluster whose age is greater than this.

### The Size of H II Regions as a Distance Indicator

Bright blue stars, especially of O type, are intense sources of ultraviolet radiation. The radiation ionizes nearby interstellar hydrogen, producing a sphere of ionized matter around the star with a radius of 150 light-years or more in the case of the hottest O stars. The sphere

[3] Bright red stars are now also being used by Sandage as distance indicators.

of ionized hydrogen is known as an H II Region.

Some of the ionized hydrogen atoms in the H II Region recapture electrons, and the electrons cascade down to the ground state in several steps, emitting a part of their recombination energy in the form of photons with energies in the visible part of the spectrum. This radiation creates a luminous sphere around the star, which makes the H II Region detectable.

The sizes of H II Regions in our galaxy have been measured by determining the distances to O-type stars in clusters, using the method of spectroscopic parallax. H II Regions are seen in some other galaxies also, and their angular dimensions can be measured. Assuming that the correlation between the size of an H II Region and the hot star at its center is the same for H II Regions in all galaxies, the measurement of the angular diameter of an H II Region in another galaxy permits the distance to this galaxy to be calculated. If $\Theta$ is the angular diameter, measured in radians, of an extragalactic H II Region and $L$ is the linear diameter of a standard H II Region in galaxy of known distance, the distance to the external galaxy is $d = L/\Theta$. This method can be used out to roughly 50 million light-years.

## Supernovas as Standard Candles

None of the methods described thus far is useful for distances greater than 50 million light-years. In general, the methods are limited by the luminosities of the astronomical objects involved. All these objects have been stars or clusters with luminosities no greater than roughly $10^6 L_\odot$. To extend the range of distance measurements beyond 150 million light-years, it is necessary to find considerably brighter objects that can be calibrated and used as standard candles.

Only two objects are known that have the necessary brightness. These are supernovas and entire galaxies. The calibration of galaxies as standard candles is discussed below. The use of supernovas for this purpose depends on the fact that the peak luminosity of a supernova is about equal to that of a bright galaxy. If we assume that the supernovas in our galaxy have the same peak brightness as supernovas in other galaxies, an observation of the peak brightness of a supernova in another galaxy immediately supplies a measure of the distance to the galaxy.

## Galaxies as Standard Candles

Following the familiar reasoning, this method assumes that the brightest galaxies in our neighborhood are typical of very bright galaxies anywhere in the Universe. That is, a maximum brightness is assumed for galaxies, just as a maximum brightness was assumed for stars. This assumption is supported by observational evidence. Measurements of the absolute luminosities of galaxies, in the various clusters of galaxies around us for which distances have been determined, show that within each cluster of galaxies there is approximately the same cutoff in maximum luminosity. The most luminous galaxy in each cluster is almost invariably a giant elliptical galaxy (page 220) and luminosities of the largest giant elliptical galaxies are roughly the same (within a factor of two or so) from one cluster to another. These luminosities of the largest giant elliptical galaxies are roughly $10^{45}$ ergs/second, which is about 10 times higher than the luminosity of the Milky Way Galaxy and other large spirals.

The observation of a uniform maximum luminosity for giant elliptical galaxies makes these objects useful standard candles for distance determination. The method is effective out to the range of visibility of the giant elliptical galaxies, which is two billion light-years. A major cause of uncertainty is the possibility that the evolution of the distant ellipticals has produced large changes in their intrinsic luminosities over the course of the one or two billion years required for their light to reach us. This effect would invalidate the calibration of the distant galaxies in terms of the close ones.

**The Red Shift**

How can distances greater than two billion light-years be determined? The only method available at the present time is based on the red shifts of galaxies. As discussed in Chapter 11, distant galaxies show a shift to the red in their spectra, which has been found to be proportional to distance for all galaxies whose distances have been determined with reasonable precision. The limit on such distance determinations for galaxies is about two billion light-years. If it is assumed that the same relationship between red shift and distance holds for distances greater than two billion light-years, a measurement of the red shift for a galaxy may be used to determine its distance.

Quasars (See Chapter 10) are the only visible objects detected thus far—with a few exceptions such as 3C295—that have red shifts suggesting a distance in excess of two billion light-years. The red shifts measured for these peculiar objects have been interpreted as an indication that the radius of the observable universe is roughly 10 billion light-years.

However, the use of the red shift in determining quasar distances rests on two assumptions. The first is the validity of the Hubble Law beyond two billion light-years. The second is that quasars are similar to galaxies.

**The Hierarchy of Astronomical Yardsticks**

The relationships among the methods of distance determination are shown schematically in Figure A.9. Spectroscopic parallax and Main-

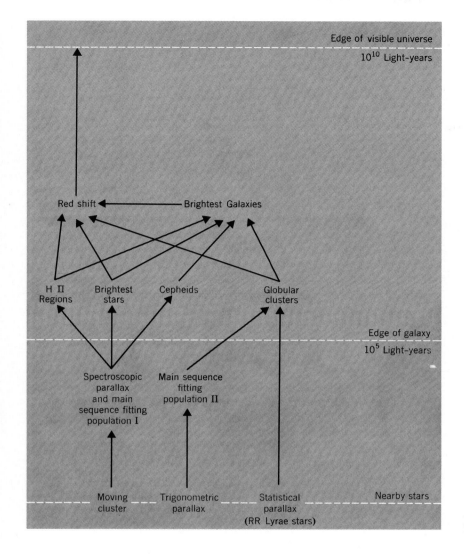

Figure A.9   *The hierarchy of astronomical yardsticks.*

Sequence fitting, both resting on the moving-cluster method, provide the most important base for measuring distances beyond our galaxy. These methods involve population I stars, which are the most common stars in galaxies like ours. Trigonometric parallax and the RR Lyrae stars provide useful independent checks on the measurement of extragalactic distances because they permit Main-Sequence fitting to be carried out for the population II stars that make up the globular clusters. In this way, these latter two methods lead to the calibration of the globular clusters in the Galaxy as standard candles. All the methods for distance determination outside the Galaxy, extending from neighboring galaxies to the boundary of the observable Universe, depend on the three fundamental methods—the moving-cluster method, trigonometric parallax and the RR Lyrae stars—used for relatively close stars. Thus, the observed dimensions of the Universe are determined by measurements of the distances to a handful of stars in the neighborhood of the sun.

## MEASUREMENT OF APPARENT MAGNITUDES[4]

In the discussion of methods for determining distance or absolute magnitude, we assumed a knowledge of the apparent magnitude. Measuring the apparent magnitudes that provide the basis for this discussion would seem to present no difficulties. One need only observe the brightness of a star with the eye, or record it on a photographic plate, to secure a value for $m$. However, the measurement of $m$ actually involves serious problems, because this quantity means the energy reaching earth *summed over all wavelengths*. The eye cannot make this measurement because it is sensitive only to a limited band of wavelengths. Also, the earth's atmosphere does not transmit fully in the infrared and the ultraviolet. In the case of very cool or very hot stars, radiating most of their energy in the infrared or ultraviolet, the atmospheric effect is large. Thus, the *visual apparent magnitude,* that is, the apparent magnitude as perceived by the eye, may be very different from the true apparent magnitude.

If a photographic plate is used to record the apparent magnitudes in place of the eye, the situation is improved for hot stars because photographic emulsions are sensitive to ultraviolet radiation. However, plates are less sensitive to red and infrared light. Also, the photographic measurement of apparent magnitudes is equally affected by absorption in the earth's atmosphere.

It is not possible to obtain the missing information about the energy radiated from stars in the ultraviolet and infrared regions unless detectors sensitive to all wavelengths are sent up above the atmosphere in rockets or satellites. Measurements of this kind have been carried out, but they are still too few in number and too recent to have had a major impact on astronomy, although they will undoubtedly make enormous contributions to future work.

As far as ground-based astronomy is concerned, no purely observational solution exists to the problem of determining apparent magnitudes, and it is necessary to rely on theoretical estimates of the energy radiated from stars in the missing bands of wavelengths. These estimates are described and tabulated below under the heading of Bolometric Corrections.

### Visual and Photographic Magnitudes

Before the advent of photography, the eye was the only practical instrument for measuring the brightness of stars. But we know that the brightness of an object depends on the color, or color range, we use to observe it. A red-hot poker, for example, may appear bright to the eye and extremely bright to an infrared detector. However, it is a very weak emitter in the ultraviolet. The human eye is most sensitive to yellow-green light, and thus is best suited for observation in this range of radiation. We employ the term *visual magnitude* to denote the brightness of a star as estimated by the human eye in the yellow-green, or the region around $\lambda = 5500\text{Å}$.

The early photographic emulsions, although sensitive to violet and ultraviolet light, did not respond to radiation in the visible region with wavelengths greater than about 5000Å. Some of the sensitivity in the ultraviolet was wasted, because most glass lenses absorb light of wavelengths shorter than about 3700Å, and silvered mirrors lose reflectivity at about 3400Å. Hence, the early photographic plates recorded the brightness of the stars in the blue-violet region, and the adopted term, *photographic magnitude,* defined a system of magnitudes for wavelengths between roughly 4000Å and 5000Å.

Photographic and visual magnitudes differ by an amount that depends on the color of the star. Astronomers have arbitrarily set the zero points of the two magnitude scales to be equal for white stars of spectral class A, such as Sirius. Since the eye is relatively more sensitive to red light than the photographic emulsion, a red star, which radiates relatively more energy in the red than in the blue-violet, appears brighter visually than it does photographically. Stars bluer than Sirius, on the other hand, are brighter photographically than visually.

*Photovisual Magnitudes.* Ordinary photographic plates are always sensitive in the blue and ultraviolet regions. They can also be made sensitive to other wavelengths by treatment with dyes that absorb certain colors. Thus it is now possible to take photographs in yellow and red light, and even in the infrared beyond the range of the eye. If an "orthochromatic" emulsion, sensitive to green and yellow light (as well as blue, violet and ultraviolet), is combined with a yellow filter (that blocks the violet and ultraviolet), the combination closely simulates the color sensitivity of the eye. Magnitudes measured with this arrangement are known as *photovisual* magnitudes.

## Color Index

The *color index* of a star is defined as the photographic magnitude minus the visual or photovisual magnitude. The color index is a quantitative measure of the color of a star. For greater accuracy, the photo-

[4] Adapted from *A Survey of the Universe,* by D. H. Menzel, F. Whipple and G. deVaucouleurs, pages 438–441. By permission of Prentice-Hall, Inc.

graphic magnitude is replaced by measurement with a photometer and filter constructed to transmit radiation with wavelengths mostly in the blue region between 3800 and 4800 angstroms. The visual magnitude refers to the intensity of radiation mostly in the spectral band between 4900 and 5900 angstroms. Because the magnitude scale is such that smaller numbers denote brighter stars, the color index is *positive* for all stars redder than Sirius and *negative* for stars bluer than Sirius. The blue supergiant Rigel has a visual magnitude 0.14, a photographic magnitude −0.03 and, therefore, a color index of −0.17; the red Supergiant Betelgeuse has visual magnitude 0.70, photographic magnitude 2.14, and color index +1.44.

A star's color index depends on the temperature of its surface. The index provides a useful measure of temperature as long as passage through interstellar dust has not reddened the starlight.

The photographic magnitudes of large numbers of stars can be measured on one photograph. In practice, the stellar images are compared with those of a few stars on the same photograph whose magnitudes have been carefully measured beforehand. The comparison can be made most simply through visual estimates of the relative sizes of the star images (the brighter the star, the larger the blackened area on the plate); with a little training a good observer can easily intercompare stars to within 0.1 mag., or about 10 percent in relative luminosity.

Precise measurements require a more objective device than the eye, such as a photoelectric photometer, to measure the blackness or density of the photographic images. Here again the magnitudes must be referred to a series of stars of known magnitudes on the photographs, which have been carefully determined beforehand. Photoelectric photometers today provide the most precise measurements of individual stellar magnitudes, but the photographic plate is still unsurpassed for the rapid, efficient measurement of the magnitudes of large numbers of stars.

*The U, B, V System.* Photographic magnitudes are accurate within a few hundredths of magnitude at best, but photoelectric photometry is capable, in principle, of a precision of about a thousandth of a magnitude. The most commonly used photometric magnitudes have been established in three colors: *near-ultraviolet (U); blue (B);* and *yellow or visible (V).* This *U, B, V* system of standard magnitudes, as it is called, and the relations between the color indices *B − V* and *U − B,* have provided precise apparent magnitudes of stars and valuable data on stellar evolution. Approximate shapes of the *U, B,* and *V* bands are shown in Figure A.10.

*Relation Between Color Index and Spectral Type or Temperature.* Table I on page 496 shows the relation between color index and spectral type for stars of various spectral and luminosity classes. It is not surprising that such a relationship exists, since the spectral type of a star depends on its temperature, and the temperature has a strong influence on the star's color.

However, conditions other than surface temperature can also affect the color. If, for example, the starlight has passed through a region of space that contains considerable interstellar dust, scattering of the stellar

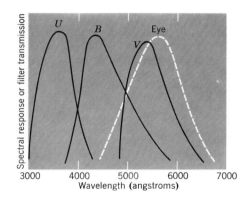

*Figure A.10    U, B, and V bands compared with the response of the eye (from Clayton, Principles of Stellar Evolution and Nucleosynthesis, McGraw Hill, 1968).*

photons by the dust may reduce their energy and redden the light appreciably.

Recently, observations of the colors of stars have been greatly refined by the precise techniques of photoelectric photometry. If, instead of determining a simple color index by comparing the star's light at two wavelengths only, we use the three-color $U, B, V$ system or similar systems, the effect of interstellar reddening often can be detected.

## Bolometric Magnitude

Some detectors of electromagnetic radiation respond to all wavelengths. These detectors measure the total rate of energy radiated by a star, that is, its absolute luminosity. Expressed in magnitudes, the total rate of energy radiated by a star at all wavelengths, including wavelengths shorter and longer than those in the visible region, is called the *bolometric magnitude*. Instruments that detect radiation over a large range of wavelengths are frequently called bolometers, from the Greek word *bole,* meaning ray.

The bolometric magnitude is substantially different from the visual absolute magnitude for stars that radiate most of their energy in the blue or red, outside the visible band of wavelengths. However, it is close to the visual magnitude for stars like the sun, whose peak radiation is in the middle of the visible region. The difference between the visual magnitude and the bolometric magnitude is called the *bolometric correction* (B.C.). It is large and negative for stars distinctly more blue or red than the sun, and negligible for stars with temperatures in the neighborhood of the sun's temperature (Table 1, page 496).

A reference to the bolometric magnitude of a star implies that the observations of the radiation from the star have been corrected for the part of the energy curve cut out by absorption in the earth's atmosphere. Most of the radiation of hot stars is in the ultraviolet part of the spectrum, which the earth's atmosphere does not transmit. For these stars we cannot measure the bolometric magnitude directly, except from space vehicles.

The earth's atmosphere introduces many inaccuracies into the measurement of stellar magnitudes by any method. Haze, dust, clouds, molecular absorption, and scattering in the atmosphere all contribute their share of the uncertainty. An observatory, to be useful for photometric work, must possess both a uniform atmosphere and high atmospheric transparency. An apparently clear sky can sometimes be almost useless photometrically because of variable haze too faint to be detected visually. The best photometric sites are generally on top of mountains, in arid, dry regions, as in the southwest of the United States or in the Andes.

*Table I*

**The visual and bolometric magnitudes, bolometric correction (B.C.) and color index (B-V) for Main-Sequence stars, giants, supergiants and for a range of spectral types.**

| | $T_E$ | Spectral Type | $M_{VIS}$ | $M_{BOL}$ | B.C. | B – V |
|---|---|---|---|---|---|---|
| **Main Sequence** | 50000 | O5 | −6 | −10.6 | −4.6 | −0.32 |
| | 30000 | B0 | −3.7 | − 6.7 | −3.0 | −0.30 |
| | 9700 | A0 | +0.7 | + 0.5 | −0.2 | 0.00 |
| | 7200 | F0 | +2.8 | + 2.8 | 0.0 | +0.30 |
| | 6000 | G0 | +4.6 | + 4.5 | −0.1 | +0.57 |
| | 4700 | K0 | +6.0 | + 5.7 | −0.3 | +0.84 |
| | 3800 | M0 | +8.9 | + 7.8 | −1.1 | +1.39 |
| **Giants, Luminosity Class, III** | 5400 | G0 | +0.8 | + 0.7 | −0.1 | +0.65 |
| | 4700 | K0 | +0.8 | + 0.5 | −0.3 | +1.06 |
| | 3700 | M0 | −0.4 | − 1.5 | −1.1 | +1.55 |
| **Supergiants, Luminosity Class, I** | 25000 | B0 | −6.4 | − 8.4 | −2 | −0.27 |
| | 9400 | A0 | −6.6 | − 6.8 | −0.2 | 0.00 |
| | 7400 | F0 | −6.6 | − 6.6 | 0.0 | +0.30 |
| | 5800 | G0 | −6.4 | − 6.5 | −0.1 | +0.76 |
| | 5000 | K0 | −6.4 | − 6.7 | −0.3 | +1.10 |
| | 3600 | M0 | −5.4 | − 6.6 | −1.2 | +1.70 |

## METHODS FOR MASS DETERMINATION

Five methods are widely used for the determination of steller masses, in addition to the binary-star method (Chapter 5). They vary in importance, depending on the number of stars whose masses can be measured with a given method, and the accuracy of the measurement.

*Empirical Mass-Luminosity Relationship.* Mass and luminosity have been measured independently for many stars extending over a broad range of masses. Figure A.11 shows measurements obtained from these stars — visual and spectroscopic binaries — and a dashed line representing an approximate fit to the data. The scatter in the data reflects the observational uncertainties in the mass and luminosity values, and also the effect of variations in chemical composition from star to star. If a star's distance and, therefore, absolute luminosity have been determined, its mass can be read off this curve.

*Theoretical Mass-Luminosity Relationship.* Computations on stellar structure provide a theoretical relationship between mass and luminosity for stars on the Main Sequence, which can be used to determine the mass of any star whose absolute luminosity is known. The theoretical relationship is shown in Figure A.12. The width of the shaded area represents the effect of compositional differences in the population of stars with respect to helium and metals abundance.

*Stellar Spectra.* The analysis of the line intensities in a stellar spectrum

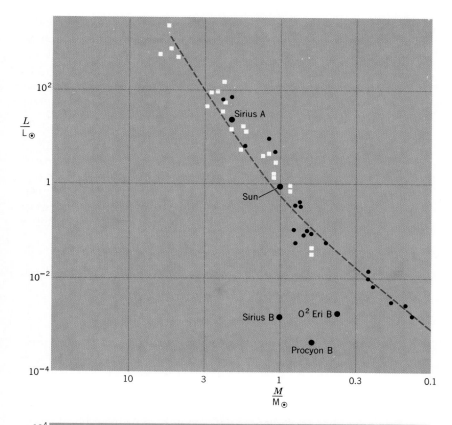

Figure A.11 Empirical mass-luminosity curve. Visual and spectroscopic binaries are indicated by circles and squares, respectively.

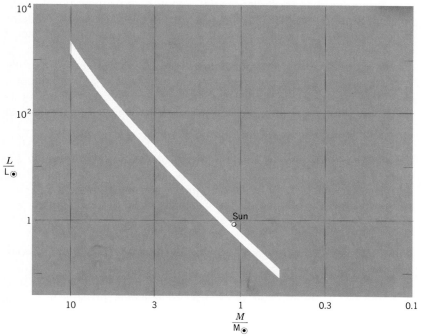

Figure A.12 Theoretical mass-luminosity curve.

yields the star's temperature, and an analysis of the spectral line widths yields the surface acceleration of gravity. From this information an observational value for $M/L$ can be obtained. We use the expressions for the acceleration of gravity at the surface of a star, $g = MG/R^2$, and its luminosity, $L = 4\pi R^2 \sigma T^4$, where $M$ and $R$ are the star's mass and radius, $G$ is the gravitational constant and $\sigma$ is the Stefan-Boltzmann constant. Dividing the first relation by the second, we obtain

$$M/L = g/4\pi G\sigma T^4$$

For stars of known distance and, therefore, known $L$, the value of $M$ follows.

*Pulsating Stars.* The period $P$ of a pulsating star depends on the mean density of the star, just as the period of vibration of a spring depends on the mass per unit length of the spring. The observed relationship is

$$P = 0.06\sqrt{\bar{\rho}_\odot/\bar{\rho}} \text{ days}$$

where $\bar{\rho}_\odot$ and $\bar{\rho}$ are the mean densities of the sun and the pulsating star, respectively. The mean density, $\bar{\rho}$, is defined by the formula,

$$M = (4\pi/3)R^3\bar{\rho}.$$

$R$ is determined from the luminosity and temperature of the star, using the relation

$$L = 4\pi R^2 \sigma T^4 \quad \text{or} \quad R = \sqrt{L/4\pi\sigma T^4}$$

$T$ is known from the spectral type. If the variable is a cepheid, $L$ is known from the period-luminosity law. If not, $L$ must be determined by an independent measurement of the distance. Assuming that $L$ has been determined, the mass is finally computed from these relationships after $R$ has been eliminated.

*Gravitational Red Shift.* Relativity theory predicts a red shift in the wavelength of the radiation emitted by a massive body. The amount of the shift depends on the mass of the star and its radius. If the gravitational red shift can be measured, and the radius is known, the mass of the star can be determined.

The relation between the gravitational red shift and the mass can be derived heuristically as follows. Suppose that a photon of energy $E$ is emitted from an atom at the surface of a star. An amount of mass, $m_{ph}$, can be assigned to the photon according to Einstein's relation,

$$m_{ph} = \frac{E}{c^2}$$

In escaping to space against the pull of the star's gravity, the photon does work against the gravitational pull of the star and loses energy. The photon's energy is related to its wavelength by an inverse proportion: the smaller the energy, the longer the wavelength. Thus, when the photon is pulled back by the star's gravity and loses energy, its wavelength increases. I.e., it is shifted toward the red. From Newton's law of gravity it can be shown that the formula for the fractional red shift is

$$\frac{\Delta\lambda}{\lambda} = \frac{MG}{Rc^2}$$

where $M$ is the mass of the star, $G$ is the universal constant of gravity, $R$ is the star's radius, and $c$ is the velocity of light.

For a star of solar mass and radius, the fractional red shift is

$$\frac{\Delta\lambda}{\lambda} = \frac{M_\odot G}{R_\odot c^2} = \frac{2 \times 10^{33} \times 6.7 \times 10^{-8}}{7 \times 10^{10} \times (3 \times 10^{10})^2} = 2 \times 10^{-6}$$

which amounts to $10^{-2}\text{Å}$ for a line in the middle of the visible spectrum at 5000Å. A red shift of roughly the predicted magnitude has been measured in the sun's spectrum, and it is also barely detectable in other Main-Sequence stellar spectra. In white dwarfs, however, the gravitational red shift is much larger, because $R$ is smaller. For a white dwarf of solar mass, with a typical radius of 20,000 km,

$$\frac{\Delta\lambda}{\lambda} = 7 \times 10^{-5}$$

which corresponds to a shift of 0.35Å for a 5000Å line. A shift of this magnitude can be measured with good accuracy in white dwarf spectra. Thus, the gravitational red shift method of mass determination is useful for white dwarfs, but not for most other stars.

In applying the method to white dwarfs, an obvious question arises: How can the gravitational red shift be separated from the Doppler shift caused by the motion of the white dwarf relative to the sun? The two effects cannot be separated for a single white dwarf, but if measurements are made on several white dwarfs, a statistical analysis can be performed to eliminate the effect of the Doppler shift. The idea of the statistical analysis is that if many stars are considered, on the average half will be moving toward the sun, and half away from it. The two groups of stars produce Doppler shifts of opposite signs. Therefore, if the wavelength shifts are averaged for several white dwarfs, the Doppler effect should cancel out, leaving only the gravitational shift.

On this basis the average mass of white dwarfs has been determined to be about $0.7M_\odot$. Most white dwarf masses are believed to be close to this value, although some, determined by independent methods, have turned out to be as low as $0.4M_\odot$.

# Appendix B
# Nuclear Reactions in Stars

Chapter 6 described a sequence of three reactions by which protons combine directly to form helium nuclei. These reactions are the main source of energy release for Main Sequence stars of solar mass or smaller, but in more massive stars a different sequence of reactions, involving carbon and nitrogen nuclei in addition to protons, constitutes the principal energy source. The discussion below adds some important information to the discussion of the proton-proton cycle, and provides a corresponding description of the alternate sequence involving carbon and nitrogen nuclei.

## THE PROTON-PROTON REACTION

The main reactions in which four protons combine in sequence to form a helium nucleus are tabulated below.

| Reaction | Energy | Reaction time |
|---|---|---|
| $p + p \rightarrow d + e^+ + \nu$ | +1.44 | $1.4 \times 10^{10}$ yr. |
| $d + p \rightarrow He^3 + \gamma$ | +5.5 | 0.6 sec |
| $He^3 + He^3 \rightarrow He^4 + 2p$ | +12.85 | $10^6$ yr. |

The energy released in each reaction of the sequence is indicated in units of *MeV*. (1 *MeV* = one million electron volts = $1.6 \times 10^{-6}$ ergs.) In adding up the total energy release, the first two reactions, leading to formation of $He^3$, must be counted twice, since two $He^3$ nuclei must be produced before the last reaction can occur. The overall energy yield is $2(1.44 + 5.5) + 12.85 = 26.7$ *MeV*. However, in the first reaction the energy that is carried off by the neutrino must be subtracted, because neutrinos have a very small interaction with matter and escape from the star without energy loss. Each neutrino carries off 0.26 *MeV*. With allowance for the two escaping neutrinos, the total energy available to the stars per helium nucleus formed, is 26.2 *MeV*.

In the table the number in parentheses is the reaction time, defined as the time required for half the constituents involved to undergo that particular reaction. The reaction time is, of course, very sensitively dependent on the density and temperature of the particles. The times in the table are calculated for density and temperature in the range of those prevailing at the center of the sun, namely 150 g/cm³ and $13 \times 10^6$ °K.

The table shows that the slowest reaction, pacing all the others, is the one in which two protons unite to form a deuteron. As soon as the deuteron appears, it is snapped up almost instantly by another proton to form $He^3$. The reactions between $He^3$ nuclei proceed at a relatively measured pace, although ten thousand times more rapidly than the initial proton-proton reaction.

## THE CARBON-NITROGEN CYCLE

The sequence of reactions that starts with two protons fusing into a deuteron was undoubtedly the primary energy source when the Universe was young and made up largely or entirely of hydrogen. Other elements built up steadily in the Universe during the course of time, as a result of stellar evolution, and their proportions in the interstellar medium gradually increased. As a result, an alternate way of transmuting hydrogen into helium became possible, in which carbon and nitrogen nuclei, made out of protons earlier in the history of the Universe, played the role of catalysts. This sequence of reactions called the carbon-nitrogen cycle, or sometimes the carbon cycle, is tabulated below.

| Reaction | Energy release (MeV) | Reaction time |
|---|---|---|
| 1) $C^{12} + p \rightarrow N^{13} + \gamma$ | $+1.95$ | $1.3 \times 10^7 y$ |
| 2) $N^{13} \rightarrow C^{13} + e^+ + \nu$ (0.7 MeV) | $+2.22$ | 7 min. |
| 3) $C^{13} + p \rightarrow N^{14} + \gamma$ | $+7.54$ | $3 \times 10^6 y$ |
| 4) $N^{14} + p \rightarrow O^{15} + \gamma$ | $+7.35$ | $3 \times 10^5 y$ |
| 5) $O^{15} \rightarrow N^{15} + e^+ + \nu$ (1 MeV) | $+2.7$ | 82 s |
| 6) $N^{15} + p \rightarrow C^{12} + He^4$ | $+4.96$ | $10^5 y$ |

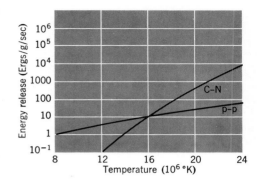

*Figure B.1 Comparison of temperature dependence for proton-proton and carbon-nitrogen cycles.*

The total energy release in the carbon-nitrogen reaction is 26.7 *MeV*. When allowance is made for the loss of 1.7 *MeV* carried off by the two neutrinos created in the sequence, the energy gain to the star becomes 25.0 *MeV* per cycle.

Note that each time the sequence of reactions occurs the net effect is to convert four protons into one helium nucleus, just as in the proton-proton reaction sequence. The carbon and nitrogen nuclei play the role of catalysts in the cycle; they are consumed in the early stages of the sequence but emerge at the end, leaving the original number $C^{12}$ and $N^{14}$ nuclei unchanged.

The electrical barriers acting in the carbon-nitrogen cycle are much higher than those acting in the proton-proton cycle, the barrier in the basic carbon-proton reaction being six times higher than the proton-proton barrier. As a result, the proton-proton cycle dominates at low temperatures, and the carbon-nitrogen cycles does not become important until higher temperatures are reached. Figure B.1 indicates the separate nuclear energy release from both cycles in ergs/g/sec for various temperatures. The proton-proton cycle is dominant up to a temperature of 16,000,000°K—slightly higher than the sun's central temperature— and the carbon-nitrogen cycle takes over above that level. About 10 percent of the sun's energy is contributed by the carbon-nitrogen cycle and the remainder by the proton-proton cycle.

The Main Sequence is divided into two segments, depending on whether the proton-proton or carbon-nitrogen cycle is the dominant energy source. The segment in which the proton-proton dominates corresponds to the *lower Main Sequence;* the other segment corresponds to the *upper Main Sequence.* The sun lies near the dividing point between the two segments of the Main Sequence.

# Photo Credits

Solar Observatory  **Fig. 12.6** Courtesy Martin Schwarzschild, Princeton University Observatory  **Fig. 12.7(a)** Courtesy NASA  **Fig. 12.7(b)** Courtesy Martin Schwarzschild, Princeton University Observatory  **Fig. 12.7(c)** Courtesy W. M. Chiplonkar, University of Poona, Poona, India  **Fig. 12.8** Courtesy Robert B. Leighton, California Institute of Technology  **Fig. 12.9** Courtesy Hale Observatories  **Fig. 12.15** Courtesy M. Kanno, Hida Observatory, Kyoto, Japan  **Fig. 12.18** Courtesy Harvard College Observatory  **Fig. 12.19** Courtesy Richard B. Dunn, Sacramento Peak Observatory, Air Force Cambridge Research Laboratories  **Fig. 12.20** Courtesy Richard B. Dunn, Sacramento Peak Observatory, Air Force Cambridge Research Laboratories  **Fig. 12.21** Courtesy Gordon Newkirk, Jr., High Altitude Observatory, Boulder Colorado  **Fig. 12.23** Courtesy Martin Schwarzschild, Princeton University Observatory  **Fig. 12.24** Courtesy Hale Observatories  **Fig. 12.25** Courtesy R. J. Bray, C. S. I. R. O., Australia. From *Sunspots* by R. J. Bray and R. E. Loughhead, Chapman and Hall Ltd., London, 1964.  **Fig. 12.27** Courtesy Hale Observatories  **Fig. 12.31** Courtesy Sara F. Martin and Harry E. Ramsey, Lockheed Solar Observatory  **Fig. 12.32** Courtesy Sara F. Martin and Harry E. Ramsey, Lockheed Solar Observatory  **Fig. 12.33** Courtesy NOAA  **Fig. 12.34** Courtesy NOAA  **Fig. 12.35** Courtesy NOAA  **Fig. 12.36** Courtesy Richard B. Dunn, Sacramento Peak Observatory, Air Force Cambridge Research Laboratories  **Fig. 12.37** Courtesy Observatoire de Paris, Meudon  **Fig. 12.38** Courtesy Yerkes Observatory  **CHAPTER 13 — Opener** Courtesy Hale Observatories  **Fig. 13.4** Courtesy Royal Society  **Fig. 13.8** Courtesy Lowell Observatory  **Fig. 13.9** Courtesy Hale Observatories  **Fig. 13.12** Courtesy Yerkes Observatory  **Fig. 13.13** Courtesy Hale Observatories
**CHAPTER 14 — Opener** Fred Ward-Black Star  **Fig. 14.8** Barrett Gallagher
**CHAPTER 15 — Opener** Courtesy NASA  **Fig. 15.1** Courtesy NASA  **Fig. 15.2** Courtesy NASA  **Fig. 15.3** Courtesy U.S. Air Force  **Fig. 15.4** Courtesy NASA  **Fig. 15.5** Courtesy NASA  **Fig. 15.6** Courtesy NASA  **Fig. 15.7** Courtesy NASA  **Fig. 15.8** Courtesy NASA  **Fig. 15.9** Courtesy NASA  **Fig. 15.10** Courtesy NASA  **Fig. 15.12** Courtesy Hale Observatories  **Fig. 15.13** Courtesy NASA  **Fig. 15.14** Courtesy NASA
**CHAPTER 16 — Opener** Courtesy NASA  **Fig. 16.1** Courtesy Lowell Observatory  **Fig. 16.2** Courtesy NASA  **Fig. 16.7** Courtesy M. Ya. Marov, USSR Academy of Sciences  **Fig. 16.8** Helmut Wimmer, The American Museum of Natural History  **Fig. 16.9** Courtesy Lowell Observatory  **Fig. 16.10** From *La Planete Mars* by E. M. Antoniadi  **Fig. 16.11** Courtesy NASA  **Fig. 16.12** Courtesy NASA  **Fig. 16.13** Courtesy NASA  **Fig. 16.14** Courtesy NASA  **Fig. 16.15** Courtesy NASA  **Fig. 16.16** Courtesy NASA  **Fig. 16.17** Courtesy NASA
**CHAPTER 17 — Opener** From *Conquest of Space* by Chesley Bonnestel and Willy Ley. Courtesy Chesley Bonnestel and Viking Press
**CHAPTER 18 — Opener** Courtesy J. William Schopf, Elso S. Barghorn, Morton D. Maser and Robert O. Gordon  **Fig. 18.2** Courtesy the British Information Service  **Fig. 18.5** Courtesy C. A. Knight and R. C. Williams, Virus Laboratory, University of California, Berkeley  **Fig. 18.6** Courtesy Carl T. Mattern and Herbert duBuy, National Institutes of Health, Bethesda, Maryland

# Color Plates

**Plates 4, 5, 6, 7, 8, 9, 11, 14, and 15** Copyright by the California Institute of Technology  **Plates 10 and 12** Courtesy U.S. Naval Observatory  **Plates 13 and 17** Courtesy NASA  **Plate 16** Courtesy Kitt Peak Observatory  **Plate 18** Courtesy William C. Atkinson  **Plate 19** Courtesy John Free  **Plates 20, 23, 24, and 25** Courtesy Richard B. Dunn, Sacramento Peak Observatory, Air Force Cambridge Research Laboratories  **Plates 21 and 22** Courtesy Sara F. Martin and Harry E. Ramsey, Lockheed Solar Observatory  **Plate 26** Courtesy H. Zirin, Big Bear Solar Observatory  **Plate 27** Courtesy Hale Observatories  **Plate 28** Courtesy Sacramento Peak Observatory, Air Force Cambridge Research Laboratories  **Plate 29** Courtesy Gordon Newkirk, Jr., High Altitude Observatory, Boulder, Colorado

506

# Index

Hydrogen atoms, excited states of, 83-84
    radiation from at 21-centimeters,
        205-207
    spectrum of, 86-88
Hygiea, 256
Hyginus, Rille, 392

**Icarus**, 356
Igneous rocks, 364
Ilmenite, 398-399
India-Australia Plate, 376-378
Infrared light, 28
Infrared radiation from galaxy M 82, 233
Infrared telescopes, 62-63
Inheritance, mechanism for, 458-460
Intelligence, origin of on the earth,
    469-472
Interstellar dust, 203
Interstellar formaldehyde, radio emission
    from, 207-208
Interstellar medium, 200-203
Inverse-square law of gravity, 345
Irregular galaxies, 224
Irregular variable stars, 135
Island arcs, 378

**Jansky, Carl**, 271
Jeans, James, 340
Jupiter, 351, 354, 438-440
    atmosphere of, 438-439
    life on, 439-440
    spacecraft exploration of, 440

**K stars**, 100-101
Kepler, Johannes, 345-346
Kiloparsec, 478
Kirchhoff, Gustav Robert, 80, 92

**Lancelet**, 470
Laurasia, 381, 382
Lava, 368, 381, 395, 405
Law of conservation of mass, 158
Lemaître, Father, 261, 263
Lenses, 41-42
Life, building blocks of, 456-461
    creation of from atmospheric gases,
        460-461
    in the cosmos, 455-473

on earth, history of, 466-472
on earth compared to Venus,
    421-422
on Jupiter, 439-440
on Mars, 436-438, 466
    tests for, 437-438
on Mars compared to earth, 436-437,
    451
on the moon, 396, 399
origin of on the earth, 456-462
on other planets, 464-466
on Venus, 450-451
    compared to earth, 421-422
Life span of stars, 172-173
Light, absorption of by atoms, 85
    diffraction of, 43-45
    Dopper shift of, 32-35
    as an electric vibration, 24-27
    emission of from excited atoms, 84-85
    focusing of, 41
    frequency of, 28
    infrared, 28
    nature of, 23-35
    refraction of, 40
    scattering of, 119-120
    speed of, 23, 28
    ultraviolet, 28
Light-gathering power of telescopes, 49-50
Light-year, 11
Limb-darkening of the sun, 292-294
Local Group, 15-16
Logarithmic scales, 125
Low, Frank, 62, 233
Lowell, Percival, 425
Luminosity, absolute, versus surface
    temperature of stars, 125-130
    apparent, 123-125
    of quasars, 250
    of stars, 123-125
Luminosity-empirical mass relationship,
    mass determination with, 496
Luminosity-theoretical mass relationship,
    mass determination with, 496
Lunar eclipses, I-12
Lunar highlands, 385
Lunar orbiters, 396
Lunar tides, I-15–I-16, I-18
Lyman series, 87

**M82 (galaxy)**, 232-234
M stars, 101

detection of in other solar systems, 336-337
giant, 351-356
   moons of, 354
life on, 464-466
motion of, 347-350
   laws of, 345-346
properties of (table), 346-347
terrestrial, 350-351
   moons of, 351
without stars, 339
Plates, 374
   African, 375
   India-Australia, 376-378
   map of, 376-377
   Pacific, 378
   South American, 378
Pluto, 351
Polaris, I-30–I-31
Population I stars, 215
Population II stars, 215-216
Populations of stars, 125-134, 213-216
Pores, 316
Positrons, 152, 153, 255
Potassium, 372
Precession of the axis, I-5
Primates, 471-472
Primordial fireball radiation, 270-271
Prism spectroscopes, 70-72
Prominences, 323-324
Proper motion, 478
Proteins, 457-459
Protogalaxies, 225-226
Proton-proton reactions in stars, 500-501
Protons, 82, 152, 153
Protostars, 164-172
   evolution of to a star, 171-172
   on the Hertzsprung-Russell diagram, 171
   rising temperatures of, 167-170
Pulsars, 190-192
Pulsating stars, mass determination with, period of, 498
Pyrex glass, 54
Pyroxene, 368, 369

**Quartz**, 368, 369
Quasars, 245-255
   discovery of, 245
   location of, 247-249
   luminosity of, 250

radio images of, 249-250
red shift in, 245-247
relation to antimatter, 254-255
relation to galaxies, 252-253
relation to supernovas, 253-254
size of, 250-252
spectra of, 252
visible images of, 249

**Radar**, 29, 415
Radar astronomy, 56
Radiation from stars and wavelengths of, 120-122
Radio astronomy, and the Milky Way Galaxy, 203-208
   and Venus, 415-416
Radio galaxies, 238-243
Radio images, of galaxies, 240-242
   of quasars, 249-250
Radio signals, intelligent, from outer space, 61-62
Radio telescopes, 55-62
   advantages of, 55-57
   comparison with optical telescopes, 57-61
Radio waves, 29
   emission of by galaxies, explanation of, 243
   emitted from Venus, 415
   from interstellar formaldehyde, 207-208
   21-centimeter line, 205-207
Radioactive carbon in tests for Martian life, 437
Radioactivity and heating of the earth's center, 366-374
Radium, 372
Reber, Grote, 238
Red giants, 121, 130
   evolution of and Hertzsprung-Russell diagram, 174-177
   meaning of, 133-134
   structure of, 177
Red shift, cosmological, 35
   of galaxies, distance measurements with, 490
   gravitational, mass determination with, 498-499
   in quasars, 245-247
Reflecting telescopes, 50-55
Refraction, 40

Urey, Harold, 460

van de Kamp, Peter, 337-338
Variable stars, 134-138
    Cepheid, 135-137
    irregular, 135
    Novas, 137-138
Venera flights, 416, 418-420
Venus, 350, 412-423
    atmosphere of, 420-423
      greenhouse effect, 422-423
    atmosphere of compared to earth,
      445-451
      accumulation of $CO_2$ in, 446-447
    dryness of, 423
    and earth, 443-451
    exploration of, 416-420
    greenhouse effect on, 422-423
    life on, 421-422, 450-451
    planetary properties of, 412-414
    radio astronomical study of, 415-416
    spacecraft exploration of, 416-420
    surface of, 414-416
    water vapor scarcity of, 423, 449-450
Vernal Equinox, I-19
Vertebrates, 470
Vesta, 356
Vibrating electric force, 24
Vibrational nature of light, 24-27
Viking spacecraft, 437
Viruses, 456, 461-464
Visible region of electromagnetic spectrum,
    28
Visibility of quasars, 249
Visual apparent magnitudes, 492
Visual binary stars, 140
Visual magnitudes of stars, 125, 493
Volcanism, on Mars, 428-429
    on the moon, 401, 402, 405

Water, absence of on moon, 396, 399
    on Mars, 428, 429, 433, 435, 436
    on Venus, 423, 449-450
Watson, James D., 459
Wavelengths, formula for, 29
    relation of to energy radiated by
      bodies, 120-122
Waves, electric, 26-27
    electromagnetic, 28-31
    gravity (See Gravity waves)
Weakness, zone of, 374-375
Weber, Joseph, 68, 255
Weber experiment, 69-70
Wegener, Alfred, 378
White dwarfs, 130
    death of, 185-187
    gravitational red shift for mass
      determination of, 499
    meaning of, 133-134
Wien's law, 32
Wilson, Robert, 271
Woolaston, William Hyde, 91

X-ray astronomy, 65-66
    from Skylab, 69-70
X-rays, 29
    and cosmology, 273

Year, length of, 348
Ylem, 261-262

Zeeman effect, 113-114, 318
Zero-Age Main-Sequence, 484
Zone of convection in the sun, 287-288
    effect of on temperature profile, 288
      relation to granules, 290-291
Zone of weakness in the earth, 374-375